Successful Agricultural Innovation in Emerging Economies

World population is forecast to grow from 7 to 9 billion by 2050, one in six is already hungry and food production must increase by 70–100% if it is to feed this growing population. No single solution will solve this problem but recent developments in the genetic technologies of plant breeding can help to increase agricultural efficiencies and save people from hunger in a sustainable manner, particularly in African nations where the need is greatest. These advances can rapidly incorporate new traits and tailor existing crops to meet new requirements and also greatly reduce the time and costs taken to improve local crop varieties. This book provides a collected, reliable, succinct review which deals expressly with the successful implementation of the new plant genetic sciences in emerging economies in the context of the interrelated key regulatory, social, ethical, political and trade matters.

DAVID J. BENNETT has long-term experience in the relations between the biosciences, industry, government, education, law, the public and the media. He works with the European Commission, government departments, companies, universities, public interest organisations and the media, having worked in universities and companies in the UK, USA, Australia and Europe.

RICHARD C. JENNINGS is an Affiliated Research Scholar at the University of Cambridge. His research interests are in the ethics of science and technology. He pioneered the university's teaching of ethics in science and continues to run graduate ethics workshops. He has developed, with others, a framework for assessing ethical issues in new technologies.

This publication was made possible through the support of a grant from the John Templeton Foundation. The opinions expressed in this publication are those of the author(s) and do not necessarily reflect the views of the John Templeton Foundation.

Successful Agricultural Innovation in Emerging Economies

New Genetic Technologies for Global Food Production

EDITED BY

DAVID J. BENNETT
St Edmund's College, Cambridge

RICHARD C. JENNINGS
University of Cambridge, UK

CAMBRIDGE
UNIVERSITY PRESS

CAMBRIDGE
UNIVERSITY PRESS

University Printing House, Cambridge CB2 8BS, United Kingdom

Published in the United States of America by Cambridge University Press, New York

Cambridge University Press is part of the University of Cambridge.

It furthers the University's mission by disseminating knowledge in the pursuit of education, learning and research at the highest international levels of excellence.

www.cambridge.org
Information on this title: www.cambridge.org/9781107675896

© Cambridge University Press 2013

First published 2013
First paperback edition 2014

A catalogue record for this publication is available from the British Library

Library of Congress Cataloguing in Publication data
Successful agricultural innovation in emerging economies : new genetic technologies for global food production / edited by David J. Bennett, Richard C. Jennings.
    p. cm.
    Includes index.
    ISBN 978-1-107-02670-4 (Hardback)
    1. Agriculture–Developing countries. 2. Crop improvement–Developing countries. 3. Food crops–Biotechnology–Developing countries.
    I. Bennett, David J. II. Jennings, Richard C.
    HD1417.S83 2013
    338.10917204–dc23
    2012031145

ISBN 978-1-107-02670-4 Hardback
ISBN 978-1-107-67589-6 Paperback

Contents

Authors' biographies

ALFREDO AGUILAR

Alfredo Aguilar has been at the European Commission, Directorate-General DG Research, since 1986. His job involves preparation, implementation and follow-up of the EC Research Framework Programmes and being at the interface between research and other Community policies. He is the Head of Unit for Biotechnologies, and his main objective is to promote the so-called KBBE (Knowledge-Based Bioeconomy in Europe) through the R&D activities in Biotechnology in FP7. Prior to this, Aguilar was the Head of the INCO (International Cooperation) Programme, Head of Unit for Biotechnology and Applied Genomics, Head of Unit 'The Cell Factory' and Head of Unit of Demonstration Projects in Life Sciences.

Aguilar has a PhD in chemistry (biochemistry). He is Associate Professor of Microbiology at the University Complutense of Madrid and prior to this was at the University of León. He has carried out research at postdoctoral level for several years at the John Innes Centre (Norwich, UK) and at the Institut für Zellbiologie, Eidgenössiche Technische Hochschule (Zurich, Switzerland). He also worked as manager at the Chemistry R&D Department, Lever Ibérica (Unilever), Aranjuez, E. He has three Japanese patents, more than 100 publications on science management and science policy in Europe and around 85 scientific publications in international journals, books, patents, etc. on experimental research prior to his incorporation to the European Commission.

KLAUS AMMANN

Professor Klaus Ammann is Professor Emeritus of Biodiversity at the University of Bern, and formerly Director of the Bern Botanical Garden (1996–2006). He gained his PhD in 1972 at the University of Bern, for a thesis on vegetation and glacier history, *summa cum laude*. His research interests include biodiversity, vegetation ecology, lichens and mosses, bioindication of air pollution, ethnobotany, plant biotechnology, and biosafety and ecology of transgenic crops. He has been a guest lecturer at Delft, The Netherlands and Istanbul, Turkey, and has researched at Duke University, North Carolina, USA. He is the author of over 350 publications. He was involved in several international research projects focusing on chemotaxonomy of macro-lichens, ecological monitoring and ecological risk assessment of vertical gene flow. He is also involved in two EU projects on gene flow and plant conservation. Professor Ammann is active as editor or co-editor of journals published by Elsevier, Springer and Landes, and sits on numerous scientific committees in Switzerland and the EU.

TINA BARSBY

Dr Tina Barsby was appointed Director and CEO of the National Institute of Agricultural Botany (NIAB), Cambridge in September 2008. She has significant experience in the agricultural crop sector and is internationally recognised for scientific achievements in plant biotechnology. Her earlier work laid the foundation for the application of somatic hybridisation in commercial plant breeding and the development of cytoplasmic male sterility in vegetable breeding. She has led and managed multifaceted, product-driven R&D programmes in North America and UK/mainland Europe.

Tina has a first degree in agricultural botany from the University of Wales at Bangor, and a PhD from the University of Nottingham. She spent a postdoctoral period at Kansas State University, and worked at Allelix Inc., Ontario, Canada for several years before returning to the UK in 1989. She joined Nickerson UK (now part of the LG Group) where she remained until joining NIAB in 2006. Under her leadership NIAB and The Arable Group (TAG) combined operations and merged to form a single entity offering impartial science-based research and information. The company is active at all points along the crop development pipeline, with partnerships in place to ensure first-class research and service provision.

DAVID BAULCOMBE

Professor Sir David Baulcombe is the Regius Professor of Botany and Royal Society Research Professor at the Department of Plant Sciences of the University of Cambridge. He is a molecular biologist and his research interest in plants focuses on how genes can be silenced. He has now moved into the field of epigenetics – the science of how nurture can influence nature. This research links to disease resistance in plants and understanding of hybrids. David is a poor field botanist. Extramural activities include membership of the Biotechnology and Biological Sciences Research Council and in 2009 he chaired a Royal Society policy study on the contribution of biological science to food crop productivity.

ROGER BEACHY

Dr Roger Beachy was appointed by President Obama as the first Director of the National Institute of Food and Agriculture (NIFA) in the US Department of Agriculture, serving from October 2009 to May 2011. He served as Chief Scientist of the US Department of Agriculture from January to October 2010. Dr Beachy was founding president of the Donald Danforth Plant Science Center in St Louis, Missouri, 1999–2002, and headed the Division of Plant Biology at The Scripps Research Institute as the Scripps Family Chair in Cell Biology from 1998 to 2001. He was Professor in the Biology Department at Washington University in St Louis, and Director of the Center for Plant Science and Biotechnology from 1978 to 1991. Dr Beachy is a member of the US National Academy of Sciences, and was awarded the Wolf Prize in Agriculture (2001). He is a Fellow of the American Association for the Advancement of Science and the American Academy of Microbiology, Foreign Associate of the National Academy of Science India, the Indian National Science Academy, and The Third World Academy of Sciences, and is a Fellow in the Academy of Science of St Louis. He received the Bank of Delaware's Commonwealth Award for Science and Industry (1991) and the Ruth Allen award 1990 from the American Phytopathological Society (1990), among other awards.

DAVID J. BENNETT

Dr David Bennett is now a Visitor to the Senior Combination Room of St Edmund's College, Cambridge, UK and a Guest at the Department of Biotechnology of the Delft University of Technology in The Netherlands. He has a PhD in biochemical genetics and an MA in science policy studies with long-term experience, activities and interests in the relations between science, industry, government, education, law, the public and the media. He works with the European Commission, government departments, companies, universities, public interest organisations and the media in these areas, having worked in universities and companies in the UK, USA, Australia and, most recently, The Netherlands. He was a founder member and secretary of the European Federation of Biotechnology Task Group on Public Perceptions of Biotechnology established in 1991 and one of the first organisations in this field. For the last 20 years or so he has been running many international network-based, multidisciplinary projects, courses, conferences, workshops, etc. funded by the European Commission and other bodies in biotechnology and, of late, nanobiotechnology. He co-edited with Richard Jennings *Successful Science Communication: Telling It Like It Is* published in 2011 by Cambridge University Press.

JACK A. BOBO

Jack Bobo is the Senior Advisor for Biotechnology in the Bureau of Economic and Business Affairs (EB) at the US Department of State. He also serves as the Chief of the Department's Biotechnology and Textile Trade Policy Division. He works on trade policy, food security, climate change and development issues related to agricultural science and technology, including agricultural biotechnology. He is responsible for developing and implementing US trade policy related to new technologies and working with foreign governments to address regulatory barriers to US agricultural exports. Prior to joining the State Department, Jack Bobo practised law at the Washington, DC firm of Crowell & Moring, LLP. His education includes a degree in law and an MSc in environmental science, as well as degrees in chemistry, biology and psychology. Previously, he received a research fellowship in international law at the University of Cambridge, served as an advisor to the President's Information Technology Advisory Committee, and taught science in Mekambo, Gabon in Central Africa as a Peace Corps volunteer.

GRAHAM BROOKES

Graham Brookes is an agricultural economist and consultant with 26 years' experience of examining economic issues relating to the agricultural and food sectors. He is a specialist in analysing the impact of technology, policy changes and regulatory impact. He has undertaken several evaluations of EU regulations including economic impact assessments of pesticide approvals, GMO regulations, novel foods and health claims. He has also, since the late 1990s, undertaken a number of research projects relating to the impact of agricultural biotechnology and written widely on this subject in peer-reviewed journals. This work includes annual updates of the global economic and environmental impact of GM crops since 1996, the impact of insect-resistant maize in Spain and herbicide-tolerant soya beans in Romania, the impact of GMO labelling in Europe, the economic impact of GMO zero-tolerance legislation and the cost to the UK economy of failure to embrace agricultural biotechnology.

SAMUEL BURCKHARDT

Samuel Burckhardt graduated at the University of Zurich with an MSc in business information systems. He also holds an MBA from the London Business School where he did various engagements in the social sector. At the time of writing this chapter, he was employed as an external consultant to the Cambridge Malaysian Education and Development Trust, c/o Trinity College, Cambridge, UK. Samuel currently works as a Senior Associate with McKinsey & Co. in Zurich, focusing on various industries on technology as an enabler.

CLAUDIA CANALES HOLZEIS

Dr Claudia Canales Holzeis is an Associate Scientist at the Department of Plant Sciences, University of Oxford. She has 9 years' experience in plant genetics research. She worked for 2 years as a Senior Project Officer for the International Service for the Acquisition of Agri-Biotech Applications (ISAAA), based in the Philippines, on science communication and technology transfer projects. She is a graduate of the University of Reading in environmental biology and holds a PhD in plant genetics from the University of Oxford.

MARK F. CANTLEY

Mark Cantley graduated in mathematics (Cambridge, 1963), economics (London School of Economics, 1965), operational research (London School of Economics, 1965) and accounting and finance (1980s). His initial career was in the operations research department of BISRA (British Iron and Steel Research Association), then he went to the University of Lancaster 1967–77 as lecturer in operations research and project leader. Mark detached to the Management and Technology Department of the International Institute for Applied Systems Analysis (IIASA) in Vienna, 1978–9. In September 1979, he moved to Brussels, as a member of the FAST team (Forecasting and Assessment in Science and Technology) in Directorate-General XII (Research and Innovation) of the European Commission, switching interests to biotechnology. There he headed a new unit, CUBE (Concertation Unit for Biotechnology in Europe), 1982–92. Then he went to the Organisation for Economic Co-operation and Development (OECD), 1993–8, heading the biotech unit, and drafting reports. In 1999–2005, Mark was back at DG XII as Adviser to Director, Life Sciences – he was involved in policy discussions, reporting. His activities have included frequent public speaking, reporting, etc. on modern biotech, with extensive travel in Europe, and occasionally further afield. Mark retired in 2005.

EUGENIO J. CAP

Dr Eugenio J. Cap is currently Director of the Institute of Economics and Sociology, with Argentina's National Institute of Agricultural Technology (INTA). He graduated from the University of Buenos Aires, did graduate work both at the University of California, Davis (MSc in vegetable crops) and at the University of Minnesota (PhD in agricultural and applied economics). His research work includes the development of quantitative tools for the analysis of the agricultural sector, with emphasis on technological variables and the impact assessment of agricultural research. He has authored numerous publications on those subjects. He has also worked as a consultant for the Food and Agriculture Organization, The World Bank, the Interamerican Development Bank and the Interamerican Institute for Cooperation in Agriculture, among other national and international organisations.

DANUTA CICHOCKA

Danuta Cichocka graduated in environmental protection at Warsaw University and environmental resources at the University of Salford. She carried out her PhD research on microbial degradation of environmental contaminants and the application of stable isotope techniques for environmental monitoring at the Helmholtz Centre for Environmental Research in Leipzig. She was awarded a PhD in environmental microbiology from the Technical University in Freiberg in 2008, and completed postdoctoral training at the Catholic University in Leuven (KUL). In 2009 she joined the Research and Innovation Directorate-General of the European Commission. As a Research Programme Officer, she worked on both projects and policies in the area of Knowledge-Based Bioeconomy, with a major focus on environmental biotechnology. She has represented the EC in both the European Federation of Biotechnology and the EU–US Task Force on Biotechnology Research, and led the preparation of the EC publication *A Decade of EU-Funded GMO Research (2001–2010)*, which led to her interest in the European GMO research and regulatory framework.

GORDON CONWAY

Professor Gordon Conway FRS trained in agricultural ecology, attending the Universities of Bangor, Cambridge, West Indies (Trinidad) and California (Davis). In the 1960s he was a pioneer of sustainable agriculture developing an integrated pest management programme for the State of Sabah in Malaysia. He joined Imperial College in 1970, setting up the Centre for Environmental Technology in 1976. In the 1970s and 1980s he lived and worked extensively in Asia and the Middle East, for the Ford Foundation, the World Bank and USAID. He directed the Sustainable Agriculture Programme at the International Institute for Environment and Development (IIED) and then became representative of the Ford Foundation in New Delhi. Subsequently he became Vice-Chancellor of the University of Sussex and Chair of the Institute of Development Studies (IDS). From 1998–2004 he was President of the Rockefeller Foundation and from 2004 to 2008 Chief Scientific Adviser to the Department for International Development and President of the Royal Geographical Society. He is a KCMG, Deputy Lieutenant of East Sussex, and an Honorary Fellow of the Royal Academy of Engineering. He holds five honorary degrees and fellowships. He is the author of *The Doubly Green Revolution: Food for All in the Twenty-First Century*, and currently Director of the advocacy group, Agriculture for Impact.

ADRIAN DUBOCK

Adrian Dubock has a PhD in vertebrate zoology from Reading University. After 2 years working for the UK Ministry of Agriculture Fisheries and Food he joined ICI, Plant Protection Division in 1977. Following company demergers and mergers in 2001 Adrian moved to Switzerland as head of Syngenta's global function responsible for Mergers, Acquisitions, Ventures and Intellectual Property Licensing. Adrian has lived in four and worked in more than 90 countries with a broad range of business development, strategic and operational responsibilities. As a sideline, for 15 years he also owned and worked a 52-hectare UK grass farm with 320 sheep. He retired from Syngenta after 30 years' service at the end of 2007. Since 2008 he has been a member of the Advisory Board of the Freiburg Institute of Advanced Studies, School of Life Sciences and School of Soft Matter Research, Albert-Ludwigs-Universität, Freiburg, Germany, one of the German universities' excellence cluster. In 2000 he negotiated with the inventors of Golden Rice, the first purposefully created biofortified crop, and has worked with them closely and others since in trying to bring their humanitarian not-for-profit vision to fruition. Currently he is Executive Secretary of the Golden Rice Humanitarian Board, and also Golden Rice Project Manager.

JIM M. DUNWELL

Professor Jim M. Dunwell graduated in botany from the University of Oxford, and worked for 16 years at the John Innes Institute (Norwich, UK) where he obtained a PhD in plant physiology. He then spent 10 years at ICI Seeds, later Zeneca Plant Sciences, where he was responsible for an international programme on transgenic crops. He moved to the University of Reading where he is Professor of Plant Biotechnology and has research interests in plant breeding, gene expression and protein evolution. He was a member of the Royal Society Working Group on food-crop production, and at present is a member of the Advisory Committee for Releases to the Environment, the committee that advises the UK government on the field testing of GM plants.

IOANNIS ECONOMIDIS

Dr Ioannis Economidis is a graduate of the Agricultural University of Athens and holds a PhD in biochemical genetics from the University of Texas at Austin. His research was concentrated on protein/nucleic acid interactions related to gene expression. He has worked as a research associate in the USA, in Belgium and in Greece. In 1987, he joined the Research Directorate-General of the European Commission where he developed research sectors in the interface of biotechnology and the environment such as the safety of GMOs, topics of environmental biotechnology and issues on molecular diversity. In 2005, he joined the team which developed the concept of the Knowledge-Based Bioeconomy and developed sectors such as environmental biotechnology, molecular diversity, metagenomics, GMO safety and emerging technologies serving biotechnology such as synthetic biology. He represented his service in the European Federation of Biotechnology, the Working Party on Biotechnology of the OECD and the EU–US Task Force on Biotechnology

Research. Dr Economidis is a Visiting Professor at the University of Athens and Crete on topics of bioeconomy, biosafety and bioethics. He is also a Member of the Advisory Boards of the Centre for Environmental Risk Assessment of the International Life Sciences Institute Research Foundation, of the International Symposia of Biosafety Research and of the Working Group on Synthetic Biology of the EU–US Task Force on Biotechnology Research.

CLAUDE FISCHLER

Claude Fischler is a "Directeur de recherche" at CNRS, the national research agency of France, and heads IIAC (Institut Interdisciplinaire d'Anthropologie du Contemporain, Interdisciplinary Institute for Contemporary Anthropology), a research unit of the Ecole des Hautes Etudes en Sciences Sociales in Paris. His main area of research has been a comparative, social science perspective on food and nutrition. His main current research is on commensality - eating together - its forms and functions, and its possible impact on public health. He served on the Scientific Committee and the Expert Committee on Human Nutrition of AFSSA, the French Agency for Food Safety and on its board of directors. He has been a member of the steering committee of the French National Program on Nutrition and Health and of the Executive Committee of the European Sociological Association. He serves on the the the Strategic Committee on Sustainable Agriculture and Development advising the French Minister of Food and Agriculture and on the Advisory Group on Risk Communication of the European Food Safety Authority.

GEORGE GASKELL

Professor George Gaskell BSc PhD, Professor of Social Psychology, is Pro-Director of the London School of Economics and Political Science. His research focuses on science, technology and society; in particular the issues of risk and trust; how values influence people's views about technological innovation, and the governance of science and technology. His research competencies include survey methodology and both quantitative analysis and qualitative inquiry. Since 1996 he has coordinated the series of Eurobarometer surveys on *Biotechnology and the Life Sciences* for the EC Directorate-General for Research and Innovation. He was principal investigator of *Life Sciences in European Society*, a European comparative study of biotechnology in the public sphere funded by the European Commission, and is now leading a project on sensitive technologies and European public ethics. He is a member of the Expert Group on Risk Communication of the European Food Standard Authority and chairs the International Advisory Group of the Centre for Genomics and Society in the Netherlands. He was Vice-Chair of the European Commission's Science and Society Advisory Committee and was also a member of the Science in Society Committee of the Royal Society.

IAN GRAHAM

Professor Ian Graham holds the Weston Chair of Biochemical Genetics and is the Director of the Centre for Novel Agricultural Products (CNAP) at the University of York. His research interests focus on the metabolic regulation of gene expression in higher plants and metabolic engineering of novel oils and other high-value chemicals. Current projects range from the development of novel oilcrops such as *Jatropha curcas* to the production of new varieties of *Artemisia annua* that deliver higher yields of the antimalarial compound artemisinin. Funding for Ian's research comes from a range of sources including industry, UK government, EU, UK and overseas charities.

JULIAN GRAY

Julian Gray is a double graduate of Imperial College London, with a BSc in cellular and molecular biology and a PhD in proteomics. Julian also holds an MBA from the London Business School and The Wharton School, University of Pennsylvania (Exchange Program). At the time of writing this chapter, he was employed as an external consultant to The Cambridge Malaysian Education and Development Trust, c/o Trinity College, Cambridge, UK. Julian currently works as a consultant at The Boston Consulting Group with a special interest in the social impact, healthcare and public sector practices.

JONATHAN GRESSEL

Professor Jonathan Gressel is a plant scientist devoted to issues of world food security and how it can only be obtained by increasing crop productivity and the introduction of new crops, using all available tools, especially genetic engineering. He has performed extensive research on issues of transgene flow between crops and weeds, as well as developing crops resistant to weeds, especially the parasitic weeds that plague Africa. He has authored or co-authored over 300 scientific publications including eight books. His most recent book is *Genetic Glass Ceilings: Transgenics for*

Crop Biodiversity. He is presently a Professor Emeritus at the Weizmann Institute of Science in Rehovot, Israel, and recently spent 3 years as chief scientist at TransAlgae Ltd, a company he co-founded.

BRIAN HEAP

Professor Sir Brian Heap CBE FRS is President of the European Academies Science Advisory Council. He was Master of St Edmund's College, Cambridge, Vice-President and Foreign Secretary of the Royal Society, President of the Institute of Biology and President of the International Society for Science and Religion. He has doctorates from Nottingham and Cambridge with publications in physiology, biotechnology, sustainability and science advice for policy. He was Director of Research at the Institute of Animal Physiology and Genetics Research (Cambridge and Edinburgh) and at the Biotechnology and Biological Sciences Research Council (Swindon). He served as UK Representative at the European Science Foundation at Strasbourg and the NATO Science Committee at Brussels. With the Nuffield Council on Bioethics, the Department of Health's Expert Group on Cloning, the Parliamentary Select Committee on Science and Technology and the President's Advisory Group on Biotechnology he has been engaged in public issues of biotechnology, population growth, sustainability and science policy working with the World Health Organization, the UK–China Forum and the European Commission. He was scientific consultant for several international pharmaceutical companies and is Special Scientific Adviser for ZyGEM Co. Ltd, New Zealand. He was knighted in 2001 for services to science internationally.

T. J. V. HIGGINS

T. J. V. Higgins is an Honorary Fellow at CSIRO Plant Industry. He works on protecting food legumes from insect damage, researching the application of gene technology for plant improvement. His current research is focused on international agriculture with particular emphasis on Africa and India and he has a special interest in public awareness of science.

JENS HÖGEL

Jens Högel graduated in 1992 from the Technical University of Berlin as engineer for fermentation technology and expert for quality management systems based on the Hazard Analysis Critical Control Points (HACCP) concept in the food and beverage industry. His professional career is marked by more than 10 years of experience in international sales and product management in the food and beverage and pharmaceutical industry. After joining the European Commission in 2002 as Food and Veterinary Inspector, he specialised further in the areas of food safety and genetically modified organisms. In this position he reviewed the governance of food safety systems in EU Member States and countries outside the European Union, in particular in the North and South Americas and Asia. After working in the Directorate-General for Environment as Policy Officer in the area of biotechnology he joined the Directorate-General for Research and Innovation in July 2008, where he is working as Research Programme Officer in the area of Knowledge-Based Bioeconomy. Apart from project management, he remains heavily involved in policy making in relation to genetically modified organisms in the European Union. He is currently also involved in the preparation of Horizon2020 and related issues, and in the preparation and management of the annual work programmes in the area of biotechnologies under the Seventh Framework Programme for Research.

RICHARD C. JENNINGS

Dr Richard C. Jennings is an Affiliated Research Scholar in the Department of History and Philosophy of Science in the University of Cambridge. His research interests are focused on the Responsible Conduct of Research, and the ethical uses of science and technology. He is a member of BCS, the Chartered Institute for IT, has worked with the BCS Ethics Forum defining and refining the BCS Code of Conduct, and with four other members has developed a Framework For Assessing Ethical Issues in New Technologies. He teaches philosophy of science to undergraduate natural science students and philosophy to undergraduate philosophy students. He lectures on ethics in science and runs graduate workshops on ethical conduct and ethical practice in science. He has a long-standing interest in the history of science and the history of philosophy of science. He is a member of Queens' College, Cambridge. He co-edited with David Bennett *Successful Science Communication: Telling It Like It Is* published in 2011 by Cambridge University Press.

DREW L. KERSHEN

Professor Drew L. Kershen is the Earl Sneed Centennial Professor of Law at the University of Oklahoma, College of Law. Professor Kershen has been teaching at the University of Oklahoma since 1971. His academic work has been primarily in the areas of agricultural law and water law. His degrees are: BA Notre Dame (1966), JD Texas (1968) and LLM Harvard (1975).

Since 1998, Professor Kershen has focused his teaching, writing and speaking on agricultural biotechnology law and policy issues. He has written extensively on legal liability, intellectual property and regulatory issues in agricultural biotechnology. In addition to many presentations in the USA, he has been a speaker on agricultural biotechnology to conferences in 11 other nations. He is also a member of the Public Research Regulation Initiative, a public service organisation, through which he has participated in international negotiations concerning agricultural biotechnology.

CHRISTOPHER J. LEAVER

Professor Christopher J. Leaver CBE FRS FRSE was Sibthorpian Professor (1990–2007) and Head of the Department of Plant Science (1991–2007) at the University of Oxford. He is now Emeritus Professor of Plant Science and a Fellow of St John's College. His main research interests were the molecular and biochemical basis of plant growth and differentiation. This included an investigation of energy metabolism and the regulation of mitochondrial biogenesis during plant growth and differentiation, coupled with an investigation of the molecular and biochemical basis, and phenotypic consequences, of mitochondrial genetic diversity in plants in general and cytoplasmic male sterility in important crop plants in particular. He also worked on the metabolic and genetic regulation of genes encoding glyoxylate cycle enzymes and factors regulating the biosynthesis of glutathione and its function during oxidative stress. In recent years he has made important contributions to our understanding of senescence and programmed cell death in plants.

Professor Leaver has a strong interest in the public understanding of science and has been actively involved in the debate on genetically modified crops both nationally and internationally. He was elected a member of the European Molecular Biology Organisation in 1982, Fellow of the Royal Society in 1986, Royal Society of Edinburgh 1987, Academiae Europaeae 1989. In 2000, he was awarded the CBE for services to plant sciences, elected a corresponding member of the American Society of Plant Biologists (2003) and Inaugural Fellow of the American Society of Plant Biologists (2007) and has served on the councils of the Agricultural and Food Research Council and the Biotechnology and Biological Sciences Research Council, and was Chair of the UK Biochemical Society, the Council of the European Molecular Biology

Organisation and a Trustee of the Natural History Museum, London. He has been a member of the Governing Body of the John Innes Centre since 1984 and a Trustee of the John Innes Foundation since 1987. He is a founding Trustee of Sense about Science and visiting Professor at the University of Western Australia and serves on a number of international advisory boards and committees.

LU BAO-RONG

Professor Lu Bao-Rong is the Chair of the Department of Ecology and Evolutionary Biology in Fudan University, Shanghai, PR China. He received a PhD from the Swedish University of Agricultural Sciences in 1993. He was awarded as a Distinguished Young Scholar in 2001 by the Natural Science Foundation of China (NSFC). He is actively involved in research and teaching of botany, evolutionary biology and biodiversity conservation. He has published more 260 articles which have been cited more than 3500 times. Currently, he focuses on research of the molecular evolution and environmental biosafety of GMOs, particularly transgene flow and its ecological impacts. He is the President of the International Society for Biosafety Research (ISBR), and a Member of the Chinese National Biosafety Committee. He sits on the editorial board of a number of national and international journals.

DIRAN MAKINDE

Dr Diran Makinde is Director of the NEPAD Planning and Coordinating Agency, African Biosafety Network of Expertise (ABNE) based in Ouagadougou, Burkina Faso. He is the immediate past Director of the NEPAD West African Biosciences Network in Dakar, Senegal. He earned the degrees of Doctor of Veterinary Medicine in 1976, and a PhD in veterinary physiology from the University of Ibadan in 1986. Prior to his current appointment, he was Professor of Animal Science at the University of Venda, South Africa where he also served a 5-year term as Dean of the School of Agriculture, Rural Development and Forestry (1997–2001). In addition, he taught at the universities of Ibadan (Nigeria), 1977–91 and Zimbabwe, 1989–95. His research interest was in the field of gastrointestinal physiology of monogastrics, which includes such areas as digestibility and intestinal transport. As a result he developed and applied this in risk-assessment studies and became involved in advocacy for agricultural biotechnology and biosafety. He has well over 45 publications in peer-reviewed journals locally and internationally, as well as several contributions in the form of books/chapters in books. He is a C-rated scientist as evaluated (1998) by the South Africa National Research Foundation.

CAREL DU MARCHIE SARVAAS

Carel du Marchie Sarvaas is Director for Agricultural Biotechnology at EuropaBio, the European Association for Bioindustries. Carel has worked as a public affairs and communications advisor in Brussels, The Hague and Washington DC over some two decades. He leads a team of six specialists at EuropaBio and coordinates biotechnology outreach programmes across Europe. In agriculture, EuropaBio represents all main producers of biotechnology seeds as well as developers of new breeding techniques.

NATHALIE MOLL

Nathalie Moll is Secretary-General of EuropaBio, the European Association for Bioindustries. Nathalie is an honours graduate in biochemistry and biotechnology from St Andrews University, UK and has spent the past 16 years working for the biotechnology industry at European Union and national level focusing on improving awareness of the importance and benefits of biotechnology for society.

LARRY MURDOCK

Larry Murdock is Professor of Entomology at Purdue University, Indiana. His interests range from basic research in insect physiology, biochemistry and molecular biology to biotechnology and simple technology. He hopes his work benefits food production and availability in developing nations, especially Africa.

MARTIN PORTER

Martin Porter is Managing Director of Edelman | The Centre and was a co-founder of The Centre, Brussels' first think–do tank, before it joined the Edelman network in March 2010. Since April 2012, he has also taken on the role of overseeing Edelman's European Medicines Agency (EMEA) Public Affairs practice. He is closely involved and highly active in consulting activities, where he specialises in strategic advice and high-level public engagement support, especially in areas related to the low carbon economy, sustainability and risk-management issues.

His track record during his 15 years of European Union public affairs activity in Brussels includes advocacy and communications campaign design and support to companies, associations, coalitions, NGOs and governments involved in the above areas to created shared-value outcomes that deliver public goods as well as private enterprise benefits.

With a PhD in political science from the University of Bath and a first-class joint honours degree in modern languages and European studies from the same university, he is also a Visiting Fellow of the School of Public Policy at University College London, a Board Member of its European Institute and a Senior Associate of the Cambridge University Programme for Sustainability Leadership.

WAYNE POWELL

Professor Wayne Powell trained in plant breeding and cytogenetics at the University College Wales, Aberystwyth (H. Rees, FRS) and in quantitative genetics (Birmingham, School of Biometrical Genetics, J.L. Jinks FRS). His earlier work laid the foundations for the use of doubled haploids and cross prediction in plant breeding. He was awarded the Broekhuizen Prize in 1990 for outstanding contribution to European cereal science and obtained his DSc (plant genetic manipulation) in 1993 (University of Birmingham). His personal research interests are at the interface of plant genetics, genome science, plant breeding and conservation of genetic resources with a strong emphasis on the delivery of 'public good' outcomes. He has published over 250 refereed scientific journal papers which have attracted 9300 citations, with an H index of 50, and a further 120 book chapters and conference proceedings, and he has presented numerous invited papers at international meetings and successfully supervised 20 PhD students and numerous visiting workers. His professional career has spanned academia, the public sector and commerce (Dupont). He is currently Professor and Director of the Institute of Biological, Environmental and Rural Sciences (IBERS) at Aberystwyth University. This was formed in 2008 by merging the Institutes of Biological and Rural Sciences of the university with the Institute of Grassland and Environmental Research (IGER) to form one of the largest groups of academics and support staff in the UK focusing on plant and land-based sciences, with more than 350 staff, 1200 undergraduates and a complex range of stakeholders. In his current role he has led and managed a significant organisational and cultural change programme to create a fully integrated institute that embraces mission diversity: education, research and enterprise. This has involved attracting more than £80 million of investment to support new faculty appointments, capital infrastructure and national capability to underpin world-class research.

TIM RADFORD

Tim Radford is a freelance journalist. He was born in New Zealand in 1940. He joined the *New Zealand Herald* as a reporter at the age of 16, and moved to the UK in 1961. Apart from a brief spell as a Whitehall information officer, he has spent all his life in weekly, evening or daily newspapers. He worked for the *Guardian* for 32 years, becoming – among other things – letters editor, arts editor, literary editor and science editor. He won the Association of British Science Writers award for Science Writer of the Year four times, and a lifetime achievement award in 2005. He has served on the UK committee for the UN International Decade for Natural Disaster Reduction; on the council of Copus, the Royal Society's committee for the public understanding of science; and on the external relations committee of the Geological Society of London. He is an Honorary Fellow of the British Science Association and a Fellow of the Royal Geographical Society. He has lectured on, or taken part in debates about, science and the media in many cities in Europe, the USA, China and Russia. He has also written for *Nature*, *The Lancet*, *New Scientist*, the *London Review of Books*, *Geographical* and many other journals. He has written two books – *The Crisis of Life on Earth* (1990) and *The Address Book: Our Place in the Scheme of Things* (2011) – and edited two books of science writing for the *Guardian*.

CHAVALI KAMESWARA RAO

Professor Chavali Kameswara Rao, born 1937, received his BSc (Hons), MSc and PhD in botany from the Andhra University, Waltair and his DSc *(hon. Cos.)* from the Open International University for Complementary Medicines, Colombo. His university teaching and research career spanned from 1965 to 2003. He was Professor and Chairman, Department of Botany and Department of Sericulture, Bangalore University, Bangalore, India. On a Commonwealth Academic Staff Fellowship and a Royal Society and Nuffield Foundation Bursary, he worked at the Natural History Museum, London, the Royal Botanic Gardens, Kew, and Royal Botanic Gardens, Edinburgh on computer applications in plant systematics. He was awarded a Certificate of Merit by the Lama

Gangchen Peace Foundation (affiliated to the UN), Beijing, for services rendered to alternative and complementary systems of medicine. He is an author of four books on flora and medicinal plants and co-author of three international reports on biosecurity and modern biotechnology. He served on several international committees and special working groups of the US National Academies of Science, US National Institutes of Health, World Health Organization, etc., and several national policy committees of the Department of Biotechnology, Ministry of Environment and Forests, and Department of Science and Technology, Government of India. He is a Member of the European Federation of Biotechnology, and European Association of Pharma Biotechnology, Delft, the Netherlands and is on the Expert Panel on Agricultural Biotechnology, Council for Biotechnology Information, Washington DC. For the past 12 years he has been the Executive Secretary, Foundation for Biotechnology Awareness and Education (FBAE), Bangalore (www.fbae.org). For over a decade he has been regularly writing critical articles on various issues of bio-security and agricultural biotechnology in developing countries, posted at www.plantbiotechnology.org.in, www.fbae.org and AgBioView. Some of these articles have been translated into other languages such as German, French, Belorussian, etc. He was an invited speaker at inter-national conferences/workshops/seminars on biotechnology and bio-security, held at Washington DC, Bethesda, Cornell University, Princeton University, Brussels, Basel (Syngenta Foundation, Basel University), Geneva, Brighton, Berlin, Budapest, Vienna, Cuernavaca (Mexico), Beijing, Tripoli, Bangkok, Des Moines (US) Cambridge (UK), and Den Hague, besides at numerous places in India.

PAMELA RONALD

Pamela Ronald is Professor of the Department of Plant Pathology and the Genome Center at the University of California, Davis. Ronald's laboratory has engineered rice for resistance to disease and tolerance to flooding which seriously threaten rice crops in Asia and Africa. In 1996, she established the Genetic Resources Recognition Fund, a mechanism to recognise intellectual property contributions from less developed countries. She and her colleagues were recipients of the US Department of Agriculture 2008 National Research Initiative Discov-ery Award for their work on rice submergence tolerance. Ronald was awarded a Guggenheim Fellowship, the Fulbright–Tocqueville Distin-guished Chair and the National Association of Science Writers Science in Society Journalism Award. She is an elected Fellow of the American

Association for the Advancement of Science. Ronald has written opinion pieces for the *Boston Globe*, *The Economist* and the *New York Times* and is a blogger for *National Geographic*'s ScienceBlogs. She is co-author with her husband, Raoul Adamchak, an organic farmer, of *Tomorrow's Table: Organic Farming, Genetics and the Future of Food*. Bill Gates calls the book 'a fantastic piece of work'. In 2011, Ronald was selected as one of the 100 most creative people in business by *Fast Company Magazine*.

PIET SCHENKELAARS

Piet Schenkelaars finished his studies in molecular sciences and philosophy of science in 1984 at the Agricultural University Wageningen in The Netherlands. Thereafter he worked for many years for civil society organisations in The Netherlands. From 1990 to 1993 he worked as coordinator of the Clearinghouse on Biotechnology at Friends of the Earth Europe in Brussels. Subsequently, he was employed by an environmental consultancy and a consultancy in societal communication both located in The Netherlands. In 1998 he founded Schenkelaars Biotechnology Consultancy. Since then he has conducted many studies on regulatory and sustainability aspects of biotechnology. He has also coordinated European Union research projects, organised conferences and moderated stakeholder dialogues.

IDAH SITHOLE-NIANG

Idah Sithole-Niang is an Associate Professor in the Department of Biochemistry, University of Zimbabwe. She has worked on cowpea crop improvement for two decades, and maintains a keen interest in agricultural biotechnology and biosafety issues in sub-Saharan Africa.

SALLY STARES

Dr Sally Stares is a Research Fellow in the Methodology Institute at the London School of Economics and Political Science. Her research is based around the methodological theme of social measurement, with a focus on the measurement of social psychological concepts in cross-national social surveys, and an emphasis on the use of latent variable models to address common challenges encountered in this field of work, such as capturing complex constructs, taking account of 'don't know' responses and exploring the comparability of measures between

groups – particularly between countries. She works on public perceptions of various aspects of science and technology, as well as on public
opinion and social attitudes and values more broadly, particularly as
they relate to aspects of civil society. She works on a European Commission project 'Sensitive technologies and European public ethics', and on
an ESRC-funded project 'Latent variable models for categorical data:
tools of analysis for cross-national surveys'.

EDUARDO J. TRIGO

Dr Eduardo J. Trigo has a PhD in agricultural economics from the
University of Wisconsin. He is Director of Grupo CEO (www.grupoceo.
com.ar), a consulting and research group specialising in agricultural
development policy and management issues in Buenos Aires, Argentina.
He also sits on the Academic Council of the Alberto Soriano Graduate
School of the Faculty of Agriculture of the University of Buenos Aires
and is Scientific Adviser for International Relations at the Ministry of
Science, Technology and Innovation of Argentina. In the past he has
served as a member of the Global Authors Team within the Global
Conference on Agricultural Research for Development (GCARD 2010),
Director of Science and Technology at the Interamerican Institute
for Cooperation on Agriculture (IICA) and Director of Research at the
International Service for National Agricultural Research (ISNAR). He has
extensive academic and consulting experience in the areas of science
and technology policy, organisation and management with emphasis on
agricultural research and biotechnology applications to the agricultural
and food sectors. He has worked with a number of national governments in Latin America and other parts of the world and different
international organisations, and published extensively.

PIERO VENTURI

Dr Piero Venturi has a degree in agricultural science and a PhD in
agricultural engineering from the University of Bologna, Italy. He
worked for several years as a researcher at the Polytechnic University
of Madrid, at Wageningen Agricultural University and at the University
of Bologna, dealing with energy crops production and non-food-chains
modelling. He was lecturer in agricultural mechanisation and energy
integrated systems at the University of Bologna, and worked on the
development of non-food-chains for several uses. Presently he works at
the Directorate-General for Research and Innovation of the European

Commission, as Policy Officer dealing with biotechnologies and international cooperation.

KATY WILSON

Katy Wilson has a background in biological sciences and an MSc in environmental technology from Imperial College, London. She has worked in research in New York and at Harnas Wildlife Foundation in Namibia. Since 2010 she has worked for Agriculture for Impact, a Bill and Melinda Gates Foundation-funded initiative, in a research and editorial role and has contributed to Gordon Conway's book *One Billion Hungry: Can We Feed the World?*

Part 1   The issues of plant science and food security

Introduction

DAVID BAULCOMBE

The underpinning fundamental challenge of food security is to match supply and demand. In a simple world this challenge would be met by straightforward technology that balances these two market forces. However, we do not live in a simple world. Innovations to save labour in production, for example, would be damaging if they eliminate the only source of personal income for farmworkers. In other settings the same innovation could release time for people to participate in education or business activities and lead to growth of the local economy. Any new technology should, therefore, be appropriate to the environment and society in which it is to be applied.

However, even setting aside the context, the balance of supply with demand is a complex topic. The crop technology which is fundamental to all aspects of food security needs to address more than simple accumulation of calorie reserves for consumption by animals and people – although yield is important. It needs to address sustainability through reduced greenhouse gas emissions and carbon sequestration. Soil erosion, aquifer depletion and impact on other ecosystem services including biodiversity should all be minimised and there should be a net benefit of the new technology on farm income through a combination of direct and indirect effects.

Successful Agricultural Innovation in Emerging Economies: New Genetic Technologies for Global Food Production, eds. David J. Bennett and Richard C. Jennings. Published by Cambridge University Press. © Cambridge University Press 2013.

The following chapters set out the background to the food security challenge and they describe appropriate and diverse technologies based on progress in plant science. These technologies aid sustainable production, improve the nutritional quality of the product and they help reduce the level of waste that would otherwise occur through pre- and post-harvest deterioration. The writers of these chapters are all proponents of biotechnology at least to some extent but, reflecting the complexity of the challenge, the writers do not advocate a GM single solution. They describe a range of solutions in which biotechnologies of various types are an important component.

Conway and Wilson categorise crop technologies as being 'traditional', 'intermediate', 'conventional' and 'new platform' and they point out that innovation at all levels has a place in the global effort to achieve food security. They refer, for example, to a traditional *Zai* system in Burkino Faso and adjacent countries. The method involves crops being planted in manure-filled pits in which termites make porous tunnels that store water. Similarly I describe an intermediate companion cropping approach in which crops are fertilised and protected from insect and parasitic weeds by legumes and forage grasses that are cultivated either between or around the main crop.

An illustration of innovation in the 'conventional' category is the development of New Rices for Africa (NERICAs) through tissue-culture-assisted hybridisation of a traditional African species with Asian rice (Conway and Wilson) and by a new generation of agrochemicals that enhance endogenous plant defence pathways rather than having components of the pathogen as their direct target (Baulcombe).

Many of the innovations in the 'new platform' category are dependent on DNA sequence data. These advances are in a continuing state of flux because there is a continuing revolution in DNA sequencing technology. As a result it is now easy to link genomic DNA sequences with traits (Graham, Baulcombe). This capacity is new because, until recently, a state-of-the-art research laboratory could identify the DNA sequence affecting defined traits only in model plant species rather than crops. It would take years of work for each trait. Now, as a result of these new technologies, gene identification is relatively routine in crops as well as model species.

The consequence of this new capacity is more precise breeding of improved major crops. Similarly, breeding minor or orphan crops including *Artemisia* and *Jatropha* (Graham), for example, can be accelerated and there is the opportunity to diversify global agriculture due to the application of this new technology. Powell and Barsby describe the principles of conventional plant breeding and how they vary depending on the system of propagation – vegetative, inbreeding or outbreeding – and how the approaches are influenced by genomics. A particularly exciting opportunity is derived from genome sequencing in which genome-wide panels of genetic markers are used to predict performance and estimated breeding value of the new variety. This new approach will allow more precise breeding of complex traits affected by multiple loci.

The new sequencing technology also facilitates molecular isolation of genes associated with traits so that they can be transferred between varieties by GM. It is likely, therefore, that many of the GM varieties to be developed in the near future will involve the transfer of plant genes to plants. In contrast, the first generation of GM crops were improved with bacterial or viral genes.

It has been suggested that the use of plant rather than alien genes in GM crops should be described as cis- rather than trans-genesis (Schouten et al., 2006). However, the non-plant genes are similar to plant genes in that they have the same nucleotide composition and use the same genetic code. There is, therefore, no rational reason why genes of plant or non-plant origin should be differentiated or subject to distinct regulation or risk assessment. In both instances there should be appropriate safeguards for the farmer, the consumer and the environment but they may not need to be as restrictive as those in the European Union (Dunwell). Brookes describes how the transfer of non-plant genes can have both direct and indirect benefits for the farmer and the environment. In this light the differentiation of cis- and trans-genesis should perhaps be discouraged because it implies wrongly that there is a hazard associated with a useful technology.

Biotechnology is often presented as being inevitably linked to multinational corporations and as being inconsistent with the interests of small farmers and less developed countries. However, these chapters illustrate the diversity of the opportunities from modern biotechnology. Such diversity may be inconsistent with the business priority of large companies in which the scale of operation requires focus on large targets that can be applied over large areas in many regions. Ironically it may be that biotechnology and GM is most useful when linked with traditional and intermediate technology. Once we have identified a framework to support this linkage we will be well on the way towards sufficiency of food supply in a sustainable manner over the peak of global population and as we feel the first major effects of anthropogenic climate change. We will have achieved a doubly Green Revolution (Conway and Wilson).

REFERENCE

Schouten, H. J., Krens, F. A., and Jacobsen, E. (2006). Do cisgenic plants warrant less stringent oversight? *Nature Biotechnology* **24**, 753.

DAVID BAULCOMBE

1

Reaping the benefits of plant science for food security

Global food security can be achieved by reducing demand for food and by increasing sustainable crop production. Both approaches are necessary. To increase sustainable production it will be necessary to harness recent developments in plant science for both genetic improvement of crops and their agronomy. The technological innovations will be most effective if they can be developed as integrated components of agricultural systems. This chapter presents four case histories to illustrate the potential of new developments in plant science. It illustrates how new technology can help improve existing crop production systems and, through grand challenge projects, produce radical innovations.

FOOD SECURITY AND CROP PRODUCTION

The recent upheaval of global economies illustrates how quickly the illusion of sufficiency can translate into a catastrophe. At the start of 2008 most financial commentators were optimistic although there were some indications that global economies were not sustainable. The general view was that 'fundamentals' were sound and that there could be growth in many stockmarkets during the year (Barber, 2008). Unfortunately the optimism was not justified and the subsequent market collapse and squeeze on credit will affect us all in the Western world for some time to come.

The economy in the recent past could well serve as a lesson for global food supplies over the next generation because, as with the economy before 2008, there is an illusion that the fundamental systems

Successful Agricultural Innovation in Emerging Economies: New Genetic Technologies for Global Food Production, eds. David J. Bennett and Richard C. Jennings. Published by Cambridge University Press. © Cambridge University Press 2013.

for production are sound. However, as with the economy in 2008, there are indications from various studies and reports that warn of insufficient capacity to meet demand for food over a 30–50-year period (IAASTD, 2009; Royal Society, 2009; Foresight, 2011).

The proposed solutions to food security place emphasis to differing degrees on production or demand. Those addressing demand involve control of population growth, distribution mechanisms and reduced consumption of grain-fed animals or on minimising the massive waste of food.

Of course crop production would be much less of a challenge than it seems now if these approaches to reduce demand could be successful. Unfortunately we cannot be confident that the global community will succeed if the emphasis is only on demand. If the Millennium Development Goals aiming to free people from extreme poverty and multiple deprivations are a guide then we should not be optimistic. The goal of halving the number of undernourished people between 1990 and 2015 is a long way behind schedule and the trend may even be in the wrong direction (United Nations, 2010).

Failure to meet demand for food would have catastrophic humanitarian and political consequences and it would be irresponsible to rely on any single solution. Population growth, elimination of waste and moderated consumption should all be addressed. However a prudent strategy requires that we look not only to solutions aimed at reducing demand but that we also attempt to increase supply via improvements in crop production.

SUSTAINABILITY AND YIELD

Current crop production is not always based on sustainable practice (Foley et al., 2011). Croplands cover 12% of the available land and they have a massive environmental impact. Natural resources are depleted, ecosystem services are degraded and there may be pollution of groundwater with pesticide and fertiliser residues or the atmosphere with nitrous oxide which is a potent greenhouse gas. Future climate change may also make it difficult to sustain high levels of crop production in regions where rainfall is reduced or crops are subject to high temperature stress at critical stages in their life history. The prudent strategies for improved crop production will, therefore, have to generate an increase in yield but using more sustainable production methods than those in present use. Unfortunately there are very few regions where additional land is available for cultivation without adverse environmental impacts

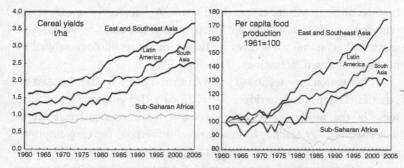

Figure 1.1 For the past four decades, cereal yields in sub-Saharan Africa have been stagnant and per capita food production has declined. The right-hand panel shows the percentage increase or decrease from 1961 which was assigned 100. (From Toenniessen *et al.*, 2008)

(Tilman *et al.*, 2011). Sustainable and productive agriculture needs to operate to a large extent on existing agricultural land.

The strategies will need to be tailored to different regions. This need is illustrated clearly by the variation in yield growth in different continents over the last 50 years (Toenniessen *et al.*, 2008) (Figure 1.1).

In regions with industrialised crop production the yields can be greater than 10 tonnes per hectare and the future focus will need to be on sustainability and environmental impact as much as yield. In the parts of Central and South America and much of Asia that benefited from the first Green Revolution there may be scope for further yield increase although sustainability is an important consideration. However, in sub-Saharan Africa, there has been no overall yield increase (Figure 1.1) and future strategies will need to focus on both yield and sustainability. Clearly an increase of only 1 or 2 tonnes per hectare provides a large proportional increase in African productivity and would add greatly to global supplies

CROP PRODUCTION AND THE ROLE OF PLANT SCIENCE

Improvements in crop production can be achieved in various ways including those that do not involve new technology. Yield increases in Malawi, for example, were achieved by farmer subsidy so that fertilisers and pesticides could be purchased (Denning *et al.*, 2009): no new technology was required. Agri-environment schemes have been used in the UK to encourage farmer practice that promotes environmental sustainability (Stevens and Bradbury, 2006) and again new technology was not required. However these and other examples are not evidence that technology advances are irrelevant. A balance of measures is required and, in connection with production, the balance will involve social and economic structures and appropriate technology as well as new science-based technology.

Discussion over technical innovation and crops is often focused on genetics (Tester and Langridge, 2010). In part this emphasis is because some of the most spectacular progress in basic science has been in molecular biology leading to new powerful methods for crop improvement through conventional breeding and genetic manipulation. However crop production can also be improved through innovation in the ways that crop plants are grown or the chemicals that are applied to them (Royal Society, 2009). These agronomic advances have an advantage over genetic improvements in that they can be applied to existing varieties of crop and, once developed, applied much more rapidly than genetic improvements that normally take many years.

Plant science is the key to improvements in crop genetics and agronomy and, as with many other areas of biology, it is in the throes of a revolution. The emerging methods in plant science differ from the traditional approaches in that they involve a much larger component of computing and the use of very large datasets. Imaging of cellular structure, for example, is no longer with a simple microscope but it may be linked to confocal or two-photon systems enabling much deeper tissue penetration and computational analysis of the data. Combined with immunodetection of different proteins it is now possible to monitor the changes to the subcellular structures that are well below the limits of detection of normal light microscopy.

Chemical analysis of plant extracts is similarly more sophisticated than in the last century. Complex extracts can be characterised using mass spectrometry so that previously uncharacterised proteins or small-molecule components of cells can be monitored during cellular transitions during development or associated with responses to external stimuli (Kopka et al., 2005). Computing again features prominently in these chemical analyses.

New methods for sequencing of DNA or RNA also illustrate the increased power of new technology (Lister et al., 2009). We can generate sequence data for an entire genome, the organism's genetic information, or transcriptome, all the different types of RNA molecules in its cells, for relatively low cost and very quickly. The challenge therefore is in the computational analysis rather than the generation of the data. This 'next-generation' sequencing is useful for large projects involving the characterisation of new species' genomes. It is also useful for more specific projects in which a mutant or genetic variant is identified or in which differentially expressed genes are characterised (Lister et al., 2009).

The following sections illustrate how recent progress in plant science either has or will generate technologies for improved crop production. In selecting these examples the aim is to illustrate how new science could enhance many different approaches to crop production and, in particular, how new science links not only with biotechnology including genetic modification (GM): it also links with approaches that are classically associated with organic or other low-input approaches. There is no contradiction in these examples because the aim of biotechnology and low-input agriculture is the same: to achieve the highest possible yield of the crop with the lowest possible impact and the greatest sustainability.

CASE HISTORY I: 'PUSH–PULL' SYSTEMS AND COMPANION CROPPING

Many crops are damaged by insects either directly by feeding or because the insect is a vector for virus disease. Control of these pests in industrial agriculture is typically by application of systemic insecticides or resistant varieties of crops. However, the insecticides may target insects other than the pest or even the farmer and for that reason are subject to increasingly stringent control regulation. Resistant varieties of crop are not always available or may take a long time to develop.

An alternative strategy is a component of integrated pest management strategies in low-input agricultural systems and is based on the production of chemicals by plants that can affect the behaviour of insect pests. These chemicals are referred to as semiochemicals – semeion is signal in Greek – and they influence the mating behaviour or feeding of insects and can be either attractants or repellents. One of the best-known strategies to exploit semiochemicals is known as 'push–pull' and it is used, for example, in the control of stem-borer moths on maize in East Africa (Cook et al., 2007) (Figure 1.2). The term push–pull is used because the strategy uses a push plant grown between the maize that produces a volatile repellent and a trap crop (the pull) on the outside of the plots that produces a volatile attractant of the stem-borer moth. The approach is also referred to as companion cropping. Under field conditions in East Africa, push–pull has increased maize yields by 100% or more without additional inputs (Khan et al., 2008).

There are many advantages to push–pull as an alternative to the use of insecticides. The push plant used as the intercrop is a

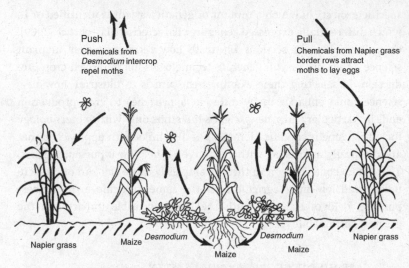

Push Chemicals from *Desmodium* intercrop repel moths

Pull Chemicals from Napier grass border rows attract moths to lay eggs

Napier grass   Maize   *Desmodium*   *Desmodium*   Napier grass

Maize   Maize

Figure 1.2 Push–pull in maize cultivation. The maize field is surrounded by a border of the forage grass *Pennisetum purpureum* (Napier grass). Napier grass is more attractive to the moths than maize for laying their eggs (the 'pull' aspect). The Napier grass produces a gum-like substance which kills the pest when the stem-borer larvae enter the stem. Napier grass thus helps to eliminate the stem-borer in addition to attracting it away from the maize. In addition, rows of maize are intercropped with rows of the forage legume silverleaf (*Desmodium uncinatum*). *Desmodium* releases semiochemicals which repel the stem-borer moths away from the maize (the 'push' aspect). *Desmodium* has the additional benefit of fixing atmospheric nitrogen, thereby contributing to crop nutrition. Remarkably, *Desmodium* has also been found to be toxic to *Striga* (witchweed), so has an additional crop protection benefit. (*Source*: The Gatsby Charitable Foundation, *The Quiet Revolution: Push–Pull Technology and the African Farmer*)

legume (*Desmodium* species) that fixes nitrogen and so fertilises the main maize crop. It produces a diffusible compound in the soil that benefits the maize crop. This compound suppresses the African witchweed that parasitises maize and causes major reductions in yield. Finally, the ground cover provided by *Desmodium* helps with soil and water conservation. The pull crop may also promote parasites of the stem-borer moth and be a forage grass for livestock.

An insect pest in industrial agriculture is often under strong selection pressure to develop resistance to an insecticide. In contrast, in push–pull the pest is not eliminated and the selection pressure is minimised. This example of integrated pest management is, therefore, more suited than pesticide use as a component of sustainable agriculture: it is likely to be durable.

There are probably three reasons to explain why, despite the evident benefits of push–pull, companion cropping has not been applied widely to intensive or industrial agriculture. The first is scientific understanding of how plants and their pests interact; the second involves engineering and farm machinery; and the third is related to economic and other incentives.

The economic considerations arise because, even when fully optimised, a companion cropping system would produce less than an intensively cultivated crop with fertiliser inputs. However, as oil and other fossil fuel costs increase, there is an increased incentive to use less energy in crop production. At some future point the lower yield of a companion cropping system will be offset by the reduced costs of crop husbandry and there will be an incentive to develop the biotechnology and agricultural engineering that would be compatible with companion cropping.

Biotechnology could focus on varieties of the main and companion crops that are adapted to companion cropping. The main crop might perform better, for example, if it has an architecture and growth dynamics that are compatible with the companion plants. It would be important, for example, to avoid shading and root competition between these plants. The companion crops could be selected similarly and also for optimal production of the semiochemicals. Case history III, described below, illustrates new methods that could be used to accelerate the identification of varieties that would be adapted to companion cropping.

Other companion cropping systems include agroforestry in which trees or shrubs are grown together with a main crop. Although such systems involve different plants and do not necessarily have the same benefits as the push–pull system, the principle is the same: exploiting the agricultural system rather than the individual components is an effective approach. The application of science to generate better understanding of agricultural systems could lead to improved mixed cropping systems that are consistent with sustainable practice and that can compete over an extended period with intensive monocultures and industrial agriculture.

CASE HISTORY II: PRIMING FOR DEFENCE

Crops are often protected against pests or pathogens by chemicals. Crop protection chemicals are extremely useful because they can be used on crop genotypes that are susceptible to pests and diseases if there is no genetic resistance. However there are disadvantages to the current generation of crop protection chemicals: their utility is restricted to invertebrates (for example insects or nematodes) or fungi rather than bacteria or viruses; they are expensive and not affordable to subsistence farmers who are likely to suffer crop disease; and there may be unintended effects that affect the environment or the farmer. There is also the problem of resistance: many pests and pathogens acquire resistance to crop protection chemicals.

However there could be an alternative chemical strategy in which the target mechanism is in the plant rather than the pest or pathogen. The aim is to harness the endogenous, inbuilt plant defence systems rather than kill the pest or pathogen directly. Until now the understanding of these host signalling mechanisms has been rudimentary and harnessing host defence is an aspiration rather than common practice.

One of the oldest observations to support the possible use of host defence involves an effect known as systemic acquired resistance (SAR) (Mur et al., 2008) in which a plant infected with one pathogen becomes resistant to infection with a second pathogen. In a classic experiment the inducing pathogen was tobacco mosaic virus (TMV) inoculated to tobacco. The virus triggered necrotic lesions, dead tissue, on the local inoculated leaves but it did not spread throught the plant because the plant carried a gene known as N that confers resistance against TMV. However, there was an effect that spread through the plant: systemic acquired resistance against fungi and bacteria was transmitted to the upper leaves of the inoculated plant and it persisted for weeks after the initial virus inoculation.

This simple experiment illustrated several key features of defence systems in plants. First, there must be a pathogen recognition system that initiates the process. Second, there must be a signalling mechanism that transmits the defence signal out of the inoculated leaves and third, there is a memory mechanism that allows the induced defence system to persist for some time.

Recent advances have shed some light on the various components of this system. The recognition systems, for example, involve extra- or intracellular receptors that perceive the pathogen either by direct interaction with a pathogen-derived elicitor or through a less direct method. The signalling systems involve both intra- and intercellular mechanisms in which jasmonate fatty acid derivatives and salicylic acid derivatives play a role (Figure 1.3). The memory mechanisms are less clear but there is some indication that DNA methylation and histone modification may play a role (Conrath et al., 2006). This molecular memory is persistent even through cell divisions and, as it does not involve changes to the DNA sequence, it is referred to as epigenetic rather than genetic.

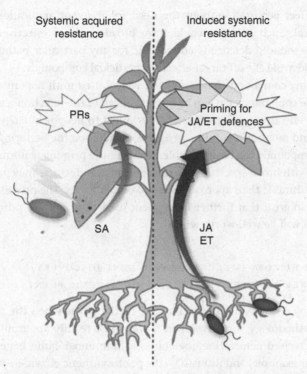

Figure 1.3 Systemic acquired resistance (SAR) is typically activated in healthy systemic tissues of locally infected plants. Upon pathogen infection, a mobile signal travels through the vascular system to activate defence responses in distal tissues. Salicylic acid (SA) is an essential signal molecule for the onset of SAR, as it is required for the activation of a large set of genes that encode pathogenesis-related proteins (PRs) with antimicrobial properties. Induced systemic resistance (ISR) is typically activated upon colonisation of plant roots by beneficial microorganisms. As in SAR, a long-distance signal travels through the vascular system to activate systemic immunity in above-ground plant parts. ISR is commonly regulated by jasmonic acid (JA)- and ethylene (ET)-dependent signalling pathways and is typically not associated with the direct activation of PR genes. Instead, ISR-expressing plants are primed for accelerated JA- and ET-dependent gene expression, which becomes evident only after pathogen attack. Both SAR and ISR are effective against a broad spectrum of virulent plant pathogens. (From Pieterse et al., 2009)

In principle the exploitation of these endogenous defence systems would involve application of a chemical that mimics either the eliciting or the endogenous signalling molecules. This chemical would then activate the defence signalling system and the memory mechanism would ensure that the effect persisted without the need for frequent application of the chemical. Such systems would have broad-spectrum effectiveness because the induced defence is not specific for any particular pathogen and it would avoid the off-target effects of pesticidal compounds.

Priming compounds have not been widely used until now in agriculture. The compounds have to be applied ahead of the infection and so their most effective application would be linked to new technology for detection and monitoring of disease agents. In addition the performance of these compounds has been variable. However the priming is normally associated with broad-spectrum disease resistance and its use may prove to be more durable than approaches based on pesticides. The promise of priming is so great that further investment to identify suitable priming compounds will be well worth while.

CASE HISTORY III: ACCELERATED GENE DISCOVERY, MARKER-ASSISTED SELECTION AND CISGENIC PLANTS

Gene cloning has revolutionised biology. In the 1970s when the gene cloning methodology was developed it was most readily applicable to well-characterised genes. Haemoglobin and immunoglobulin genes in animals, for example, and RuBisCO – the photosynthetic enzyme – were some of the first genes to be cloned. However, since the 1990s, there have been methods for gene cloning without prior knowledge of the protein product. These methods allowed cloning of genes that had been known previously from the wild-type or mutant phenotype. Included in this category were regulatory genes affecting flowering or other developmental transitions and disease-resistance genes.

Initially the cloning of genetic loci was a major task. In my laboratory in 1999, for example, it took 3 years to isolate a virus resistance gene from potato (Bendahmane et al., 1999). We had to track many DNA markers in large populations of plants so that the interval surrounding the gene of interest could be defined. Progress was limited by the number of available markers and the size of available mapping populations.

However genetic mapping is now greatly facilitated by next-generation DNA sequencing (Lister et al., 2009). A feasible strategy would involve the use of small populations of plants – fewer than 200 – with the gene of interest (Schneeberger et al., 2009). The approach would involve sequencing DNA of two samples. One sample would be bulked DNA from 100 plants with the wild-type phenotype, i.e. characteristics, and the second would be from 100 plants with the phenotype with the lost function. The sequences would be aligned to the genome sequence of the two parents and the gene of interest would be mapped to a region in which all of the sequenced DNA resembled the genome carrying the loss-of-function gene. The approach can be successful with such small populations because the resolution of the mapping is not limited by the marker density.

The ability to characterise the DNA corresponding to genes of interest has application in both conventional plant breeding and GM. In conventional breeding the benefit is not only related to speed or expense of breeding programmes for simple traits or characteristics. It additionally means that selection can be carried out more directly for quantitative traits that vary continuously rather than being present or absent and traits that are affected by multiple genes. Such a facility might be particularly useful with drought resistance, for example, for which it has been difficult until now to identify single major genes conferring drought tolerance.

The molecular characterisation of genes affecting agronomic traits will allow GM to be used more flexibly than until now. The first generation of GM crops mostly involved genes transferred from alien species or organisms (James, 2010). Herbicide or insect resistance, for example, is from bacterial genes, virus resistance from viral genes and Golden Rice with improved vitamin A content is from transfer of maize and bacterial genes. There are relatively few traits that can be improved by this foreign gene transfer. However, now that plant genes associated with useful traits can be isolated as DNA, it is possible to use GM to tackle all of the traits that were previously associated with conventional plant breeding. The new approach involves first the identification of a useful gene or set of genes in a crop or a crop relative. These genes can then be transferred into a crop plant using standard GM technology. The value of plant genes transferred as such a 'transgene' is illustrated by the development of a disease-resistant potato by transfer of a late blight-resistance gene from a wild potato relative (van der Vossen et al., 2003). The term 'cisgenic' has been proposed for use when GM is used to transfer genes between sexually compatible species (Jacobsen and Schouten, 2007). However I do not favour that term because there is no fundamental difference between GM of native or alien genes.

There are several advantages to the use of GM rather than crossing to transfer genes between plants (Figure 1.4). In conventional or 'introgression' breeding it is often necessary to carry out repeated backcrosses, that is, crossing a plant with its parent, in order to retrieve a variety with the desired additional trait. These backcrosses obviously take time and there may be 'linkage drag' if the target trait is linked

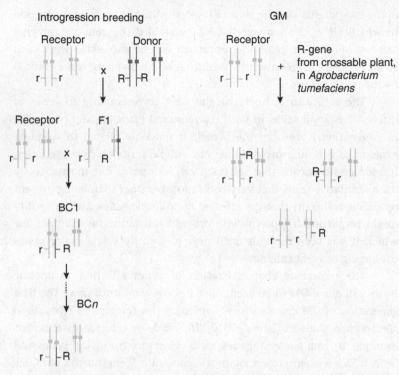

Figure 1.4 A comparison of introgression breeding and GM. Introgression breeding of a disease-resistance gene (for example) involves crossing a receptor variety of the crop with a donor of the R gene that could be a different variety or an alien species. The hybrid is then backcrossed repeatedly in each generation with selection for the presence of the R gene and the absence of other donor markers. The finished variety has the agronomic characteristics of the receptor except that it now carries the R gene and any other adjacent genes from the donor genome. In a GM approach a single manipulation introduces the R gene without any other flanking DNA. However the new R gene insert is positioned randomly in the recipient genome and several different GM lines have to be evaluated because the transgene would not be expressed correctly or stably from all positions. (From Jacobsen and Schouten, 2007)

to other genes that adversely affect the recipient variety. The conventional breeder has to assess the trade-off between the new trait and the cost of linkage drag but the biotechnologist can use GM to transfer the gene affecting the desired trait more quickly and without a penalty from linked genes.

Other advantages to the use of GM to transfer plant genes between plants are the ability to transfer one gene into several varieties at once and so help the diversity of varieties used in agriculture. In addition, now that genetic mapping can be applied to complex traits, GM can be used to introduce quantitative or multigenic traits into new varieties. As with simple traits, the qualities of the recipient variety are retained and several varieties can be improved simultaneously.

CASE HISTORY IV: HOMOLOGOUS GENE TARGETING AS A NOVEL APPROACH TO CROP IMPROVEMENT

Different varieties of a crop are descended from a common ancestor and many of their genetic differences are due to mutation. If there were no crop plant relatives with useful variation in a target trait it seemed likely that artificial mutation could be useful. A parental line would be exposed to a mutagen and mutants with the desired trait could be selected in the second or subsequent generations. The mutant line would then be backcrossed repeatedly to an elite, or already improved, variety and the progeny selected for the new mutant trait. However this process is as long-winded as introgression breeding and there is no guarantee that the desired mutants could be identified. Most random mutations are neutral in their effects or not useful for the breeder and, until now, a mutation breeding approach has not been used as often as introgression breeding (Ahloowalia et al., 2004). This situation may change following the recent developments in biotechnology for targeting changes to genomes.

The scientific discovery that underpins targeted modification of genomes involves a set of transcription factors that control whether a gene is switched on or off with DNA binding 'motifs' or sites. These proteins, known as 'TAL effectors', are remarkable because there is a simple relationship between the sequence of the protein and the sequence of the bound DNA. The natural versions of these proteins bind to motifs upstream of plant genes involved in disease and disease resistance but modified versions can be designed to bind to almost any region of the DNA (Bogdanove and Voytas, 2011).

Targeted DNA binding can be harnessed for genome modification if a DNA binding motif from one of these proteins is fused to a nuclease domain, a section of the DNA sensitive to an enzyme that cleaves the DNA (Figure 1.5). Expression of this fusion protein introduces double-stranded breaks into DNA at the targeted site and repair of the break by a process known as 'non-homologous end joining' may result in mutation at the targeted site. Such mutations may be useful in crop improvement if the targeted gene confers, for example, production of toxins or has a positive influence on growth and development of the crop.

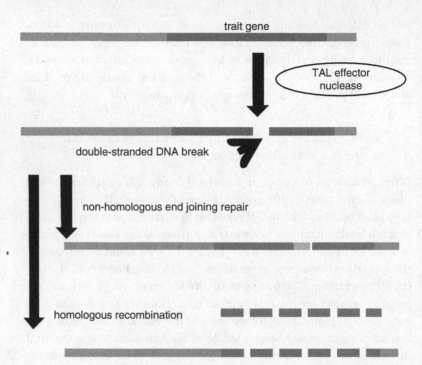

trait gene

TAL effector nuclease

double-stranded DNA break

non-homologous end joining repair

homologous recombination

Figure 1.5 Gene targeting in plants. A TAL effector nuclease is designed to target a trait gene. Expression of the nuclease in a crop then results in a double-stranded DNA break in the trait gene. Non-homologous end joining repair may introduce a mutation (hatched) into the trait gene that will often have a loss of function phenotype that would be useful if the trait is deleterious for the crop. Repair of the double-stranded break by homologous recombination results in replacement of the native trait gene with a similar DNA sequence with a beneficial phenotype (dashed line).

Double-stranded breaks in DNA can also be repaired by homologous recombination if there is DNA in the same cell with sequence identity to the regions that flank the site of the break. In natural situations this DNA will be from the second undamaged allele. However the recombination process will also use exogenous DNA from another source that could be modified to incorporate specific mutations with beneficial effects. Such changes could involve just a single base unit of the DNA that would, for example, influence the pathogen specificity of a disease-resistance gene (Farnham and Baulcombe, 2006). Alternatively the modification could include domains or even whole genes that could be inserted at defined sites in the genome.

Targeted genetic modifications based on TAL effectors can be developed as part of a transgenic strategy when part of another species' genome is transferred. However a more flexible approach would involve transient introduction of the engineered nuclease gene and the recombination-target DNA either as naked DNA or integrated into viral genomes. If plants can be regenerated from cells that have been exposed to these two types of DNA or if pollen or eggs can be formed from those cells then the targeted modification can be propagated as part of the genetic material of a new variety (Bogdanove and Voytas, 2011).

GRAND CHALLENGES FOR THE FUTURE

These four case histories are representative of technological development due to recent developments in biological science. Together with other examples of science-based innovation they illustrate how new technologies could contribute to sustainable crop production with adequate yield to meet demand. Ideally these innovations would be considered as components of integrated agricultural systems and, as with push–pull, there will be as much effort on design of the system as on the individual components. However it is important to emphasise that no single innovation or approach will be sufficient: success in developing sustainable crop production methods relies on exploitation of all available technologies.

In the examples given above the advances would be refinements, to varying degrees, of existing crop production technology. Agriculture with these innovations would be productive and sustainable but it would be fundamentally similar to present practice. However there are some grand challenges in plant science that could be linked to more radical innovations that would change agriculture beyond recognition.

One of these 'grand challenge' innovations would lead to making cereals and other crops that are currently annuals into perennials (Glover et al., 2010). At the end of the growing season the aerial part of the plant would be cut or allowed to die back but the root systems would be undisturbed. The perennial crop would not have to invest photosynthate and nitrogen each year in developing a new root system and the undisturbed soil would store water and carbon and be resistant to erosion. Weeds would be outcompeted by the standing crop and the persistent roots would be colonised by mycorrhizal fungi that provide mineral nutrients for the plant in exchange for photosynthate.

Perennialisation of annual crops is not readily achievable by conventional breeding but with gene discovery methods described in Case history III it will be possible to transfer perennialisation genes from wild crop relatives by accelerated breeding or by GM. As a grand challenge I estimate that, given the acceleration in gene discovery methods, it could be achieved within 10 years. Like the use of companion cropping, the introduction of perennial grains and other crops would drive a radical change to dominant agricultural systems in developed and developing country agriculture.

At the other end of the grand challenge spectrum is the target of improving the efficiency of photosynthesis. In many major crops, including rice and wheat, the type of photosynthesis involves the 3-carbon pathway which is inefficient because the photosynthetic enzyme ribulose bisphosphate carboxylase (RuBisCO) catalyses a non-productive and energy-consuming pathway of photorespiration, the oxidation of carbohydates (Langdale, 2011). Some plants, including maize, suppress photorespiration using a metabolic shunt, an alternative pathway – the 4-carbon pathway – to increase the level of carbon dioxide at the RuBisCO active site. Alternatively there are algae that use an aggregate of RuBisCO known as the pyrenoid to achieve the same end.

The hunt is now on to find the master regulators of 4-carbon pathways (Hibberd et al., 2008) and the determinants of pyrenoid formation (Ma et al., 2011) so that they can be transferred to 3-carbon plants. There is good progress with both approaches and there is also evidence that an artificial metabolic shunt could also be successful (Kebeish et al., 2007). Notwithstanding this progress there could be 10 years of basic research before new photosynthetic pathways could be designed and transferred to crops. However the yield increase could be as much as 50% if this grand challenge could be met and it is surely worth the effort.

Other grand challenges might involve crops other than wheat, rice and maize which collectively provide more than 75% of global calorie consumption (Naylor et al., 2004). These orphan crops have not benefited from the spectacular enhancements that have caused major crop yields to increase several-fold since the start of the last century. The lack of effort on these species may explain at least in part why the first Green Revolution failed to have a significant impact on African agriculture. It also leaves a tremendous opportunity for the application of basic science.

REFERENCES (KEY REFERENCES INDICATED BY *)

Ahloowalia, B. S., Maluszynski, M., and Nichterlein, K. (2004). Global impact of mutation-derived varieties. *Euphytica* 135, 187–204.
Barber, T. (2008). Europe cites US policies as reason for turbulence. *Financial Times*, 23 January.

Bendahmane, A., Kanyuka, K., and Baulcombe, D. C. (1999). The *Rx* gene from potato controls separate virus resistance and cell death responses. *Plant Cell* 11, 781–791.

Bogdanove, A. J., and Voytas, D. F. (2011). TAL effectors: customizable proteins for DNA targeting. *Science* 333, 1843–1846.

Conrath, U., Beckers, G. J. M., Flors, V., et al. (2006). Priming: getting ready for battle. *Molecular Plant–Microbe Interactions* 19, 1062–1071.

Cook, S. M., Khan, Z. R., and Pickett, J. A. (2007). The use of push–pull strategies in integrated pest management. *Annual Review of Entomology* 52, 375–400.

Denning, G., Kabambe, P., Sanchez, P., et al. (2009). Input subsidies to improve smallholder maize productivity in Malawi: toward an African green revolution. *PLoS Biology* 7, e1000023.

Farnham, G., and Baulcombe, D. C. (2006). Artificial evolution extends the spectrum of viruses that are targeted by a disease-resistance gene from potato. *Proceedings of the National Academy of Sciences USA* 103, 18828–18833.

*Foley, J. A., Ramankutty, N., Brauman, K. A., et al. (2011). Solutions for a cultivated planet. *Nature* 478, 337–342.

*Foresight (2011). *The Future of Food and Farming Final Project Report*. The Government Office for Science, London. www.bis.gov.uk/assets/foresight/docs/food-and-farming/11-546-future-of-food-and-farming-report.

*Glover, J. D., Reganold, J. P., Bell, L. W., et al. (2010). Increased food and ecosystem security via perennial grains. *Science* 328, 1638–1639.

Hibberd, J. M., Sheehy, J. E., and Langdale, J. A. (2008). Using C-4 photosynthesis to increase the yield of rice: rationale and feasibility. *Current Opinion in Plant Biology* 11, 228–231.

IAASTD (2009). *Agriculture at a Crossroads: Global Report*. www.agassessment.org/reports/IAASTD/EN/Agriculture

*Jacobsen, E., and Schouten, H. J. (2007). Cisgenesis strongly improves introgression breeding and induced translocation breeding of plants. *Trends in Biotechnology* 25, 219–223.

James, C. (2010). A global overview of biotech (GM) crops: adoption, impact and future prospects. *GM Crops* 1, 8–12.

Kebeish, R., Niessen, M., Thiruveedhi, K., et al. (2007). Chloroplastic photorespiratory bypass increases photosynthesis and biomass production in *Arabidopsis thaliana*. *Nature Biotechnology* 25, 593–599.

Khan, Z. R., Midega, C. A. O., Amudavi, D. M., Hassanali, A., and Pickett, J. A. (2008). On-farm evaluation of the 'push–pull' technology for the control of stemborers and striga weed on maize in western Kenya. *Field Crops Research* 106, 224–233.

Kopka, J., Schauer, N., Krueger, S., et al. (2005). GMD@CSB.DB: the Golm Metabolome Database. *Bioinformatics* 21, 1635–1638.

Langdale, J. A. (2011). C(4) cycles: past, present, and future research on C(4) photosynthesis. *Plant Cell* 23, 3879–3892.

Lister, R., Gregory, B. D., and Ecker, J. R. (2009). Next is now: new technologies for sequencing of genomes, transcriptomes, and beyond. *Current Opinion in Plant Biology* 12, 107–118.

Ma, Y. B., Pollock, S. V., Xiao, Y., Cunnusamy, K., and Moroney, J. V. (2011). Identification of a novel gene, *CIA6*, required for normal pyrenoid formation in *Chlamydomonas reinhardtii*. *Plant Physiology* 156, 884–896.

Mur, L. A. J., Kenton, P., Lloyd, A. J., Ougham, H., and Prats, E. (2008). The hypersensitive response: the centenary is upon us but how much do we know? *Journal of Experimental Botany* 59, 501–520.

*Naylor, R. L., Falcon, W. P., Goodman, R. M., *et al.* (2004). Biotechnology in the developing world: a case for increased investments in orphan crops. *Food Policy* 29, 15–44.

*Royal Society (2009). *Reaping the Benefits: Science and the Sustainable Intensification of Global Agriculture.* http://royalsociety.org/Reapingthebenefits/.

Pieterse, C. M. J., Leon-Reyes, A., Van der Ent, S., and Van Wees, S. C. M. (2009). Networking by small-molecule hormones in plant immunity. *Nature Chemical Biology* 5, 308–316.

Schneeberger, K., Ossowski, S., Lanz, C., *et al.* (2009). SHOREmap: simultaneous mapping and mutation identification by deep sequencing. *Nature Methods* 6, 550–551.

Stevens, D. K., and Bradbury, R. B. (2006). Effects of the Arable Stewardship Pilot Scheme on breeding birds at field and farm-scales. *Agriculture, Ecosystems & Environment* 112, 283–290.

*Tester, M., and Langridge, P. (2010). Breeding technologies to increase crop production in a changing world. *Science* 327, 818–822.

*Tilman, D., Balzer, C., Hill, J., and Befort, B. L. (2011). Global food demand and the sustainable intensification of agriculture. *Proceedings of the National Academy of Sciences USA* 108, 20260–20264.

*Toenniessen, G., Adesina, A., and DeVries, J. (2008). Building an alliance for a green revolution in Africa. *Annals of the New York Academy of Sciences* 1136, 233–242.

United Nations (2010). *The Millennium Development Goals Report 2010.* www.un.org/millenniumgoals/pdf/MDG%20Report%202010%20En%20r15%20-low%20res%2020100615%20-.pdf#page=13.

van der Vossen, E., Sikkema, A., and Hekkert, B. T. L. (2003). An ancient R gene from the wild potato species *Solanum bulbocastanum* confers broad-spectrum resistance to *Phytophthora infestans* in cultivated potato and tomato. *Plant Journal* 36, 867–882.

2

Global population growth, food security and food and farming for the future

HISTORICAL POPULATION GROWTH

At the start of agriculture 10 000 to 15 000 years ago, the estimated world population was between 1 and 10 million. It then grew to 300 million about 2000 years ago, and after a further 1600 years it doubled to 600 million; it has since risen from about 1.5 billion at the end of the nineteenth century to the present level of about 7 billion (Figure 2.1) (OECD, 2011). This recent rapid growth in population began in 1950, with reductions in mortality in the less developed regions; this resulted in an estimated population of 6.1 billion in the year 2000, nearly two-and-a-half times the population in 1950. However, with the recent declines in fertility in most of the world, the growth rate of the global population has been decreasing since its peak of 2.0% in 1965–70.

There are significant international differences in the historic and predicted population trends according to the state of each country's economy. During the transition from a subsistence rural economy to that of more developed economy, most countries' populations go through a 'demographic transition', a shift from high fertility and mortality rates to lower mortality, followed by declining fertility and a stable or even shrinking population. Global data show that people under the age of 25 now comprise 43% of the world's population, reaching as much as 60% in some countries (Population Action International, 2011).

It is generally recognised that as populations progress through the demographic transition fertility tends to decline earlier and at a more rapid pace in urban areas. For example, surveys conducted in sub-Saharan Africa since 2000 show that urban fertility rates are lower than

Successful Agricultural Innovation in Emerging Economies: New Genetic Technologies for Global Food Production, eds. David J. Bennett and Richard C. Jennings. Published by Cambridge University Press. © Cambridge University Press 2013.

23

Millions   **Billions**

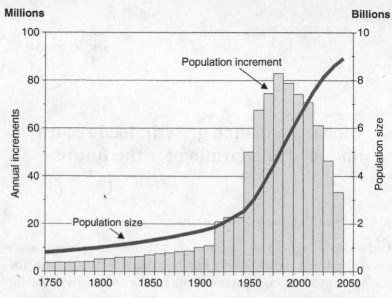

Figure 2.1 Long-term population growth, 1750–2050. (United Nations)

those in rural areas by at least one child per woman in 22 of 23 countries. In cities, the factors that contribute to lower fertility rates include higher costs of raising children, a more educated population, higher age at marriage and greater access to contraception (Lutz and Samir, 2010).

AGRICULTURAL PRODUCTION

The rapid increase in population that took place since 1950 has been accompanied in many countries by a corresponding increase in agricultural production, due to an increase in area devoted to agriculture and a genetic and agronomic improvement of the crops grown (Figure 2.2).

Selected data for the annual percentage rate of growth in population, agricultural land area and value of agricultural production show values of 0.82, 0.87 and 1.62, respectively, over the period from 1908 to 1930. The corresponding values from 1930 to 1960 were 1.21, 0.88 and 1.68 and from 1960 to 2008 were 1.68, 0.20 and 2.16. Food and Agriculture Organization of the United Nations (FAO) data from 1961 to 2003–4 show total food production increased by about 150%, and production per capita has also increased. Food prices tended to go down, although there was a spike in the 1970s and another in 2008–9. Global data, however, obviously obscure regional variation. For example, specific data from China over the last half century show the pattern of grain production per person increasing from about 200 kg in 1949 to about

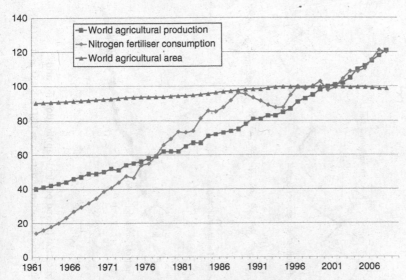

Figure 2.2 Indices of world agricultural production, nitrogen fertiliser use and world agricultural area. (100 = 1999–2001). (FAO, 2011)

400 kg in the early 1990s, during a period when the land available per person has been reduced by half to about 1000 m². Such progress has not occurred in other regions, most obviously in sub-Saharan Africa.

A retrospective analysis of some aspects of this recent success in improving crop yields is given in Figure 2.3, which shows the reduction in agrichemical inputs and the increases in nitrogen use and water use efficiency of maize in the USA. It can be assumed that these trends, based on improved genetics and agronomy, will continue in the future, although the continuing role of agrichemicals may be limited by environmental concerns, and fertiliser prices are linked to energy costs, which may increase significantly over the coming decades.

The most recent FAO data for agricultural output are given in Table 2.1. It has been estimated by the FAO that in order to meet the world's growing needs, food production will have to double by 2050, with the required increase mainly in developing countries where the majority of the world's rural poor live, and where 95% of the population increase during this period is expected to occur.

GLOBAL HUNGER INDEX

Despite the great improvement in agricultural production over recent decades the associated benefits are not distributed equally. Today, approximately 75% of the world's 7 billion people live in the developing

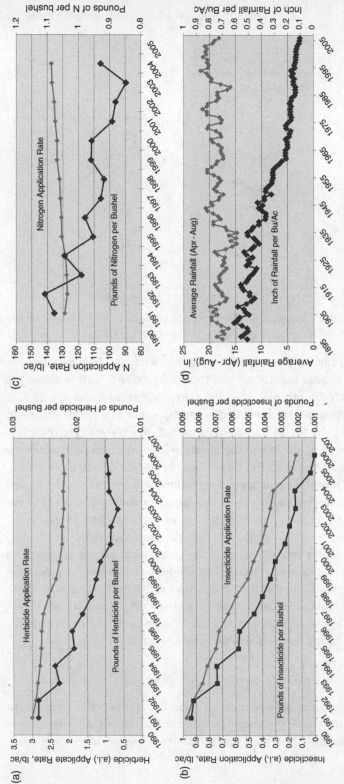

Figure 2.3 Maize yields in USA (3-year running averages). Evidence of declining herbicide application (a) and pesticide application (b), and improvements in nitrogen use efficiency (c) and water use efficiency (d). (Amended from Monsanto presentation of data from USDA, NASS and NOAA)

Table 2.1 *Global production of agricultural commodities 2010*

		Production	
Rank	Commodity	$ billion	Billion tonnes
1	Cow milk	179.1	599.4
2	Rice, paddy	174.7	672.0
3	Indigenous cattle meat	167.9	62.1
4	Indigenous pig meat	166.7	109.1
5	Indigenous chicken meat	121.6	85.4
6	Wheat	81.2	650.9
7	Soya beans	64.8	261.5
8	Maize	55.1	844.4
9	Sugar cane	53.6	1685.4
10	Tomatoes	53.6	145.7
11	Hens' eggs, in shell	52.7	63.5
12	Potatoes	44.5	324.1
13	Vegetables, fresh	40.7	240.1
14	Grapes	39.0	68.3
15	Buffalo milk, whole fresh	36.3	92.5
16	Cotton lint	33.6	23.4
17	Apples	29.2	69.6
18	Bananas	28.3	102.1
19	Indigenous sheep meat	23.4	8.6
19	Mangoes, etc.	23.2	38.7

Source: FAO (2010).

world, where most of the world's existing poverty is concentrated. It is estimated that 1 billion people live on less than a dollar a day and spend half their income on food. Some 854 million people are hungry and each day about 25000 people die from hunger-related causes. A number of different indicators can be used to measure hunger. Of these indicators, probably the best known is the Global Hunger Index (GHI). This is designed comprehensively to measure and track hunger both globally and by country and region. It is calculated each year by the International Food Policy Research Institute (IFPRI), and to reflect the multidimensional nature of hunger, it combines three equally weighted indicators in one numerical index:

1. Undernourishment: the proportion of undernourished people as a percentage of the population (reflecting the share of the population with insufficient intake of calories);

2. Underweight children: the proportion of children younger than 5 who are underweight (low weight for age reflecting wasting, stunted growth, or both); this is one indicator of child undernutrition; and

3. Child mortality: the mortality rate of children younger than 5. This partially reflects the interaction between inadequate dietary intake and an unhealthy environment.

In comparative terms, a GHI value below 4.9 is considered to be low, between 5.0 and 9.9 to be moderate, 10.0 to 19.9 to be serious, from 20.0 to 29.9 to be alarming and more than 30.0 to be extremely alarming. The 2011 Index, published jointly by the three organisations IFPRI, Concern Worldwide and Welthungerhilfe, shows that although there has been some recent progress in reducing hunger, the proportion of hungry people, as well as the absolute number, remain very high (International Food Policy Research Institute, 2011). The 2011 report focuses particular attention on the issue of food price spikes and volatility, which have played a large role in the global food crises of 2007–8 and 2010–11. Many poor people already spend a large proportion of their incomes on food, and surges in food prices leave them unable to pay for the food, health-care, housing, education, and other goods and services they need.

In summary, this report shows that although the number of undernourished people increased from the mid-1990s until 2009, the proportion of undernourished people has declined slightly during this period. The 2011 world GHI fell by 26% from the value in 1990, from a score of 19.7 to 14.6. However, such global averages do not reveal the dramatic differences among regions and countries. For example, the 2011 score fell by 18% in sub-Saharan Africa, by 25% in South Asia, and by 39% in the Near East and North Africa. Progress in Southeast Asia and Latin America and the Caribbean was particularly notable, with the scores each decreasing by 44%, although the score was already low in Latin America and the Caribbean.

It should be noted that the food and economic crises of recent years provide a particular challenge to achieving the Millennium Development Goal of a reduction by half in the proportion of people who suffer from hunger by 2015 (Fan and Pandya-Lorch, 2012).

PREDICTED POPULATION GROWTH

The Population Division of the United Nations Department of Economic and Social Affairs, in its *World Population Prospects: The 2010 Revision* (UNFPA, 2011) foresees a global population of 9.3 billion people

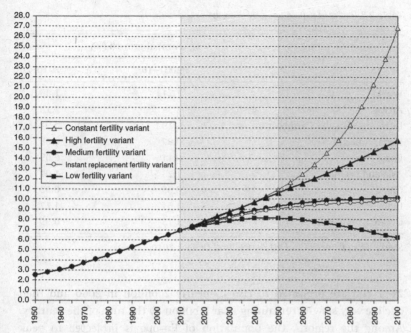

Figure 2.4 Estimated and projected world population according to different variants, 1950–2100 (billions). (United Nations, Department of Economic and Social Affairs, Population Division, 2011)

at 2050, an increase over earlier estimates, and more than 10 billion by the end of this century – and that scenario assumes lower fertility rates over time. With only a small variation in fertility, particularly in the more populous countries, the total could be higher, 10.6 billion by 2050 and more than 15 billion in 2100.

Whether these projections are realised is dependent on the continued decline of fertility in countries where this measure is above replacement level (that is, women have, on average, more than one daughter) and an increase of fertility in the countries that have below-replacement fertility. Predictions of global population, based on the various rates of fertility are given in Figure 2.4.

An analysis of regional trends suggests that Asia will remain the most populous area during the twenty-first century but Africa's population will increase from 1 billion in 2011 to 3.6 billion in 2100. Recently, this latter population has been growing 2.3% per year, a rate more than double that of Asia (1%). The population of Asia, at present 4.2 billion, is expected to peak around the middle of this century (5.2 billion in 2052) and then to start a slow decline. The populations of all other major areas combined (the Americas, Europe and Oceania) amount to 1.7 billion in

Table 2.2 *Predicted population growth in the next decade*

	Annual growth (%)	
	2001–10	2010–20
World	1.21	1.05
Africa	2.34	2.18
Latin America and Caribbean	1.19	0.91
North America	0.97	0.88
Europe	0.11	0.09
Asia and Pacific	1.23	1.01
China	0.65	0.55
India	1.51	1.17
Oceania Developed	1.13	0.99

Source: Data from *OECD–FAO Agricultural Outlook 2010*.

2011 and are projected to rise to nearly 2 billion in 2060 and then decline very slowly, remaining near 2 billion by the turn of the century. Among the regions, the population of Europe is projected to peak around 2025 at 0.74 billion and decline thereafter. These data are summarised in Table 2.2.

URBANISATION

In 2010, the absolute number of people living in urban centres worldwide overtook the number of people living in the countryside for the first time. This shift in the rural–urban balance has occurred very rapidly over recent years: of a total world population of 3.7 billion people in 1970, 2.4 billion (64.8%) were rural dwellers and 1.3 billion (35.2%) were urban. The change in the ratio of agricultural to non-agricultural population has been even more dramatic since that time. Whereas in 1970, the agricultural population was 2.0 billion and the non-agricultural population 1.7 billion, by 2010, out of a world population of 6.91 billion, only 2.62 billion (38.1%) were employed in agriculture. Moreover, the percentage of rural poor people continues to be higher than that of the urban poor; three-quarters of the world's poor today live and work in the countryside (FAO, 2010). Although the global rural population is 49%, values vary from a maximum of 89% in Burundi to 2% in Kuwait. In terms of the proportion of the economically active population working in agriculture, the values also vary significantly from 93% in Bhutan to 1% in countries that include the UK.

Currently, about 3.2 billion people are urban and require 2.4 billion tonnes of food each year. By 2050, the urban population will have grown significantly and about 3.6 billion tonnes of food will be required. An additional and important consequence of urbanisation is that it continues to modify consumption patterns towards higher-value processed products and convenience foods. Such changes include a move towards sugar, vegetable oil and livestock products. Consumption of livestock products has also been increasing dramatically in developing countries since the 1960s. Consumption of milk has almost doubled, consumption of meat has tripled and that of eggs has increased fivefold over the same period. However, this growth is very unevenly distributed. The greatest growth has occurred in East and Southeast Asia. In China, per capita consumption of meat, milk and eggs increased by a factor of four, ten and eight respectively. These increases have in turn driven an increased demand for animal feed, principally soya, such that the consumption of soya meal in China has risen 30-fold since 1990 to a total of more than 35 million tonnes per year. Indeed, the trade in soya from South America to China is the largest of any agricultural commodity.

CONSTRAINTS ON AGRICULTURE

Land

Access to appropriate land is the most fundamental limitation to agricultural production (Franck *et al.*, 2011; Lotze-Campen *et al.*, 2010). The combination of various global crises (financial, environmental, energy, food) in recent years has contributed to a notable revaluation of, and rush to control, land, especially that located in the global South. National governments in 'finance-rich, resource-poor' countries (e.g. parts of the Middle East and Asia) are looking to 'finance-poor, resource-rich' countries to help secure their own future needs of food and especially energy, the latter often provided by biofuels. These so-called 'land grab' transactions have been much criticised for their dependence on financial speculation, often linked to non-transparent agreements involving national and local governments. There is a concern that such deals may lead to dispossession when 'local communities' do not have formal, legal and clear property rights over the contested lands. This suggests that regulation of land deals is needed, whether through the Responsible Agricultural Investments (RAI) principles put forward by the World Bank, the United Nations Conference on Trade and Development (UNCTAD), the International Fund for

Agricultural Development (IFAD), and FAO, or through the Voluntary Guidelines being advocated by social movements and NGOs within the Committee on Food Security (CFS) of the FAO.

Water

If land is available, then the most important constraint is water. Given both population increase and climate change there is a high probability that Green Water (GW) (water in soil) and Blue Water (BW) (water in rivers, lakes, reservoirs and aquifers) availability will decrease in many regions, such that countries and river basins in Africa, the Near East and the Middle East that are presently water-scarce will remain so in the future (Gerten *et al.*, 2010). However, agricultural systems in most countries in the Americas and also Australia will still have enough GW/BW resources to generate sufficient calories for their populations. The World Economic Forum report in 2010 suggested that globally there will be a 40% shortfall between forecast demand for water and available supply by 2030. Such disparity in access is likely to lead to significant regional conflict, and potentially it may have an even greater influence on migration.

Climate change

By 2050 the global CO_2 concentration is likely to be approximately 550 ppm and FACE (free air CO_2 enrichment) experiments suggest that this will increase yields of C3 crops by about 13% (Jaggard *et al.*, 2010), although it will not affect yields of C4 species.[1] It will also decrease water consumption, making rain-fed crops less prone to water stress. However, by that date most regions will be warmer by 1–3 °C. This increase in temperature will speed up the development of crops, increasing the yields of indeterminate species, such as sugar beet, that do not flower before harvest and potentially decreasing the yields of determinate crops like wheat and rice. It is likely that plant-breeders will be able to select lines with increased yields in a CO_2-enriched environment, and it should remain possible to control most weeds and airborne pests and diseases so long as policy changes do not remove too many

---

[1] During the first steps of CO_2 fixation in photosynthesis, C3 plants form a pair of three carbon-atom molecules while C4 plants initially form four carbon-atom molecules. C3 plants comprise more than 95% of the plant species on Earth including trees while C4 plants include such crop plants as sugar cane and maize.

crop-protection chemicals. However, it is predicted that soil-borne pathogens are likely to be an increasing problem since warmer weather will increase their multiplication rates. Future control may depend upon transgenic approaches to breeding for resistance.

FORECASTS OF AGRICULTURAL PRODUCTION

In its most recent estimate of current production the FAO has raised its forecast for 2011 world cereal production to 2327 million tonnes, which would be an increase of 3.6% from 2010 and a new record. Following the completion of 2011 wheat harvests, the forecast for world wheat output is now established as a new record of 694.5 million tonnes, up 6.3% from 2010. However, this short-term optimism is not likely to be maintained in the longer term. The *OECD–FAO Agricultural Outlook* for 2011–20 suggests that global agricultural production is projected to grow at 1.7% annually, on average, compared to 2.6% in the previous decade (OECD/FAO, 2011). Slower growth is expected for most crops, especially oilseeds and coarse grains, which face higher production costs and slowing growth of productivity. Despite the slower expansion, production per capita is still projected to rise 0.7% annually.

In order to assess the changes that are likely in the food system by 2050, it should be noted that currently about 3.5 billion hectares (Bha) are under pastures, 1.5 Bha are cultivated and about 280 million ha are irrigated. Although improved genetics has led to an increase in resource use efficiency in crops, there is still significant use of fertilisers, namely 87 million tonnes of nitrogen and 34 million tonnes of phosphorus. Additionally, more than 3.5 million tonnes of pesticides are applied (Burney et al., 2010). Quantitative assessments show that the environmental impacts of meeting the demand for doubled production by 2050 depend on the process by which global agriculture expands (Foley et al., 2011; Foresight, 2011; Tilman et al., 2011). If the current trends of greater agricultural intensification in richer nations and greater land clearance (extensification) in poorer nations were to continue, then ~1 Bha of land would be cleared globally by 2050, with CO_2-C equivalent greenhouse gas (GH) emissions reaching ~3 Gt/y and N use ~250 Mt/y by then. In contrast, if crop demand in 2050 was met by limited intensification focused on existing croplands in countries with low yields, adaptation and transfer of the most productive technologies to these croplands, and global improvements in technology, land clearance would be only ~0.2 Bha, GH emissions ~1 Gt/y and global N use ~225 Mt/y.

NOVEL CROPS

Over the past century, agricultural development as summarised above has been based on an ambition of increasing productivity and maximising the production of cereals. This paradigm has produced an agricultural system that provides the world's primary source of calories and employs 60–80% of people in low-income countries. The increases in cereal production in the Green Revolution, for example, saved countless lives in Asia, and agricultural growth there was linked to a rapid rate of economic growth. Some experts consider, however, that such agricultural intensification has led to an overconcentration on grain production and has diminished the focus on nutrient-dense crops like pulses, fruits and vegetables. In this context, many suggest that underutilised crops have great potential to alleviate hunger directly, through increasing food production in challenging environments where major crops are severely limited, through nutritional enhancement to diets focused on staples, and through providing the poor with purchasing power and thereby helping them purchase the food that is available. One specific initiative in this area is Crops for the Future (www.cropsforthefuture. org/); this group will complement the Consultative Group for International Agricultural Research (CGIAR) centres which tackle issues relating to major crops. For the first time, underutilised crops have their own organisation to assist their research and development for greater food security. It is highly significant to note that 80% of the plant food consumed globally is provided by just 12 species of plants – the cereals barley, maize, millet, rice, rye, sorghum, sugar cane and wheat, and tubers cassava, potato, sweet potato and yam. This low number should be compared with the 7000 edible and partly domesticated plants and an estimated 30 000–75 000 edible wild species of plants on the planet (Brown, 2011).

TRADE AND RESEARCH

Other important factors that interact with agricultural production include the impact of world trade and the support for fundamental and applied research. In an analysis of the influence of trade, Schmitz et al. (2012) recently used MagPIE, a mathematical programming model which minimises global agricultural production costs. On a global scale their results demonstrate that increased trade liberalisation will lead to lower global costs of food production. Depending on the degree of liberalisation, results from the model show that between

Table 2.3 *Growth rates (%) in public agricultural research expenditures, 1981–2000*

	1976–81	1981–91	1991–2000
Latin America and the Caribbean	8.54	1.86	0.32
Asia-Pacific	7.98	4.67	3.35
High income	2.5	2.43	0.52
Sub-Saharan Africa	0.94	1.02	−0.15
West Asia and North Africa	–	4.12	2.93

Source: Adapted from Beintema and Elliott (2009).

6% (US$5.4 trillion) and 10% (US$9.4 trillion) will be saved in the period 2005–45. Moreover, their model shows that trade liberalisation leads to a much slower increase in the index of food scarcity.

Another important aspect of policy is the investment into technological change. Although the need has increased, investments into agricultural research and development (R&D) have slowed down in the past decades and thereby reduced the growth in agricultural production (Alston *et al.*, 2009; Alston, 2010; Pardey and Alston, 2010). Table 2.3 shows the extent of decreased public investment in agriculture between 1981 and 2000. This trend was not equal across all regions of the world. In the Asian region, public investment decreased but stayed quite high (around 4%), mostly because of the high growth of agriculture R&D in China and India. In contrast, spending in Africa stayed level between 1980 and 2000, and actually diminished during the 1990s. In Latin America, public spending for agricultural R&D grew slightly in the 1990s (less than 1%) after a significant reduction from the late 1970s.

Many experts recommend a significant global expansion in funding for agricultural R&D. One specific suggestion is to strengthen the current reform process of the CGIAR, along with increased support for national research systems. These changes should contribute to long-term solutions to food insecurity, especially in the context of land degradation, water scarcity and climate change (Beddington *et al.*, 2011).

ROLE OF BIOTECHNOLOGY

As reviewed recently (Dunwell, 2011) there are many opportunities provided by the application of biotechnology to crop improvement. These relate both to the response of plants to their environment and to their underlying genetic and physiological potential. In the words of Phillips (2009): '*breaking the yield barrier can be achieved in two ways: Increasing "Operational yield" such as from insect resistance, etc. and from*

"*Intrinsic yield*" *such as raising the base yield due to changes in physiological processes. Transgenics thus far appear to have raised operational yield.*' Despite the present extensive cultivation of GM crops, principally in North and South America, many other countries remain resistant to their use because governments fear that they would not be able to export their products to the European Union where the governments of some countries are strongly opposed. Unfortunately, this affects several developing countries that already face food shortages, and whose farmers might benefit from growing GM crops. At one time in the 1990s, Europe had much to offer in terms of an environmentally oriented approach to agricultural technology, but policy and stakeholder interactions related to GM crops and biotechnologies over the past 10–15 years have prevented these benefits from being realised. It seems obvious that if Europe is to meet its own food security needs and to contribute to the food requirements of the rest of the world, policy and regulatory changes will be necessary (Tait and Barker, 2011).

CONCLUSION

To summarise the conclusions of Godfray *et al.* (2010, 2011), if crop yields are to be increased in a sustainable manner, then much can be achieved using existing knowledge and skills. Most particularly, growers must be encouraged and given incentives to adopt best practice. To that end it is necessary to re-establish extension services, to improve the skills base among farmers and food producers, and to introduce better economic incentives to make it worthwhile for farmers to adopt new practices. For example, a variety of methods are available to increase the nitrogen efficiency of crop and livestock production or to reduce methane emissions from livestock or wetland rice. Much more carbon could be sequestered in farmland, both in soils and agro-forestry (combining trees and shrubs with crops and/or livestock). Moreover, in a global context it is also important that the public sector maintains its ability to exploit the most effective combination of traditional and novel approaches to plant breeding. Additionally, the regulatory framework and associated political processes that interact with the agricultural science will play a critical role.

REFERENCES

Alston, J. (2010). The Benefits from Agricultural Research and Development, Innovation, and Productivity Growth, *OECD Food, Agriculture and Fisheries Working Papers*, No. 31, OECD Publishing. http://dx.doi.org/10.1787/5km91nfsnkwg-en

Beddington, J., Asaduzzaman, M., Fernandez, A., *et al.* (2011). *Achieving Food Security in the Face of Climate Change: Summary for Policy Makers from the Commission on Sustainable Agriculture and Climate Change.* CGIAR Research Program on Climate Change, Agriculture and Food Security (CCAFS). Copenhagen, Denmark. www.ccafs.cgiar.org/commission.

Beintema, N., and Elliott, H. (2009). Setting meaningful investment targets in agricultural research and development. In *FAO Expert Meeting on How to Feed the World in 2050.* Food and Agriculture Organization of the United Nations, Rome.

Brown, A. G. (ed.) (2011). *Biodiversity and World Food Security: Nourishing the Planet and its People.* The Crawford Fund, Kingston, ACT. www.crawfordfund.org/home/view.html?publication=411&rtn=1

Burney, J. A., Davis, S. J., and Lobell, D. B. (2010). Greenhouse gas mitigation by agricultural intensification. *Proceedings of the National Academy of Sciences USA* **107**, 12052–12057.

Dunwell, J. M. (2011). Crop biotechnology: prospects and opportunities. *Journal of Agricultural Science* **149**, 17–27.

Fan, S., and Pandya-Lorch, R. (eds.) (2012). *Reshaping Agriculture for Nutrition and Health.* International Food Policy Research Institute, Washington, DC. www.ifpri.org/publication/reshaping-agriculture-nutrition-and-health

FAO (2010). *FAO Statistical Yearbook 2010.* Food and Agriculture Organization of the United Nations, Rome. www.fao.org/economic/ess/ess-publications/yearbook 2010

FAO (2011). *The State of Food Insecurity in the World: How Does International Price Volatility Affect Domestic Economies and Food Security?* Food and Agriculture Organization of the United Nations, Rome. www.fao.org/docrep/014/i2330e/i2330e.pdf

Foley, J. A., Ramankutty, N., Brauman, K. A., *et al.* (2011). Solutions for a cultivated planet. *Nature* **478**, 337–342.

Foresight (2011). *The Future of Food and Farming: Final Project Report.* The Government Office for Science, London.

Franck, S., von Bloh, W., Müller, C., Bondeau, A., and Sakschewski, B. (2011). Harvesting the sun: new estimations of the maximum population of planet Earth. *Ecological Modelling* **222**, 2019–2026.

Gerten, D., Heinke, J., and Hoff, H. (2010). Estimating green–blue water availability and needs for global food production. In *The Global Dimensions of Change in River Basins: Threats, Linkages and Adaptation,* pp. 65–72. Global Water System Project International Project Office, Bonn, Germany.

Godfray H. C. J., Beddington, J. R., Crute, I. R., *et al.* (2010). Food security: the challenge of feeding 9 billion people. *Science* **327**, 812–818.

Godfray, H. C. J., Pretty, J., Thomas, S. M., Warham, E. J., and Beddington, J. R. (2011). Linking policy on climate and food. *Science* **331**, 1013–1014.

Jaggard, K. W., Qi, A., and Ober, E. S. (2010). Possible changes to arable crop yields by 2050. *Philosophical Transactions of the Royal Society B* **365**, 2835–2851.

International Food Policy Research Institute (2011). *Global Hunger Index: The Challenge of Hunger – Taming Price Spikes and Excessive Food Price Volatility.* IFPRI, Washington, DC. www.ifpri.org/sites/default/files/publications/ghi11.pdf

Lotze-Campen, H., Popp, A., Beringer, T., *et al.* (2010). Scenarios of global bioenergy production: the trade-offs between agricultural expansion, intensification and trade. *Ecological Modelling* **221**, 2188–2196.

Lutz, W., and Samir, K. C. (2010). Dimensions of global population projections: what do we know about future population trends and structures? *Philosophical Transactions of the Royal Society B* **365**, 2779–2791.

OECD (2011). *OECD Factbook 2011–2012: Economic, Environmental and Social Statistics*. OECD Publishing, Paris. www.oecd-ilibrary.org/economics/oecd-factbook-2011–2012_factbook-2011-en

OECD/FAO (2011). *OECD–FAO Agricultural Outlook 2011–2020*. OECD Publishing, Paris. www.agri-outlook.org/pages/0,2987,en_36774715_36775671_1_1_1_1_1,00.html

Pardey, P. G., and Alston, J. M. (2010). *US Agricultural Research in a Global Food Security Setting: A Report of the CSIS Task Force on Food Security*. Center for Strategic International Studies, Washington, DC.

Phillips, R. L. (2009). Mobilizing science to break yield barriers. Background paper to the CGIAR 2009 Science Forum workshop on: 'Beyond the yield curve: exerting the power of genetics, genomics and synthetic biology'. www.scienceforum2009.nl

Population Action International (2011). *Why Population Matters*. Population Action International, Washington, DC. http://populationaction.org/wp-content/uploads/2011/10/Why_Population_Matters.pdf.

Schmitz, C., Biewald, A., Lotze-Campen, H., *et al.* (2012). Trading more food: implications for land use, greenhouse gas emissions, and the food system. *Global Environmental Change* 22, 189–209.

Tait, J., and Barker, G. (2011). Global food security and the governance of modern biotechnologies. *EMBO Reports* 12, 763–768.

Tilman, D., Balzer, C., Hill, J., and Befort, B. L. (2011). Global food demand and the sustainable intensification of agriculture. *Proceedings of the National Academy of Sciences USA* 108, 20260–20264.

UNFPA (2011). *State of World Population 2011: People and Possibilities in a World of 7 billion*. United Nations Population Fund, New York. www.unfpa.org/swp/

United Nations, Department of Economic and Social Affairs, Population Division (2011). *World Population Prospects: The 2010 Revision*. United Nations, New York.

3

New genetic crops in a global context

A CONFLUENCE OF CRISES

We are beset by global crises – economic and fiscal turmoil as well as failures of political and institutional governance, and crises of energy and water, of the environment and of food and nutrition security. This chapter is about food and nutrition security, but it is important to recognise that all these crises are increasingly interlinked. Food security is dependent on efficient and fair markets, and is hard to achieve if governments are corrupt and the rule of law is weak. Secure supplies of food are difficult to achieve when the prices of energy are high, and land and water are in short supply and degraded. Sustainable solutions thus require actions on many fronts, some technological, others political and institutional. We need to put our best minds to the tasks ahead, using reasoned analysis to understand what is happening and getting agreement on actions to be taken from ordinary people as well as political leaders. If we are to succeed we will need every tool we can lay our hands on.

We face three global challenges to food security. First is the threat of repeated food price spikes. Second is the presence of around 1 billion people in the world who are chronically hungry, and third, we need to increase food production by 70–100% by 2050 if we are to feed the world.

SPIKING FOOD PRICES

The price of basic food staples, tracked by the Food and Agriculture Organization of the United Nations, began rising in 2006, following a rise in oil prices (Figure 3.1). Protests, many associated with violence,

Successful Agricultural Innovation in Emerging Economies: New Genetic Technologies for Global Food Production, eds. David J. Bennett and Richard C. Jennings. Published by Cambridge University Press. © Cambridge University Press 2013.

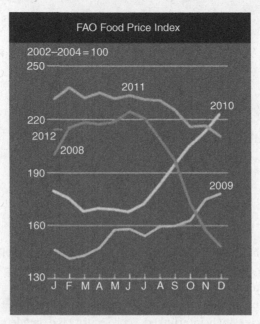

Figure 3.1 The FAO Food Price Index. (FAO, 2012)

erupted throughout Latin America, Africa and Asia, and even in Italy there were demonstrations triggered by the rise in the price of pasta.

High prices continued into 2008 when they peaked and fell, indicating a classic, short-lived commodity price spike. But they began to grow again in 2009, mimicking the 2007 rise. In 2010, a severe heat wave in Russia which forecast a 30% reduction in wheat harvest compared with the previous year triggered an export ban that extended into 2011. This prompted Egypt and other North African states to purchase large quantities of wheat. Newspaper stories, ranging from possible locust outbreaks in Australia to heavy rains in Canada, heightened perceptions of impending shortage and future price rises. The actual and perceived shortages were enough to trigger a spike which eventually peaked above the 2007/8 spike and in 2010 a new high had been reached, one that has persisted with only a small decline in 2011 (Figure 3.1).

Higher food prices coupled with economic recession has hit people in the developed world hard but the impact in developing countries, where grain prices have remained high, has been devastating. It is estimated that, globally, average crop prices will be 10–20% higher in real terms relative to 1997–2006 for the next 10 years and vegetable oils will be more than 30% higher (OECD–FAO, 2009).

The proximal causes of the 2007/08 spike have been the subject of much debate. Certainly a major factor was the fall in grain stocks to dangerously low levels as a result of steadily increasing demand and a series of poor harvests brought about by adverse weather. Global consumption of grains and oilseeds outstripped production in seven of the eight years after 2000. By 2007, stocks were only 14% of use. Prices began to spike as importing countries became concerned about their ability to meet future food needs.

HUNGER ON THE RISE

Each year, the FAO annual report *The State of Food Insecurity in the World* (SOFI), which covers 106 countries, monitors the number of hungry in the world. Over the past decade hunger has grown following a steady decline in the previous three decades, most likely as a result of population growth once again outstripping food production. In 2009, over 1 billion people suffered from hunger, the largest number in the history of the planet. The FAO recently revised its methodology estimating 868 million chronically hungry people in the world. This is acknowledged as a conservative estimate in part because it is based on a sedentary lifestyle (FAO, 2012).

In addition to chronic hunger there are periodic famines and acute periods of hunger in different localities. In 2011, the total failure of the winter 2010 rains and the poor spring 2011 rains resulted in crop failure and livestock decline in the drylands of Ethiopia, Somalia and Kenya in the 'Horn of Africa'. People were unable to respond or adapt to the drought due to a lack of livelihood alternatives, poor political support, underinvestment in rural areas and conflict. Both national and international responses to the crisis were too late, despite sophisticated early-warning systems indicating the coming crisis in August 2010. It is estimated that 50000 to 100000 people have died and that over 13 million people have been affected, although the situation has begun to improve and in February 2012 the famine was declared over in Somalia as a result of donor aid and good rains.

FOOD SECURITY IN 2050?

Food price spikes and recurrent famines are part of a series of acute crises that reflect an underlying chronic crisis. Unless that is resolved we will not prevent recurrences of these acute crises, nor will we be able to feed the world in 2050. The chronic crisis is driven by both supply and demand factors that interact with each other.

CHANGES IN DEMAND

Population

The UN estimates that the global population will rise to around 9.15 billion by 2050 after which it will begin to stabilise. The addition of 2.15 billion people to the planet, largely in developing countries, requires a corresponding increase in food production, one that could be considered relatively achievable given that production increases of food crops between 1990 and 2010 have been around 2% per annum.

Changing consumption patterns

Contributing to the scale of future food production required is the world-wide growth in per capita incomes. Recent increases have occurred in most countries of the world, for example, in high-income OECD countries growth has been over threefold in constant US dollars since the 1960s. Although from a lower base, growth in per capita incomes has been sixfold in East Asia and the Pacific and twofold in the Middle East and North Africa. Only in sub-Saharan Africa has there been little growth.

The impact of higher per capita incomes is a greater demand for better quality food in greater amounts. In combination with growing urbanisation and exposure to so-called 'diet globalisation', demand for processed and higher-value foods is increasing, often as substitutes for raw agricultural commodities. In Asia, for example, per capita consumption of rice is falling (in India, Indonesia and Bangladesh, it has been declining since its peak in the early 1990s) while consumption of such foods typical of a Western diet, wheat and wheat-based products, temperate-zone vegetables and dairy products, is growing.

A revolution in livestock production, argued to be having as significant impact on developing countries as the Green Revolution, has seen meat and milk and dairy product consumption in developing countries grow at around 5% and 3.5% per annum, respectively, in recent years (FAO, 2006) (Figure 3.2). In China, meat consumption rose from around 20 kg per person per year in 1985 to some 55 kg per person per year today (FAO, 2009).

The result of these dietary changes is the need to double global meat production from around 230 million tons in 1999/01 to over 460 million tons by 2050, the great bulk of which will be produced in developing countries.[1]

[1] N.b. 1 tonne = 1 metric ton = 1000 kg = 1 million grams.

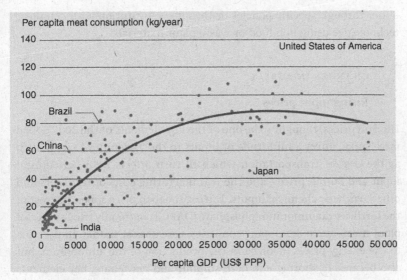

Figure 3.2 The relationship between meat consumption and per capita income. (FAO, 2009)

Corresponding increases in the production of cereals, oilseed and other feedstuffs will be needed to meet increased demand for livestock products. In general around 8 kg of grain is required to produce 1 kg of meat product although this varies considerably depending on the livestock production system, e.g. intensively raised livestock in confined forms of housing are often fed entirely on cereals and oilseeds whereas extensively raised livestock grazing on grassland can require little if any external forms of feed. Demand for feed will also increase as a result of shifts away from free-range livestock production to more intensive stall-fed systems as countries develop. By 2050 the total feed use of cereals (note this excludes oilseeds) is likely to have doubled and comprise over 40% of total cereal use.

Demand for biofuels

Biofuel production is increasing. Between 2000 and 2010 global ethanol production increased from under 20 billion litres to around 86 billion litres. Global biodiesel production increased tenfold to almost 11 billion litres between 2000 and 2008, and by 2010 had reached 19 billion litres (REN21, 2011). Despite considerable debate over the effects of a growing biofuel industry on the 2007/8 food price spike, it is clear that growing crops for biofuels will tend to amplify the prices of basic food crops. Thus biofuel production needs to be decoupled from food production either through specific policies or through substitution with cellulosic technologies using non-food or waste plant materials.

CHANGES IN SUPPLY

Rising input prices

High oil prices, thought to be one of the key elements of the 2007/8 food price spike, affect a multitude of inputs to the agricultural sector such as the cost of transportation which in turn affects both agricultural input and output prices, and the manufacturing costs of fertilisers and other synthetic chemical inputs. During the food price spike, the cost of the fertiliser diammonium phosphate (DAP), a commonly used source of plant nutrients in developing countries, rose almost sixfold, in part due to the energy prices involved in the production of the ammonium, but also because of shortages in both sulphur and phosphate, key elements in its manufacture (Figure 3.3). After falling significantly, the price of DAP fertiliser began growing again in the middle of 2009.

Slowing productivity increases

There are some data to indicate the rate of growth in major cereal and oilseed yields are slowing. One factor in their decline may be a reduction in funding of agricultural research and development by government donors, possibly as a result of low and steady food prices since the late

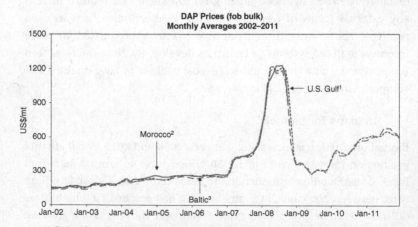

1. Derived from *Green Markets*. 2. Derived from *Fertiliser week*. 3. Derived from *FMB Weekly Report*.

Figure 3.3 Increasing fertiliser prices (diammonium phosphate, freight on board, monthly prices are all represented by a single line). (IFDC, 2011)

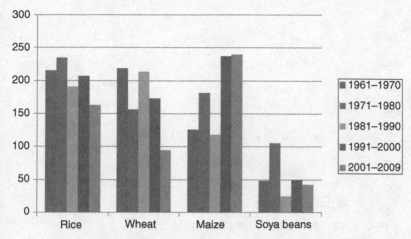

Figure 3.4 Average decadal increase in yield for rice, wheat, maize and soya bean between 1961 and 2009. (FAO, 2010)

1970s. There has also been a partly ideologically based belief that agriculture is best served by the private sector. Indeed, private funding of agricultural research has grown, but has tended to focus on the crops of developed countries where the returns have been greater. This may be the reason why yield growth in maize and soya beans has been strong in recent years (Figure 3.4).

Land and water scarcity and deterioration

More often than not, increased demand for food has previously been met by expanding agriculture onto new land. Fears over the sustainability of taking new land into cultivation have recently led to a growing interest in raising crop yields on existing agricultural land. How much potential cultivable land is in existence is contentious. When looking at area harvested over time, total global cropland has increased by only 10% over the past 50 years, in comparison with a population increase of 110%. Given the pressures to increase food production we would expect to see much greater land expansion if it were readily available. Where land expansion has occurred, e.g. soya beans and oil palms have each increased by over 300% in area and by over 700% and over 1400% respectively in production over the past 50 years (Figure 3.5), it is presumably a result of clearing the Cerrado in Brazil and rainforests in the Amazon, Africa and Southeast Asia.

A new assessment of land degradation, GLADA (Global Land Degradation Assessment), although in its early stages, has found that

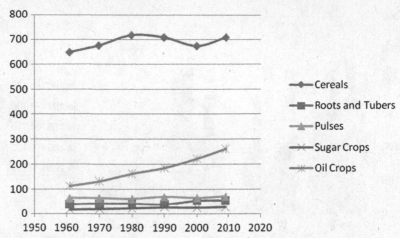

Figure 3.5 Change in harvested area for selected food crops, million hectares. (FAO, 2010)

between 1981 and 2003 a quarter of global land area has been degraded, with the most severely affected areas being Africa south of the equator, Indochina, Myanmar and Indonesia. Globally land degradation affects 1.5 billion people and over 40% of the very poor live in degraded areas.

Water is, of course, crucial to agricultural production and like land is similarly in short supply, and for similar reasons – overuse, inefficient use and degradation through pollution (Comprehensive Assessment of Water Management in Agriculture, 2007). Today, irrigated, multi-cropping systems provide the bulk of Asia's rice and wheat supplies. A large portion of this area, as well as areas in the Middle East and North Africa, maintain irrigated agricultural systems through unsustainable extractions of water from rivers or the ground. In parts of northwest India, the groundwater overdraft rate can exceed 56% (World Bank, 2006), largely thought to be due to the use of water pumps utilising subsidised or free electricity.

Sub-Saharan Africa, on the other hand, has considerable unexploited water resources for agriculture. Only around 4–5% of cultivated land is irrigated, of which two-thirds is in Madagascar, South Africa and Sudan. It is thought an additional 20 million hectares (Mha) of land could be brought under irrigation but, so far, technical, financial and socio-economic constraints have slowed this expansion. At the same time, sub-Saharan Africa has the largest number of water-stressed countries of any region with almost a quarter of the African population living in water-stressed countries, and this share is rising.

Climate change

Climate change is likely, in the longer term, to impact food supply and, in particular, agriculture, more than any other factor or economic sector.

There is convincing evidence that global climate change is occurring and is the result of man-made emissions of greenhouse gases (GHG) – primarily carbon dioxide (CO_2), methane (CH_4) and nitrous oxide (N_2O). As these gases are increasingly expelled into the atmosphere they form a layer over the Earth's surface that traps a growing proportion of the infrared radiation that would be otherwise radiated out to space, so warming the land and oceans beneath. As a consequence, the world as a whole is warming – so far by more than 0.7 °C since the Industrial Revolution.

Plant enzymes, central to photosynthetic processes, are sensitive to temperature and hence plants are particularly vulnerable to a warming climate. High temperatures can also induce physiological plant responses such as causing stomata to close in order to prevent water loss, so reducing photosynthesis, or impacting upon pollen viability and fertilisation. A lack of sufficient water, a key input to photosynthesis, is also detrimental to plant growth, leading to lower yields and, in the extreme, to death of the plant.

Many crops are already grown close to their limits of thermal tolerance. Just a few days of high temperature near flowering can seriously limit yields of crops such as wheat, fruit trees, groundnut and soya bean. In low-latitude regions, where most of the developing countries lie, even modest temperature increases of 1–2 °C can diminish harvests of major cereals and the modulating effects of adaptation are likely to be limited.

In a study by David Lobell, an environmental scientist at Stanford University, and colleagues at CIMMYT, yield data from 20 000 field trials of maize conducted in Africa between 1999 and 2007 were matched with local weather station records. The results indicated a non-linear relationship between warming and yields. Specifically, for each degree day spent above 30°C, yields declined by 1% under optimal rain-fed conditions and 1.7% under drought conditions. With a 1°C warming, around three-quarters of Africa's maize crop area would experience a 20% loss. These results indicate the necessity and urgency of improving soil moisture and breeding drought-tolerant crops (Lobell et al., 2011).

While agriculture is a major victim of climate change it is also a contributor and as such has the potential, albeit largely unrealised, to contribute to mitigating climate change by reducing GHG emissions.

The agricultural sector generates 10–12% of total global GHG emissions (12.4 gigatons of carbon), as estimated by the IPCC

(Intergovernmental Panel on Climate Change). The Global Carbon Project, however, estimates global emissions at 9.1 Gt carbon per year between 2000 and 2008. Inclusion of agricultural fuel use, fertiliser production and land-use change in these estimates increases agriculture's contribution to 30% (Smith et al., 2007). Given that agriculture's share in global GDP is only around 4%, the input of agriculture to climate change is considerably large.

Three main GHGs are produced by agriculture. They are:

- N_2O (nitrous oxide), which largely originates in the soil following applications of manure, urine and nitrogen fertiliser. It is around 300 times more potent (in terms of global warming) than CO_2.
- CH_4 (methane), which predominantly stems from ruminant digestion, rice cultivation and anaerobic soils and has 21 times as much warming potential as CO_2.
- CO_2 (carbon dioxide), which comes from microbial decay, fossil fuel combustion to supply power to machinery and construct grain silos, fuel for transport and to produce synthetic fertilisers and pesticides, as well as through land clearing and the burning of biomass, plant litter and soil organic matter.

Theoretically, using available mitigation measures and, assuming a high carbon price without economic or other barriers, the IPCC estimates agricultural emissions could be reduced or offset by around 5.5–6 Gt of CO_2 eq per year by 2030. Seventy per cent of this potential for abatement would come from developing countries, 20% from OECD countries and 10% from emerging countries. Furthermore, 89% of the mitigation potential is through soil carbon sequestration, predominantly in the developing world, while only 9% and 2% of the total potential is associated with lowering CH_4 and N_2O emissions, respectively.

THE RESPONSES OF FARMERS

In economic theory, when prices rise, suppliers increase their production of output. This is true for farmers, who will grow more of the crop in demand, and is what typically happens in the developed countries, although with the inevitable lag because of the time a crop takes to mature. Some developing country farmers can respond similarly although, in Africa in particular, the response is often weak. Farmers may lack access to inputs; fertiliser prices may be prohibitively high given the risk of failure; they may not have enough access to land or water; they may not have appropriate extension advice; the markets may be too variable and unreliable; and there may not be the roads and transport available to get their crops to market. These are formidable obstacles.

Nevertheless, African farmers are often highly innovative. This can take many forms: sometimes it may be the process of discovering something entirely new, more often it involves finding novel applications for existing technologies. David Millar, who works for the Tamale Archdiocesan Agricultural Programme, asserts that there is no farmer in northern Ghana who is not in some way experimenting.

Climate change is testing these innovative capacities. For example, in the village of Nwadhajane in southern Mozambique, the birthplace of the great Mozambique leader, Eduardo Mondlane, the villagers, already experiencing the effects of climate change, have devised a method of land sharing in order to adapt. Village-level farmer associations have been created to reassign a share of lowland and highland to each farmer. On the former, crops are very productive, but are washed out by periodic floods while the highlands produce good crops in the flood years but poor crops during the droughts. The farmer associations are also carrying out experiments with drought-resistant crops.

APPROPRIATE TECHNOLOGIES

What farmers need are technologies that are appropriate to their particular circumstances – their environment, the labour available and what they can afford. Ideally they require technologies that work, i.e. that are highly productive and add significant value. They also have to be resilient to stresses and shocks and be accessible and equitable. Farmers need to be aware of the downsides and what the counterfactual is. What do they lose if they do not adopt the technology?

On these criteria, technologies of many kinds can be appropriate:

Traditional Technologies appropriate to local conditions, which have usually evolved and been developed over an extended period of time by communities in developing countries, *e.g. home gardens, rainwater harvesting techniques.*

Intermediate Traditional technologies improved through integration with modern conventional technologies, *e.g. traditional treadle pump improved by engineers, intercropping.*

Conventional Technologies developed in industrialised countries through the application of modern physical, chemical and biological knowledge, and delivered as products in a packaged form for a regional or global market, *e.g. synthetic fertilisers and pesticides.*

New platform New scientific 'platforms' for innovation, such as biotechnology, nanotechnology and information and communication technology, which have the potential to be developed simultaneously for the needs of the industrialised and the developing world, *e.g. genetically modified organisms, drought- or pest-resistant crop varieties.*

Despite this extensive array of technologies, ascertaining those that are truly appropriate is difficult. There are often trade-offs, which can mean they transpire to be too expensive, not very productive, demanding of labour or damaging to the environment. Furthermore, as described at the beginning of this chapter, the challenges we must face are daunting. We need to increase yields, productivity and nutritive value and this entails crops that are more efficient at converting sunlight into carbohydrate and other nutrients; we need improved feed crops for livestock and biofuel crops that do not compete with food crops; we need crops that are resistant to pests and diseases, are more efficient at taking up nutrients from the soil, and reduce water degradation and, above all, we need crops that are resilient to global warming and capable of reducing GHG emissions.

This is why building desirable characteristics – high yields coupled with stability and resilience – into the seed is so attractive. The seed, in a sense, can be a 'package of desirable and appropriate technologies'. It is in this respect that the new genetics, in the form of biotechnology, becomes so relevant.

THE RISE OF BIOTECHNOLOGY

Biotechnology has traditionally been associated with the centuries-old practice of fermentation used in the making of bread, beer and spirits. Modern biotechnology is defined as encompassing any technological application that uses biological systems, living organisms or their derivatives to make useful products or processes. Driving modern biotechnology is the revolution in cellular and molecular biology that occurred in the second half of the twentieth century. In particular, it uses our knowledge of DNA and RNA to identify the genetic basis of useful traits in animals and plants.

While conventional breeding (selecting and crossing crop varieties and livestock breeds possessing desirable traits) has advanced agricultural production, it can be a slow and imprecise science. It often involves many crop or animal generations and depends on the often serendipitous discovery of new and beneficial mutations, the result of which is not always successful.

In the development of 'quality protein maize', conventionally bred improved maize varieties for developing countries, this process has been successful at producing an appropriate and useful end product but it was initiated by the discovery of maize mutants with high levels of desirable amino acids, a discovery that occurred back in the 1960s. Biotechnology's main advantage is that it makes conventional breeding much quicker, more targeted and more effective.

The new cellular and molecular techniques open up a new world in which breeders can deliberately design and engineer new plant and animal types, speedily and with much less reliance on random processes. Under the general title of biotechnology, they are already having a significant impact on both medicine and plant and animal breeding, benefiting poor people as well as those who are better off.

The application of this knowledge to crop breeding is pursued through three practical techniques.

Marker-aided selection

Marker-assisted selection (MAS) is a process whereby DNA sequences within an organism are analysed with the aim of detecting the presence and structure of genes responsible for particular traits.

The use of MAS was key to the development of submergence-tolerant rice, a potentially revolutionary rice variety able to withstand submergence in water for a number of weeks. Rice in Asia is typically grown in standing water but deep flooding for more than a couple of days is detrimental to crop growth and viability. Deep flooding affects over 25% of global rice-producing land, a proportion expected to rise as a result of global warming, and flash flooding can submerge rice plants, often at the seedling stage, for several weeks.

The development of submergence-tolerant rice began in 2006, when a team at the International Rice Research Institute (IRRI) in the Philippines identified a quantitative trait locus (QTL), a form of marker, in rice called *Submergence 1 (Sub1)* that allowed submerged plants to survive for more than 2 weeks. The resulting rice was named Scuba rice. The gene responds to flooding by limiting the elongation of the internodes that is activated by ethylene. This conserves carbohydrates so permitting regrowth as the flood recedes. The rice becomes dormant during the flooding then continues growing once floodwaters fall. Breeding based on MAS is now under way and new submergence-tolerant varieties are being produced in Laos, Bangladesh, Thailand and India. Within one year of its release one variety has been adopted by over 100000 Indian farmers.

Cell and tissue culture

This technique involves the development of whole plants from a single cell or cluster of cells in an artificial growth medium external to the organism.

Cell and tissue culture has become a powerful tool in the production of wide crosses, whereby wild relatives, often hosting desirable traits such as disease resistance, are crossed with domestic varieties. Frequently the embryo produced by a wide cross may fail to develop unless cut out and grown in a culture medium. An example is the successful production of new high-performing rices for Africa.

Demand for rice in West Africa is rising rapidly, failing to be met by local production. Population growth, rising incomes and changing consumer preferences have meant some 6 million tons of rice per year (half of the region's requirements) are imported in at a cost of over US$1 billion.

In order to address this shortfall, Monty Jones, a Sierra Leone scientist working at the Africa Rice Center (WARDA), began a programme utilising tissue culture technologies such as embryo rescue and anther culture, to develop crosses between the African species of rice (*Oryza glaberrima*) and the Asian species (*Oryza sativa*). The former typically returns low yields of around 1 ton per hectare while the latter produces yields of around 5 tons per hectare. Crossing the two species bore hundreds of new varieties but the process wasn't straightforward. At first the embryo rescue technique did not work well, but collaboration with Chinese scientists provided a new tissue culture method, involving the use of coconut oil, which proved highly successful.

The resulting rice varieties, known as the New Rices for Africa (NERICAs), grow well in drought-prone, upland conditions, are resistant to local pests and disease, tolerant of poor nutrient conditions and mineral toxicity and show early vigorous growth, crowding out weeds, all of which are characteristics of the African species of rice. Later in their development, characteristics of the Asian rice species appear:

they produce more erect leaves and full panicles of grain and are ready to harvest after 90 to 100 days, 30 to 50 days earlier than current varieties. Due to significantly higher yields than current varieties (up to 4 tons per hectare under low inputs) Uganda has been able to reduce its rice imports by half during the period 2002 to 2007. A shift from maize to NERICA production in Uganda with proper crop rotation was found to increase income by $250 per hectare.

Recombinant DNA

The technique of recombinant DNA, also known as genetic engineering (GE) technology or genetic modification (GM), involves the direct transfer of genes from one organism to another.

Recombinant DNA technology permits the extraction of genes from one DNA helix and insertion into another. Sections of DNA code are located, cleaved and reattached using a variety of naturally occurring enzymes such as restriction enzymes and DNA ligases. The nature and aim of the process are the same as for conventional breeding in that during the crossing of plants and animals, chromosomes transfer pieces from one individual to another but recombinant DNA technology has several advantages. First, combinations of chromosomes are predetermined and the process of transfer exact so the resulting offspring are precisely rather than randomly determined. Second, the process is much quicker and finally, the library with which to draw genetic material from is much larger and less restricted by geographic or biological boundaries.

Artificially transferring a gene from one organism to another can be approached in several different ways. One of the first and most successful techniques to be developed employs *Agrobacterium tumefaciens*, a bacterium that naturally invades such plants as potatoes, tomatoes and lucerne, as a vector to transport new genes to plants. It is a surprisingly easy process: first, recombinant DNA techniques are used to replace the genes *A. tumefaciens* normally integrates into plant chromosomes with genes for desired new traits (e.g. insect resistance), then pieces of leaves are dipped into a suspension of bacteria and cultured to produce whole plants containing the new gene. This technique now works effectively on cereals as well as broadleaf plants. Alternatively the new genes can be injected more forcefully. One such method is to apply a coat of gold particles to the DNA which are then fired into the plant cell with a microparticle gene gun. It is customary for millions of cells to be treated and those in which the new gene have lodged are identified by means of a marker and cultured.

The potential for transformation of the agricultural sector through the use of genetic engineering is great. New plant varieties and animal breeds could deliver higher yields but also greater tolerance and adaptability to worsening global conditions such as increasing drought, salinity and chemical toxicity. Resistance to bacterial and viral diseases and to insect pests could, once conferred, provide huge savings in terms of crop losses and potentially lessen agriculture's dependence on chemical inputs such as fungicides and pesticides.

The benefits of the practical applications of genetic engineering go beyond the economic. Indeed not only could the quantity of food produced be augmented but its quality improved, thus providing gains for both the food and nutritional insecure portions of society. An example of the latter is biofortification – the production of crops with enhanced nutritional value. In Asia, and many parts of sub-Saharan Africa, poor families consume rice as the basic staple of their diet and babies are often weaned on rice gruel but surprisingly, most cereals and other staples are deficient in a number of proteins and other micronutrients. In rice, beta-carotene, the precursor of vitamin A, is not present in the grain endosperm (it is present in the leaves and stalks, and minute amounts are contained in brown unmilled rice) and, therefore, does not provide this critical micronutrient in sufficient amounts. Golden Rice, named after its distinctive colour, has been developed by Ingo Potrykus of the Swiss Federal Institute of Technology, and his colleague Peter Beyer of the University of Freiburg to try to address the 250 million preschool children who are vitamin A deficient, a state that can lead to blindness.

Two daffodil genes and one bacterial gene were first transferred into rice as a way of increasing beta-carotene content. The biochemical pathway leading to beta-carotene is largely present in the rice grain but lacks two crucial enzymes: phytoene synthase (*psy*) – provided by a daffodil gene – and carotene desaturase (*crt1I*) – provided by a bacterium gene.

In the greenhouse this transfer gave beta-carotene levels of about 1.6 µg/g, significant but not large. Subsequently, scientists at Syngenta have found new versions of the *psy* gene in maize; which when introduced to rice increased the beta-carotene levels to 31 µg/g. Given a conversion ratio of beta-carotene to vitamin A of 4:1, the new Golden Rice (Golden Rice 2) will be able to provide the necessary boost to daily diets, even after 6 months of storage.

The second-generation Golden Rice has already been developed into locally appropriate varieties in the Philippines and India and is forecast to be available in other countries in the next 2 to 4 years.

The consequent benefits to Asia's GDP have been estimated at US$18 billion annually (Anderson *et al.*, 2005).

COMBATING DROUGHT

Biotechnology, however, is not a magic bullet. It has to be combined with the utilisation of other forms of appropriate technology and governance. A good example lies in the efforts to combat the increasing levels of drought caused by global warming.

Managing irrigation systems

Irrigation, if it is to be effective, has to be reliable, otherwise much or all of the potential benefit will be lost. Reliability depends, in turn, on an efficient and responsive organisation.

Such an organisation can be established through partnerships between government and local communities. A reliable water supply system has been in operation in the valleys of northern Thailand for some 200 years. Dams made of stone and wood maintained by local communities are linked to irrigation systems governed by representative bodies known as *muang*. In a strategy to provide year-round irrigation, the government, in the 1960s and 1970s, began the construction of large-scale diversion systems. So reliable is the community-managed arrangement that in many places the government grafted these extensions on to local structures. Indeed areas where triple-cropping is practised tend to coincide with the locations of these joint systems, most likely as a result of the reliability of the water supply, justifying the risk of planting a high-value third-season crop.

Intermediate technologies

In many areas within the least developed countries the development of large-scale irrigation systems is not feasible either physically or economically and so much attention has turned towards small-scale systems such as drip irrigation and treadle pumps. Pumps can replace the labour-intensive task of lifting water from shallow wells by bucket but such pumps were traditionally oil-powered and so expensive to run as well as purchase. New, modern treadle pumps, first developed by local people in Bangladesh in the early 1980s, are human-powered, efficient and easy to use and maintain. The machine is relatively simple:

a farmer can irrigate their field purely by standing on the apparatus and pumping their feet. This pumping action induces suction in turn raising water from a natural source or a dug well. Treadle pumps are also relatively inexpensive thanks to a combination of public subsidies with private manufactures and servicing, and community involvement. The technology has been improved upon by a number of enterprising engineers, producing a variety of effective and easy-to-maintain designs which can be used to irrigate up to 2 acres of land without any motor or fuel.

The zai system

Even in very dry situations there is usually some water available and the challenge is to harvest this in a cheap and efficient manner. Some of the most ingenious solutions are systems of so-called water harvesting. Farmers in much of the developing world are confronted by drought on an annual basis. Not surprisingly they have developed a number of drought-tolerant 'traditional technologies'.

The zai system is a technology pioneered decades ago by farmers in the parched soils of northwest Burkina Faso as a way of generating more arable land. It has since spread throughout similar environments, to the rest of Burkina Faso and in Mali and Niger.

Zais or holes 20–30 cm in diameter and 10–15 cm deep are dug in rows across fields during the dry season. Each zai is allowed to fill with leaves and sand as the winds blow across the land. Farmers add manure that, during the dry months, attracts termites, which dig an extensive network of underground tunnels beneath the holes and bring up nutrients from the deeper soils. Rainwater, when it comes, is captured in the zais, in which sorghum and millet seeds are sown, and stone and earth ridges are also constructed around the field, in order to slow run-off. Overall water capture is enhanced, even in the drought-prone environment of the Sahel, through manure, which limits water loss through drainage, and through the termites' porous tunnels, which permit deep infiltration.

Use of this technique has consistently led to increased yields. In a study conducted in Bafaloubé, Mali, between 2000 and 2003, sorghum and millet yields increased by between 80% and 170%. The labour required to build the zais in the first year is quite high, but after that farmers may reuse the holes, or dig more between the existing ones. In many cases, after around 5 years the entire land surface will be improved.

Conservation farming

Conservation farming is an alternative to the common practice of tilling soils in fields prior to sowing seeds, a practice which can make soils vulnerable to erosion and drought, harm soil structure and increase water loss. Conservation farming, an ecological, intermediate technique, is fast becoming popular as a means of protecting soils and improving their fertility. Its main principle is to reduce the use of tillage, often completely, as a way of conserving topsoil, soil structure and beneficial insects. Other benefits include the saving of labour used for ploughing and the retaining of carbon and organic matter in the soil, which not only leads to a higher microbial content but also fixes carbon and thus reduces carbon emissions from agriculture.

In southern Africa, interest in conservation agriculture has been noticeably growing as evidenced by joint investigations, by the NGO Concern Worldwide and local government bodies, into the substitution of the traditional long fallow system with conservation farming. In the former, woodland is felled and burnt before being ploughed and sown to maize. Crops are grown for only a couple of years and the land then takes several decades to return to a state where it can be felled and burnt again. Under conservation farming, on the other hand, the use of ploughing is absent and seeds are sown in small 'pockets' in the soil to which have been added two cupfuls of manure and a bottle top of fertiliser. After harvest, the soil is covered with the stems and leaves of the maize and next year's seed is sown several months later in the same holes. Despite the need to hoe weeds, the labour is much less than in the conventional systems while yields are high – some 4–5 tons of maize when growing new drought-tolerant hybrids. In addition to building carbon in the cropped soil, such a system should allow tree or shrub cover to remain unburned more or less permanently, so increasing carbon sequestration and maintaining soil carbon levels, creating a more stable and sustainable farming system.

Soda cap fertilisers

Conventional, synthetic fertilisers, if used to excess, can be costly and inefficient and may pollute water sources. For this reason there has been a move, since the 1960s, to find ways of using fertilisers in a more sparing and selective manner.

One such method, developed in Niger, is micro-dosing, which involves a soda-bottle cap filled with a 6-g mix of phosphorus and nitrogen fertiliser. The contents of the cap are poured into each hole before the seed is planted. Its advantages lie in the fact that it is an everyday household item and thus easy to obtain, and in the reduction in fertiliser usage it brings: it equates to using 4 kg per hectare, three to six times less than used in Europe and North America, but still very effective. Decreasing rates of fertiliser application to less than one-sixth leads to an average 55–70% increase in millet yield and at lower cost. ICRISAT (International Crops Research Institute for the Semi-Arid Tropics) has also shown that by reducing the amounts of nitrogen (N) and phosphate (P) fertiliser used by farmers, crops are more able to absorb water. This is a good example of where an appropriate technology is a win–win proposition.

Chaperone genes

Limited supplies of water for agriculture will increasingly test the ingenuity of engineers and agronomists but also plant-breeders. Developing crop plants (or livestock) able to make much more efficient use of water and thrive under drought conditions is perhaps the most sustainable solution to the water challenges agriculture faces.

In many instances the first step in breeding is to expose new varieties to a range of stresses, including drought, in field experiments such as mother–baby trials. These are participatory trials of crop varieties involving central 'mother' sites, managed by schools, colleges or extension agents and testing a range of varieties, and 'baby' sites, located on farmers' fields and used to investigate farmer perception of a subset of the these varieties. The biological basis of drought tolerance can then be investigated for those varieties showing a high degree of tolerance as a platform for improved breeding. Breeding for drought tolerance has so far been difficult given the 'low heritability of tolerance', the variety of effects of drought in the plant that depend on the timing of the drought and the limited understanding of drought physiology.

Recently, however, a suite of genes that regulate drought adaptation and/or tolerance have been identified and their combination with transgenic approaches has led to rapid progress in improving drought tolerance. One such gene is a so-called 'chaperone' gene. Such genes can confer tolerance to stress of various kinds, including cold, heat and lack of moisture. The product of the gene helps to repair misfolded proteins caused by stress and so the plant recovers more quickly. One such gene found in bacterial RNA has been transferred to maize with excellent results in field trials. Plants with the gene show 12–24% increase in growth in high drought situations compared with plants without the gene. Field trials are now being carried out in Africa.

Through a combination of conventional breeding, marker-assisted selection and GM technology, new drought-tolerant and royalty-free varieties of maize, named Water Efficient Maize for Africa (WEMA), are to be delivered over the next decade. The project was established in 2008 through a public–private partnership between the African Agricultural Technology Foundation, the International Maize and Wheat Improvement Center (CIMMYT), Monsanto and national agricultural research systems in the participating countries (Kenya, Tanzania, Uganda, Mozambique and South Africa). The aim is to deliver maize varieties that will increase yields by around 20–35% under moderate drought conditions in comparison to current varieties. This could result in an estimated 2 million tons of additional food with benefits to 14 to 21 million people.

CONCLUSION

In the 1960s and 1970s when many countries in Asia and Latin America were facing acute food shortages and rapidly rising populations, the advent of the so-called Green Revolution provided a partial solution. Food production grew dramatically and many millions of poor people benefited. The Green Revolution had its limitations, however. It largely passed Africa by and even in the prime, high-production Green Revolution lands poor people remained poor and hungry. There were also undesirable environmental effects resulting from overuse of fertilisers and pesticides.

Today we face both acute and chronic crises which are in many respects of greater magnitude. We need to increase food production, perhaps by as much as 100%, if we are to cope with rising populations, rising per capita incomes and changing diets, and with the growing demand for biofuels. At the same time we have to cope with the challenge of rising oil prices and the costs of fertilisers, as well as shortage of good-quality land and water, declining increases in the yields of some staples and, perhaps most alarming of all, the threats posed by global warming.

In some respects we need a repeat of the Green Revolution but with greater productivity increases, better access to the products by the poor and needy and much reduced environmental impact. It also has to be resilient to climate change. I have called this goal 'A Doubly Green Revolution' (Conway, 2012).

Appropriate technologies exist in many forms – ranging from traditional to modern genetic technologies. All have something to offer, although none is a magic bullet. Each has to be assessed on its merits, the costs and benefits, and the appropriateness to often small farmers in relatively remote situations. As part of the package of technologies available the new genetics, comprising marker-aided selection, cell and tissue culture and genetic modification, has much to offer, because so many desirable characteristics can be combined in seeds that are potentially readily available to poor farmers at prices they can afford.

REFERENCES

Anderson, K., Jackson, L. A., and Nielsen, C. P. (2005). Genetically modified rice adoption: implications for welfare and poverty alleviation. *Journal of Economic Integration* **20**, 771–788.

Conway, G. (2012). *A Billion Hungry: Can We Feed the World*. Cornell University Press, Ithaca, NY.

FAO (2006). *World Agriculture: Towards 2030/2050*, Interim report, *Prospects for Food, Nutrition, Agriculture and Major Commodity Groups*. FAO, Global Perspective Studies Unit, Rome.

FAO (2009). *The State of Food Insecurity in the World 2009: Economic Crises – Impacts and Lessons Learned*. FAO, Rome.

FAO (2012). *The State of Food Insecurity in the World 2012: Economic Growth Is Necessary but not Sufficient to Accelerate Reduction of Hunger and Malnutrition*. FAO, Rome.

Lobell, D. B., Bänzinger, M., Magorokosho, C., and Vivek, B. (2011). Nonlinear heat effects on maize as evidenced by historical yield trials. *Nature Climate Change* **1**, 42–45.

OECD–FAO (2009). *Agricultural Outlook 2009–2018*. OECD and FAO, Rome.

REN21 (2011). *Renewables 2011: Global Status Report*. REN21 Secretariat, Paris.

Smith, P., Martino, D., Cai, Z., *et al.* (2007). Agriculture. In *Climate Change 2007: Mitigation*, Contribution of Working Group III to the Fourth Assessment Report of the Intergovernmental Panel on Climate Change, ch. 8. Cambridge University Press.

World Bank (2006). *Reengaging in Agricultural Water Management: Challenges and Options*. World Bank, Washington, DC.

4

The economic and environmental impact of first-generation biotech crops

INTRODUCTION

This chapter provides insights into the reasons why so many farmers around the world have adopted crop biotechnology and continue to use it in their production systems since the technology first became available on a widespread commercial basis in the mid-1990s. The chapter draws, and is largely based on, the considerable body of peer-reviewed literature available that has examined these issues.[1] It specifically focuses on the farm-level economic effects, the production effects, the environmental impact resulting from changes in the use of insecticides and herbicides, and the contribution towards reducing greenhouse gas (GHG) emissions.

The report is based on extensive analysis of existing farm-level impact data for biotechnology crops. Whilst primary data for impacts of commercial cultivation were not available for every crop, in every year and for each country, a substantial body of representative research and analysis is available and this has been used as the basis for the analysis presented. This has been supplemented by the author's own data collection and analysis. The analysis of pesticide usage also takes into consideration changes in the pattern of herbicide use in recent years that reflect measures taken by some farmers to address issues of weed resistance to the main herbicide (glyphosate) used with herbicide-tolerant biotechnology crops. For additional information, readers should consult a detailed examination of these issues in Brookes and Barfoot (2012).

---

[1] Data from other sources, including industry, are used where no other sources of representatative data are available. All sources and assumptions used are detailed in the chapter.

Successful Agricultural Innovation in Emerging Economies: New Genetic Technologies for Global Food Production, eds. David J. Bennett and Richard C. Jennings. Published by Cambridge University Press. © Cambridge University Press 2013.

ECONOMIC IMPACTS

Farm income effects

Crop biotechnology has had a significant positive impact on global farm income derived from a combination of enhanced productivity and efficiency gains (Table 4.1). In 2010, the direct global farm income benefit from biotechnology crops was US$14 billion. This is equivalent to having added 4.3% to the value of global production of the four main crops of soya beans, maize, canola and cotton. Since 1996, farm incomes have increased by US$78.4 billion.

Table 4.1 *Global farm income benefits from growing biotechnology crops 1996–2010*

Trait	Increase in farm income 2010[a] (US$)	Increase in farm income 1996–2010[a] (US$)	Farm income benefit in 2010 as % of total value of production of these crops in biotechnology adopting countries[b]	Farm income benefit in 2010 as % of total value of global production of crop[b]
GM herbicide-tolerant soya beans	3 299.8	28 389.2	3.5	3.2
GM herbicide-tolerant maize	438.5	2 672.8	0.5	0.3
GM herbicide-tolerant cotton	148.3	1 062.4	0.4	0.3
GM herbicide-tolerant canola	472.4	2 657.8	5.7	1.4
GM insect-resistant maize	4 522.3	18 969.3	5.4	3.2
GM insect-resistant cotton	5 030.1	24 371.9	14.0	11.9
Others[c]	90.2	301.5	N/a	N/a
Totals	14 001.6	78 424.9	6.25	4.3

Notes: All values are nominal.

[a] Farm income calculations are net farm income changes after inclusion of impacts on yield, crop quality and key variable costs of production (e.g. payment of seed premiums, impact on crop protection expenditure).

[b] Totals for the value shares exclude 'other crops' (i.e. relate to the four main crops of soya beans, maize, canola and cotton).

[c] Others, virus-resistant papaya and squash and herbicide-tolerant sugar beet.

The largest gains in farm income in 2010 have arisen in the cotton sector, largely from yield gains. The US$5 billion additional income generated by GM insect-resistant (GM IR) cotton in 2010 has been equivalent to adding 14% to the value of the crop in the biotechnology growing countries, or adding the equivalent of 11.9% to the US$42 billion value of the global cotton crop in 2010.

Substantial gains have also arisen in the maize sector through a combination of higher yields and lower costs. In 2010, maize farm income levels in the biotechnology-adopting countries increased by almost US$5 billion and since 1996, the sector has benefited from an additional US$21.6 billion. The 2010 income gains are equivalent to adding 6% to the value of the maize crop in these countries, or 3.5% to the US$139 billion value of total global maize production. This is a substantial increase in value added terms for two new maize seed technologies.

Significant increases to farm incomes have also resulted in the soya bean and canola sectors. The GM herbicide-tolerant (GM HT) technology in soya beans has boosted farm incomes by US$3.3 billion in 2010, and since 1996 has delivered over US$28 billion of extra farm income (the highest cumulative increase in farm income of the biotechnology traits). In the canola sector (largely North American) an additional US$2.7 billion has been generated (1996–2010).

Table 4.2 summarises farm income impacts in key biotechnology-adopting countries. This highlights the important farm income benefit arising from GM HT soya beans in South America (Argentina, Bolivia, Brazil, Paraguay and Uruguay), GM IR cotton in China and India and a range of GM cultivars in the USA. It also illustrates the growing level of farm income benefits being obtained in South Africa, the Philippines, Mexico and Colombia.

In terms of the division of the economic benefits obtained by farmers in developing countries relative to farmers in developed countries, Table 4.3 shows that in 2010, 54.8% of the farm income benefits have been earned by developing country farmers.[2] The vast majority of these income gains for developing country farmers have been from GM IR cotton and GM HT soya beans. Over the 15 years 1996–2010, the

---

[2] The authors acknowledge that the classification of different countries into developing or developed country status affects the distribution of benefits between these two categories of country. The definition used in this chapter is consistent with the definition used by James (2009).

Table 4.2 *GM crop farm income benefits 1996–2010 selected countries (US$ million)*[a]

Country	GM HT soya beans	GM HT maize	GM HT cotton	GM HT canola	GM IR maize	GM IR cotton	Total[b]
USA[c]	12109.0	2225.0	875.4	225.5	16326.4	3267.4	35028.7
Argentina	11217.3	314.2	68.6	N/a	309.2	246.4	12155.9
Brazil	3888.3	17.8	36.4	N/a	655.5	3.8	4601.8
Paraguay	655.0	N/a	N/a	N/a	N/a	N/a	655.0
Canada	163.3	57.7	N/a	2418.9	637.8	N/a	3277.7
South Africa	7.2	3.2	2.7	N/a	769.0	27.1	809.2
China	N/a	N/a	N/a	N/a	N/a	10911.2	10911.2
India	N/a	N/a	N/a	N/a	N/a	9395.2	9395.2
Australia	N/a	N/a	31.5	13.4	N/a	362.8	407.7
Mexico	4.7	N/a	36.7	N/a	N/a	95.0	136.4
Philippines	N/a	54.6	N/a	N/a	115.7	N/a	170.3
Romania	44.6	N/a	N/a	N/a	N/a	N/a	44.6
Uruguay	76.4	N/a	N/a	N/a	8.0	N/a	84.4
Spain	N/a	N/a	N/a	N/a	113.9	N/a	113.9
Other EU	N/a	N/a	N/a	N/a	13.6	N/a	13.6
Colombia	N/a	0.3	11.1	N/a	15.6	11.4	38.4
Bolivia	223.1	N/a	N/a	N/a	N/a	N/a	223.1

Notes: All values are nominal.

[a] Farm income calculations are net farm income changes after inclusion of impacts on yield, crop quality and key variable costs of production (e.g. payment of seed premiums, impact on crop protection expenditure).

[b] Not included in the table is US$4.3 million extra farm income from GM HT sugar beet in Canada.

[c] USA total figure also includes US$296.4 million for other crops/traits (not included in the table).

cumulative farm income gain derived by developing country farmers was 50% (US$39.24 billion).

Examining the cost farmers pay for accessing GM technology, Table 4.4 shows that across the four main biotechnology crops, the total cost in 2010 was equal to 28% of the total technology gains (inclusive of farm income gains plus cost of the technology payable to the seed supply chain[3]).

[3] The cost of the technology accrues to the seed supply chain including sellers of seed to farmers, seed multipliers, plant-breeders, distributors and the GM technology providers.

Table 4.3 *GM crop farm income benefits 2010: developing versus developed countries (US$ million)*

Crop	Developed	Developing[a]
GM HT soya beans	970.8	2329.0
GM IR maize	3868.6	653.7
GM HT maize	274.3	164.2
GM IR cotton	586.0	4444.1
GM HT cotton	65.3	83.0
GM HT canola	472.4	0
GM virus-resistant papaya and squash and GM HT sugar beet	90.2	0
Total	6327.6	7674.0

[a] Developing countries: all countries in South America, Mexico, Honduras, Burkino Faso, India, China, the Philippines and South Africa.

For farmers in developing countries the total cost was equal to 17% of total technology gains, whilst for farmers in developed countries the cost was 37% of the total technology gains. Whilst circumstances vary between countries, the higher share of total technology gains accounted for by farm income gains in developing countries relative to the farm income share in developed countries reflects factors such as weaker provision and enforcement of intellectual property rights in developing countries and the higher average level of farm income gain on a per-hectare basis derived by developing country farmers relative to developed country farmers.

The analysis presented above is largely based on estimates of average impact in all years. Recognising that pest and weed pressure varies by region and year, additional sensitivity analysis was conducted for the crop/trait combinations where yield impacts were identified in the literature. This sensitivity analysis was undertaken for two levels of impact assumption; one in which all yield effects in all years were assumed to be 'lower than average' (level of impact that largely reflected yield impacts in years of low pest/weed pressure) and one in which all yield effects in all years were assumed to be 'higher than average' (level of impact that largely reflected yield impacts in years of high pest/weed pressure). The results of this analysis suggest a range of positive direct farm income gains in 2010 of +US$12 billion to +US$18.5 billion and over the 1996–2010 period, a range of +US$68.5 billion to +US$93.1 billion (Table 4.5). This range is broadly within 87% to 119% of the main estimates of farm income presented above.

Table 4.4 Cost of accessing GM technology (US$ million) relative to the total farm income benefits, 2010

Crop	Cost of technology: all farmers	Farm income gain: all farmers	Total benefit of technology to farmers and seed supply chain	Cost of technology: developing countries	Farm income gain: developing countries	Total benefit of technology to farmers and seed supply chain: developing countries
GM HT soya beans	1605.1	3299.8	4904.9	564.4	2329.0	2893.4
GM IR maize	1767.5	4522.3	6289.8	515.2	653.7	1168.9
GM HT maize	789.6	438.5	1228.1	94.6	164.2	258.8
GM IR cotton	610.3	5030.1	5640.4	400.8	4444.1	4844.9
GM HT cotton	348.0	148.3	496.3	32.8	83.0	115.8
GM HT canola	122.8	472.4	595.2	N/a	N/a	N/a
Others	72.0	90.2	162.2	N/a	N/a	N/a
Total	5315.3	14001.6	19316.9	1607.8	7674.0	9281.8

Note: Cost of accessing technology based on the seed premiums paid by farmers for using GM technology relative to its conventional equivalents.

Table 4.5 *Direct farm income benefits 1996–2010 under different impact assumptions (US$ million)*

Crop	Consistent below average pest/weed pressure	Average pest/weed pressure (main study analysis)	Consistent above average pest/weed pressure
Soya beans	28220	28389.2	28558
Maize	15772	21642.1	27618
Cotton	22065	25434.3	33520
Canola	2281	2657.8	2793
Others	159	301.5	631
Total	68497	78424.9	93120

Note: No significant change to soya bean production under all three scenarios as almost all gains due to cost savings and second-crop facilitation.

Indirect (non-pecuniary) farm-level impacts

As well as the tangible and quantifiable impacts on farm profitability presented above, there are other important, more intangible (difficult to quantify) impacts of an economic nature.

Many of the studies of the impact of biotechnology crops[4] have identified the following reasons as being important influences for adoption of the technology:

Herbicide-tolerant crops

- Increased management flexibility and convenience that comes from a combination of the ease of use associated with broad-spectrum, post-emergent herbicides like glyphosate and the increased/longer time-window for spraying. This not only frees up management time for other farming activities but also allows additional scope for undertaking off-farm, income-earning activities.

- In a conventional crop, post-emergent weed control relies on herbicide applications after the weeds and crop are established. As a result, the crop may suffer 'knock-back' to its growth from the effects of the herbicide. In the GM HT crop, this problem is avoided because the crop is tolerant to the herbicide.

---

[4] For example, relating to HT soya beans, USDA (1999), Carpenter and Ginessi (1999), Qaim and Traxler (2002), Brookes (2008); relating to IR maize, Rice (2004); relating to IR cotton, Ismael *et al.* (2002), Pray *et al.* (2002).

- Facilitates the adoption of conservation or no-tillage systems. This provides for additional cost savings such as reduced labour and fuel costs associated with ploughing, additional moisture retention and reductions in levels of soil erosion.
- Improved weed control has contributed to reduced harvesting costs – cleaner crops have resulted in reduced times for harvesting. It has also improved harvest quality and led to higher levels of quality price bonuses in some regions and years (e.g. HT soya beans and HT canola in the early years of adoption respectively in Romania and Canada).
- Elimination of potential damage caused by soil-incorporated residual herbicides in follow-on crops and less need to apply herbicides in a follow-on crop because of the improved levels of weed control.
- A contribution to the general improvement in human safety (as manifest in greater peace of mind about own and worker safety) from a switch to more environmentally benign products.

Insect-resistant crops

- Production risk management/insurance purposes – the technology takes away much of the worry of significant pest damage occurring and is, therefore, highly valued. Piloted in 2008 and more widely operational from 2009, US farmers using stacked maize traits (containing insect-resistant and herbicide-tolerant traits together) are being offered discounts on crop insurance premiums (for crop losses) equal to US$12.97/ha in 2010. Over the 3 years, this has applied to 12.7 million ha, resulting in insurance premiums savings of US$137.8 million.
- A 'convenience' benefit derived from having to devote less time to crop walking and/or applying insecticides.
- Savings in energy use – mainly associated with less use of aerial spraying.
- Savings in machinery use (for spraying and possibly reduced harvesting times).
- Higher quality of crop. There is a growing body of research evidence relating to the superior quality of GM IR maize relative to conventional and organic maize from the perspective of having lower levels of

mycotoxins.[5] Evidence from Europe (as summarised in Brookes, 2008) has shown a consistent pattern in which GM IR maize exhibits significantly reduced levels of mycotoxins compared to conventional and organic alternatives. In terms of revenue from sales of maize, however, no premiums for delivering product with lower levels of mycotoxins have to date been reported, although where the adoption of the technology has resulted in reduced frequency of crops failing to meet maximum permissible fumonisin levels in grain maize (e.g. in Spain), this delivers an important economic gain to farmers selling their grain to the food-using sector. GM IR maize farmers in the Philippines have also obtained price premiums of 10% (Yorobe, 2004) relative to conventional maize because of better quality, less damage to cobs and lower levels of impurities.

- Improved health and safety for farmers and farmworkers (from reduced handling and use of pesticides, especially in developing countries where many apply pesticides with little or no use of protective clothing and equipment).
- Shorter growing season (e.g. for some cotton-growers in India) which allows some farmers to plant a second crop in the same season (notably maize in India). Also some Indian cotton-growers have reported knock-on benefits for bee-keepers as fewer bees are now lost to insecticide spraying.

Some of the economic impact studies have attempted to quantify some of these benefits (e.g. Yorobe, 2004: see above). Where identified, these cost savings have been included in the analysis presented above. Nevertheless, it is important to recognise that these largely intangible benefits are considered by many farmers as a primary reason for adoption of GM technology, and in some cases farmers have been willing to adopt for these reasons alone, even when the measurable impacts on yield and direct costs of production suggest marginal or no direct economic gain.

Since the early 2000s, a number of farmer-survey based studies in the USA have also attempted to better quantify these non-pecuniary benefits. These studies have usually employed contingent valuation

[5] Mycotoxins are a group of toxic fungal metabolites known as fumonisins. They are produced by certain mould species of the genus *Fusarium* which thrives in the environmet of maize plants that have been damaged by corn-boring pests.

Table 4.6 *Values of non-pecuniary benefits associated with biotechnology crops in the USA*

Survey	Median value (US$/ha)
2002 IR (to rootworm) maize-growers survey	7.41
2002 soya bean (HT) farmers survey	12.35
2003 HT cropping survey (maize, cotton and soya beans) – North Carolina	24.71
2006 HT (flex) cotton survey	12.35[a]

[a] Relative to first-generation HT cotton.
Source: Marra and Piggott (2006, 2007).

techniques[6] to obtain farmers' valuations of non-pecuniary benefits. A summary of these findings is shown in Table 4.6.

Aggregating the impact to US crops 1996–2010

The approach used to estimate the non-pecuniary benefits derived by US farmers from biotechnology crops over the period 1996–2010 has been to draw on the values identifed by Marra and Piggott (2006, 2007) (Table 4.6) and to apply these to the biotechnology-crop planted areas during this 15-year period. Figure 4.1 summarises the values for non-pecuniary benefits derived from biotechnology crops in the USA (1996–2010) and shows an estimated (nominal value) benefit of US$1.02 billion in 2010 and a cumulative total benefit (1996–2010) of US$7.62 billion. Relative to the value of direct farm income benefits derived by US farmers who used crop biotechnology, the non-pecuniary benefits were equal to 18.5% of the total US direct income benefits in 2010 and 21.6% of the total cumulative (1996–2010) US direct farm income. This highlights the important contribution this category of benefit has had on biotechnology trait adoption levels in the USA, especially where the direct farm income benefits have been identified to be relatively small (e.g. HT cotton).

Estimating the impact in other countries

It is evident from the research carried out to compile the data and analysis in this chapter that GM-technology-using farmers in other countries also value the technology for a variety of non-pecuniary/

---

[6] Survey-based method of obtaining valuations of non-market goods that aim to identify willingness to pay for specific goods (e.g. environmental goods, peace of mind, etc.) or willingness to pay to avoid something being lost.

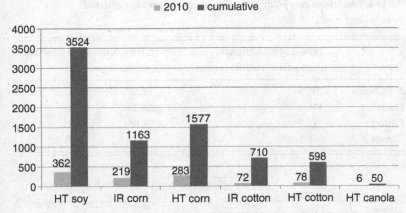

Figure 4.1 Non-pecuniary benefits derived by US farmers 1996–2010 by trait (US$ million)

intangible reasons. The most appropriate methodology for identifying these non-pecuniary benefit valuations in other countries would be to repeat the type of US farmer surveys in other countries. Unfortunately, the author is not aware of any such studies having been undertaken to date.

Production effects of the technology

Based on the yield assumptions used in the direct farm income benefit calculations presented above and taking account of the second soya bean crop facilitation in South America, biotechnology crops have added important volumes to global production of maize, cotton, canola and soya beans since 1996 (Table 4.7).

The biotechnology IR traits, used in the maize and cotton sectors, have accounted for 98% of the additional corn production and 99.4% of the additional cotton production. Positive yield impacts from the use of this technology have occurred in all user countries except GM IR cotton in Australia[7] when compared to average yields derived from crops using conventional technology (such as application of insecticides and seed treatments). The average yield impact across the total area planted to

---

[7] This reflects the levels of *Heliothis* (plant-eating insect) pest control previously obtained with intensive insecticide use. The main benefit and reason for adoption of this technology in Australia has arisen from significant cost savings (on insecticides) and the associated environmental gains from reduced insecticide use.

Table 4.7 *Additional crop production arising from positive yield effects of biotechnology crops*

Crop	1996–2010 additional production (million tonnes)	2010 additional production (million tonnes)
Soya beans	97.5	13.07
Maize	159.4	28.29
Cotton	12.5	2.06
Canola	6.1	0.65

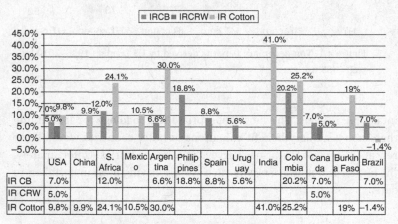

	USA	China	S. Africa	Mexico	Argentina	Philippines	Spain	Uruguay	India	Colombia	Canada	Burkina Faso	Brazil
IR CB	7.0%		12.0%		6.6%	18.8%	8.8%	5.6%		20.2%	7.0%		7.0%
IR CRW	5.0%										5.0%		
IR Cotton	9.8%	9.9%	24.1%	10.5%	30.0%				41.0%	25.2%		19%	−1.4%

Figure 4.2 Average yield impact of biotechnology IR traits 1996–2010 by country and trait. IRCB, resistant to corn-boring pests; IRCRW, resistant to corn rootworm.

these traits over the 15-year period since 1996 has been +9.6% for maize traits and +14.4% for cotton traits (Figure 4.2).

Although the primary impact of biotechnology HT technology has been to provide more cost-effective (less expensive) and easier weed control (relative to weed control obtained from conventional technology), it has delivered higher yields in some countries, e.g. HT soya beans in Romania where average yields increased by over 30% in the first years of adoption, HT maize in the Philippines where yield increases of between 5% and 15% have occurred and HT soya beans in Bolivia where average yields increased by over 15%.

Biotechnology HT soya beans have also facilitated the adoption of no-tillage production systems, shortening the production cycle. This advantage enables many farmers in South America to plant a crop of soya beans immediately after a wheat crop in the same growing season. This second crop, additional to traditional soya bean production, has

Table 4.8 *Additional crop production arising from positive yield effects of biotechnology crops 1996–2010 under different pest/weed pressure assumptions and impacts of the technology (million tonnes)*

Crop	Consistent below-average pest/weed pressure	Average pest/weed pressure (main study analysis)	Consistent above-average pest/weed pressure
Soya beans[a]	97.0	97.5	98.0
Maize	137.2	159.4	197.7
Cotton	8.8	12.5	18.2
Canola	4.6	6.1	6.5

[a] No significant change to soya bean production under all three scenarios as 99% of production gain due to second-cropping facilitation of the technology.

added 96.1 million tonnes to soya bean production in Argentina and Paraguay between 1996 and 2010 (accounting for 98.5% of the total biotechnology-related additional soya bean production).

Using the same sensitivity analysis as applied to the farm income estimates presented above to the production impacts (one scenario of consistent lower-than-average pest/weed pressure and one of consistent higher-than-average pest/weed pressure), Table 4.8 shows the range of production impacts.

ENVIRONMENTAL IMPACT FROM CHANGES IN INSECTICIDE AND HERBICIDE USE

To examine this impact, the study has analysed both active ingredient use and utilised the indicator known as the Environmental Impact Quotient (EIQ) (Kovach *et al.*, 1992) to assess the broader impact on the environment (plus impact on animal and human health). The EIQ distils the various environmental and health impacts of individual pesticides in different GM and conventional production systems into a single 'field value per hectare' and draws on key toxicity and environmental exposure data related to individual products. It therefore provides a better measure to contrast and compare the impact of various pesticides on the environment and human health than weight of active ingredient alone. Readers should, however, note that the EIQ is an indicator only and does not take into account all environmental issues and impacts. In the analysis of GM HT technology we have assumed that the conventional alternative delivers the same level of weed control as occurs in the GM HT production system.

Biotechnology traits have contributed to a significant reduction in the environmental impact associated with insecticide and herbicide use on the areas devoted to biotechnology crops (Table 4.9). Since 1996, the use of pesticides on the biotechnology crop area was reduced by 448 million kg of active ingredient (9% reduction), and the environmental impact associated with herbicide and insecticide use on these crops, as measured by the EIQ indicator, fell by 17.9%.

In absolute terms, the largest environmental gain has been associated with the adoption of GM IR cotton (23.9% reduction in the volume of active ingredient used and a 26% reduction in the EIQ indicator 1996–2010) and reflects the significant reduction in insecticide use that the technology has allowed, in what has traditionally been an intensive user of insecticides.

The volume of herbicides used in biotechnology soya bean crops also decreased by 28.8 million kg (1996–2010), a 1.4% reduction, whilst the overall environmental impact associated with herbicide use on these crops decreased by a significantly larger 16.2%. This highlights the switch in herbicides used with most GM HT crops to active ingredients with a more environmentally benign profile than the ones generally used on conventional crops.

Important environmental gains have also arisen in the maize and canola sectors. In the maize sector, herbicide and insecticide use decreased by 212.8 million kg (1996–2010) and the associated environmental impact of pesticide use on this crop area decreased due to a combination of reduced insecticide use (37.7%) and a switch to more environmentally benign herbicides (11.5%). In the canola sector, farmers reduced herbicide use by14.4 million kg (an 18.2% reduction) and the associated environmental impact of herbicide use on this crop area fell by 27.4% (due to a switch to more environmentally benign herbicides).

In terms of the division of the environmental benefits associated with less insecticide and herbicide use for farmers in developing countries relative to farmers in developed countries, Table 4.10 shows a 55:45 split of the environmental benefits (1996–2010) respectively in developed (55%) and developing countries (45%). Over three-quarters (76%) of the environmental gains in developing countries have been from the use of GM IR cotton.

It should, however, be noted that in some regions where GM HT crops have been widely grown, some farmers have relied too much on the use of single herbicides like glyphosate to manage weeds in GM HT crops and this has contributed to the development of weed resistance.

Table 4.9 Impact of changes in the use of herbicides and insecticides from growing biotechnology crops globally 1996–2010

Trait	Change in volume of active ingredient used (million kg)	Change in field EIQ impact (in terms of million field EIQ/ha units)	Percent change in active ingredient use on biotechnology crops	Percent change in environmental impact associated with herbicide and insecticide use on biotechnology crops	Area biotechnology trait 2010 (million ha)
GM HT soya beans	−28.8	−6261.7	−1.4	−16.2	71.6
GM HT maize	−169.9	−4199.2	−10.0	−11.5	27.0
GM HT canola	−14.4	−478.2	−18.2	−27.7	6.7
GM HT cotton	−12.1	−347.6	−5.2	−8.1	4.9
GM IR maize	−42.9	−1571.9	−41.9	−37.7	34.1
GM IR cotton	−170.5	−7615.1	−23.9	−26.0	17.7
GM HT sugar beet	+0.54	−2.8	+19.0	−5.0	0.46
Totals	−438.06	−20476.5	−9.0	−17.9	162.46

Table 4.10 *Biotechnology crop environmental benefits from lower insecticide and herbicide use 1996–2010: developed versus developing countries*

Crop	Change in field EIQ impact (in terms of million field EIQ/ha units): developed countries	Change in field EIQ impact (in terms of million field EIQ/ha units): developing countries
GM HT soya beans	−4571.9	−1689.8
GM HT maize	−4076.7	−122.5
GM HT cotton	−274.9	−72.7
GM HT canola	−478.2	0
GM IR maize	−1267.9	−304.0
GM IR cotton	−577.1	−7038.0
GM HT sugar beet	−2.8	0
Total	−11249.5	−9227.0

Worldwide, there are 21 weed species that are currently resistant to glyphosate (www.weedscience.org) (compared to, for example, 69 weed species resistant to triazine herbicides such as atrazine). A few of the glyphosate-resistant species, such as marestail (*Conyza canadensis*) and palmer pigweed (*Amaranthus palmeri*) are now reasonably widespread in the USA, especially marestail, where there are several million acres infested, and palmer pigweed, in southern states, where over 1 million acres are estimated to exhibit such resistance. In Argentina, development of resistance to glyphosate in weeds such as Johnson grass (*Sorghum halepense*) is also reported.

Where this has occurred, farmers have had to adopt reactive weed management strategies incorporating the use of a mix of herbicides. In recent years, there has also been a growing consensus among weed scientists of a need for changes in the weed management programmes in GM HT crops because of the evolution of these weed populations that are resistant to glyphosate. While the overall level of weed resistance in areas planted to GM HT crops is still relatively low (equal to between 5% and 10% of the total US cropping area annually planted to GM HT crops), growers of GM HT crops are increasingly being advised to be more proactive and include other herbicides in combination with glyphosate in their weed management systems even where instances of weed resistance to glyphosate have not been found. This is because proactive weed management programmes generally require fewer herbicides and are more economical than reactive weed management programmes. At the macro level, the adoption of both reactive and proactive weed management programmes in GM HT crops has already begun to influence the mix, total amount and overall environmental profile of herbicides applied to GM HT soya beans, cotton, maize and canola and this is reflected in the data presented in this chapter.

IMPACT ON GREENHOUSE GAS EMISSIONS

The scope for biotechnology crops contributing to lower levels of GHG emissions comes from two principal sources:

- Reduced fuel use from less frequent herbicide or insecticide applications and a reduction in the energy use in soil cultivation. The fuel savings associated with making fewer spray runs relative to conventional crops and the switch to conservation, reduced and no-till farming systems have resulted in permanent savings in carbon dioxide emissions. In 2010 this amounted to about 1715 million kg (arising from reduced fuel use of 642.2 million litres). Over the period 1996 to 2010 the cumulative permanent reduction in fuel use is estimated at 12232 million kg of carbon dioxide (arising from reduced fuel use of 4582 million litres).

- The use of no-till and reduced-till farming systems.[8] These production systems have increased significantly with the adoption of GM HT crops because the GM HT technology has improved growers' ability to control competing weeds, reducing the need to rely on soil cultivation and seedbed preparation as means to getting good levels of weed control. As a result, tractor fuel use for tillage is reduced, soil quality is enhanced and levels of soil erosion cut. In turn more carbon remains in the soil and this leads to lower GHG emissions. Based on savings arising from the rapid adoption of no-till/reduced tillage farming systems in North and South America, an extra 4805 million kg of soil carbon is estimated to have been sequestered in 2010 (equivalent to 17634 million tonnes of carbon dioxide that has not been released into the global atmosphere). Cumulatively, the amount of carbon sequestered

[8] No-till farming means that the ground is not ploughed at all, while reduced tillage means that the ground is disturbed less than it would be with traditional tillage systems. For example, under a no-till farming system, soya bean seeds are planted through the organic material that is left over from a previous crop such as maize, cotton or wheat.

may be higher than these estimates due to year-on-year benefits to soil quality; however, it is equally likely that the total cumulative soil sequestration gains have been lower because only a proportion of the crop area will have remained no-till/reduced-till. It is, nevertheless, not possible to confidently estimate cumulative soil sequestration gains that take into account reversions to conventional tillage because of a lack of data. Consequently, the estimate provided above of 133 639 million tonnes of carbon dioxide not released into the atmosphere should be treated with caution.

Placing these carbon sequestration benefits within the context of the carbon emissions from cars (Table 4.11) shows that:

- In 2010, the permanent carbon dioxide savings from reduced fuel use were the equivalent of removing 0.76 million cars from the road.
- The additional probable soil carbon sequestration gains in 2010 were equivalent to removing 7.84 million cars from the roads.
- In total, in 2010, the combined biotechnology-crop-related carbon dioxide emission savings from reduced fuel use and additional soil carbon sequestration were equal to the removal from the roads of 8.6 million cars, equivalent to 27.7% of all registered cars in the UK.
- It is not possible to confidently estimate the probable soil carbon sequestration gains since 1996 (see above). If the entire biotechnology crop in reduced or no-tillage agriculture during the last 15 years had remained in permanent reduced/no-tillage then this would have resulted in a carbon dioxide saving of 133 639 million kg, equivalent to taking 59.4 million cars off the road. This is, however, a maximum possibility and the actual levels of carbon dioxide reduction are likely to be lower.

CONCLUDING COMMENTS

Biotechnology has, to date, delivered several specific agronomic traits that have overcome a number of production constraints for many farmers. This has resulted in improved productivity and profitability for the 15.4 million adopting farmers who have applied the technology to 139 million hectares in 2010.

During the last 15 years, this technology has made important positive socio-economic and environmental contributions. These have

Table 4.11 *Context of carbon sequestration impact 2010: car equivalents*

Country/crop/trait	Permanent carbon dioxide savings arising from reduced fuel use (million kg of carbon dioxide)	Permanent fuel savings: as average family car equivalents removed from the road for a year (000s)	Potential additional soil carbon sequestration savings (million kg of carbon dioxide)	Soil carbon sequestration savings: as average family car equivalents removed from the road for a year (000s)
USA: GM HT soya beans	246	109	4810	2138
Argentina: GM HT soya beans	670	298	6762	3005
Brazil: GM HT soya beans	364	162	3680	1636
Bolivia, Paraguay, Uruguay: GM HT soya beans	183	81	1850	822
Canada: GM HT canola	110	49	532	237
Global: GM IR cotton	64	29	0	0
Brazil: GM IR maize	78	35	0	0
Total	1715	763	17634	7838

Note: Assumption: an average family car produces 150 g of carbon dioxide per km; a car does an average of 15000 km/year and therefore produces 2250 kg of carbon dioxide/year.

arisen even though only a limited range of biotechnology agronomic traits have so far been commercialised, in a small range of crops.

The biotechnology has delivered economic and environmental gains through a combination of their inherent technical advances and the role of the technology in the facilitation and evolution of more cost-effective and environmentally friendly farming practices. More specifically:

- The gains from the GM IR traits have mostly been delivered directly from the technology (yield improvements, reduced production risk and decreased use of insecticides). Thus farmers (mostly in less developed countries) have been able to both improve their productivity and economic returns whilst also practising more environmentally friendly farming methods.

• The gains from GM HT traits have come from a combination of direct benefits (mostly cost reductions to the farmer) and the facilitation of changes in farming systems. Thus, GM HT technology (especially in soya beans) has played an important role in enabling farmers to capitalise on the availability of a low-cost, broad-spectrum herbicide (glyphosate) and, in turn, facilitated the move away from conventional to low/no-tillage production systems in both North and South America. This change in production system has made additional positive economic contributions to farmers (and the wider economy) and delivered important environmental benefits, notably reduced levels of GHG emissions (from reduced tractor fuel use and additional soil carbon sequestration).

• Both IR and HT traits have made important contributions to increasing world production levels of soya beans, maize, cotton and canola.

In relation to GM HT crops, however, over-reliance on the use of glyphosate by some farmers, in some regions, has contributed to the development of weed resistance. As a result, farmers are increasingly adopting a mix of reactive and proactive weed management strategies incorporating a mix of herbicides. Despite this, the overall environmental and economic gains arising from the use of biotechnology crops have been, and continue to be, substantial.

REFERENCES

Brookes, G. (2008). The benefits of adopting GM insect resistant (Bt) maize in the EU: first results from 1998–2006. *International Journal of Biotechnology* **10**, 148–166. Also available at www.pgeconomics.co.uk.

Brookes, G., and Barfoot, P. (2012). *GM Crops: Global Socio-Economic and Environmental Impacts 1996–2010*. PG Economics Ltd, Dorchestor, UK. www.pgeconomics.co.uk.

Carpenter, J., and Gianessi, L. (1999). Herbicide tolerant soybeans: why growers are adopting Roundup-Ready varieties. *AgBioForum* **2**, 65–72.

Ismael, Y., Bennett, P., Morse, S., and Buthelezi, T. J. (2002). A case study of smallholder farmers in the Mahathini flats, South Africa. In *6th International ICABR Conference*, Ravello, Italy.

James, C. (2009). *Global Status of Commercialized Biotech/GM Crops*. ISAAA, Ithaca, NY.

Kovach, J., Petzoldt, C., Degni, J., and Tette, J. (1992). A method to measure the environmental impact of pesticides. *New York's Food and Life Sciences Bulletin*, New York State Agricultural Experiment Stations Cornell University, Geneva, NY. Also available at www.nysipm.cornell.edu/publications/EIQ.html

Marra, M., and Piggott, N. (2006). The value of non-pecuniary characteristics of crop biotechnologies: a new look at the evidence. In R. Just, J. Alston and D. Zilberman (eds.) *Regulating Agricultural Biotechnology: Economics and Policy*, pp. 145–178. Springer, New York.

Marra, M., and Piggott, N. (2007). The net gains to cotton farmers of a national refuge plan for Bollgard II cotton. *AgBioForum* 10, 1–10. Also available at www.agbioforum.org.

Pray, C., Huang, J., Hu, R., and Rozelle, S. (2002). Five years of Bt cotton in China: the benefits continue. *Plant Journal* 31, 423–430.

Qaim, M., and Traxler, G. (2002). Roundup-Ready soybeans in Argentina: farm level, environmental and welfare effects. In *6th International ICABR Conference*, Ravello, Italy.

Rice, M. E. (2004). Transgenic rootworm corn: assessing potential agronomic, economic and environmental benefits, *Plant Health Progress* 10, 1–10.

USDA (1999). Farm level effects of adopting genetically engineered crops: preliminary evidence from the US experience. In R. Shoemaker, J. Harwood, K. Dayn Rubenstein *et al.* (eds.) *Economic Issues in Agricultural Biotechnology*, pp. 15–17. US Department of Agriculture, Washington, DC.

Yorobe, J. (2004). Economic impact of Bt corn in the Philippines. In *45th PAEDA Convention*, Quezon City, Philippines.

5

Germplasm diversity and genetics to drive plant breeding for Africa

INTRODUCTION

The global food crisis of 2008 highlighted the necessity for innovation in agriculture to address food insecurity in the presence of a changing climate and a growing population. The world population is predicted to reach 9 billion within the next 40 years, requiring a 70–100% increase in food production relative to current levels (Beddington, 2010). A burgeoning world population is not the only threat to global food security. Changing lifestyles, population demographics, competition from subsidised biofuels, deterioration of natural resources and dwindling supplies of water will require considerable financial, intellectual and molecular investment in agriculture, particularly in the developing world. A Global Food Security Index (http://foodsecurityindex.eiu.com) has recently been created for 105 countries, with the aim of guiding the development of food security solutions. It provides a quantitative and qualitative model to identify and measure risks that drive food insecurity, based on affordability, availability and food quality, The bottom third of the table is dominated by sub-Saharan African countries. These countries include Mozambique, Ethiopia, Rwanda and Nigeria which are predicted to have the fastest growing economies over the next 2 years and are well positioned to address food security.

In these circumstances, Africa and in particular sub-Saharan Africa can lead the introduction of new agricultural technologies and in particular the deployment of the full spectrum of plant breeding technologies to address food security and sustainability of the natural resource base. Reviews outlining the deployment of new biotechnologies in plant

Successful Agricultural Innovation in Emerging Economies: New Genetic Technologies for Global Food Production, eds. David J. Bennett and Richard C. Jennings. Published by Cambridge University Press. © Cambridge University Press 2013.

breeding have been published (Lusser *et al.*, 2012) but to address the global challenges effectively will require a deeper understanding of plant breeding methods and principles together with exposure and understanding of the needs of smallholder farmers who dominate agricultural productivity in sub-Saharan Africa. We will discuss these interrelated topics through a review of the principles of plant breeding and the role of genetic diversity; a critical appraisal of crop genomics and life sciences advances and their application to plant breeding; and an analysis of the particular issues and opportunities that pertain to plant breeding in sub-Saharan Africa.

PRINCIPLES AND HISTORY OF PLANT BREEDING

During the last century genetics played a pivotal role in improving our understanding of many aspects of biology. The principles underlying heredity were first established in plants and the main practical beneficiary of these discoveries was agriculture, notably via plant breeding. The principles of plant breeding were established by Mendel who provided the first clear exposition of transmission genetics: the statistical rules governing the transmission of hereditary elements from generation to generation. Mendel introduced three fairly novel approaches to the study of inheritance: he extended observation to experimentation, he counted his plants (*Pisum sativum*), and he maintained the parent stocks of plants. At that time numerical studies were not yet part of the tradition of biology and indeed the statistical methods required to validate the observed ratios were not available. However, Mendel understood the inherent errors associated with sampling. The theory of inheritance established by Mendel is based upon the existence of hereditary particles or genes that behave in a well-defined and hence predictable manner. Arising from these principles the concepts of alleles (the alternative forms of a gene at a specific position on a specific chromosome) and the segregation and independent assortment of genes during reproduction emerged.

This we now know is a result of alleles being carried on the chromosomes, where (in general, but not always) each cell carries two paired sets of homologous chromosomes. 'Homologous' because they carry alleles for a specific characteristic at the same place on each of the chromosomes of a pair. During sexual reproduction the gametes (i.e. the sperm, pollen or eggs) each carry one or other of these homologous chromosomes. The result is that the progeny of sexual reproduction have alleles from each parent, combined, or assorted, in many different ways.

During the nineteenth century, and before the 'rediscovery' of Mendel's work in 1900, plant breeders applied the Darwinian theory, i.e. that plant characters could be improved by continuous selection. Their methods involved repeated mass selection within local varieties of plants, but no great advances were made. However, in 1890 plant breeders in Sweden and France began to make single plant selections and to grow selected progeny. In essence breeding became more systematic and careful. The Mendelian rediscovery catalysed the establishment of fundamental concepts describing the particular nature of inheritance, providing a clear distinction between genotype (the genetic make-up) and phenotype (the characteristics of the plant), and a rationale for progeny testing.

The characteristics or 'traits' described by Mendel are discontinuous, normally related to genes (alleles) at one or a few locations on the chromosome, and their gross effect on the plant or 'phenotypic expression' tends to be modestly affected by the environment. Traits that vary in a more continuous fashion are frequently controlled by more than one gene and are subject to significant modifications by the environment. However, such traits are inherited in the same way as discontinuous characters, exhibiting during reproduction segregation, recombination and linkage when on the same chromosome. The genetical approach to handling continuous variation or quantitative traits is based on the theory and predictions of biometrical genetics (Mather and Jinks, 1971). Such traits are the most economically important components of plant breeding programmes, underpinning the genetic improvement of yield, and of resistance to biotic (living) and abiotic (non-living) stresses on the plant. Successful crop improvement programmes are dependent on an understanding of Mendelian and quantitative genetics principles, and the need to capture reproducible phenotypic data.

Bernardo (2002) defined plant breeding as 'the science, art and business of improving plants for human benefit'. Progress is dependent on: the amount and type of genetic variability available and the effectiveness of the evaluation and selection techniques employed. It is important to remember that although scientific plant breeding only began in the last century, it had its origins in the dawn of agriculture many thousands of years ago. These early efforts resulted in the domestication of our major cultivated crops. For any particular species, breeding methods are largely conditioned by the stage of evolution, development and domestication of that species. Continued selection and isolation by humans are likely to have influenced the amount of genetic variability, the life cycle, mating system, the history of selection and type of gene action, all of which are interdependent and together determine the applicability of any breeding method.

The prime determinant of the methods practised in plant breeding is the crop's reproductive biology and breeding system. Consideration of the breeding system focuses attention on the relationship between the reproductive cells or gametes at reproduction and their control of the genetic structure of populations. Crops can be classified according to their system of propagation into:

(1) Vegetatively propagated crops, e.g. potato, sweet potato, cassava and yam

(2) Inbreeding (autogamous) crops, e.g. wheat, barley, rice, soya bean, cotton

(3) Outbreeding (allogamous) crops, e.g. maize, forage grasses

(4) A variable, intermediate group which will partly cross-pollinate and partly self-pollinate, e.g. *Brassica napus*, sorghum.

Approximately two-thirds of the world's food crops are inbreeders. For these self-pollinating crops individual genotypes (i.e. the alleles, or different forms of a gene, which an individual has with respect to a particular characteristic) are pure lines or homozygous lines, meaning that both chromosomes have the same allele for that particular characteristic. For inbreeding crops the variability which plant breeders need for selection of desired characteristics must be generated. The normal way in which breeders introduce new characteristics is by making sexual crosses, that is, by crossing the pure line with another line which has a different allele for that characteristic. The starting point for the production of a new variety of a self-pollinating species is a hybrid, called the F_1 hybrid, produced from a pair of different homozygous lines, each having identical alleles for a single characteristic. Self-pollination of the F_1 generates the F_2 progeny which is subsequently allowed to self-pollinate in the field. The result is a very diverse population which theoretically could contain all the recombinants from the parental gametes if sufficiently large numbers of plants could be grown. In the early generations the genetic variation is not fixed, so the progeny of any single plant continues to segregate, i.e. the different alleles are distributed throughout the population, leading to considerable diversity within the population. However, with each succeeding generation of selection the individual plant becomes progressively more inbred and true breeding so that a new uniform variety can usually be based on a single sixth (F_6) generation because virtually all other variation is eliminated through selection of plants with the desired characteristic or characteristics.

Traditionally, breeding methods for inbreeders have been based on variations of the pedigree system where the ancestors of the selections can be traced back to individual F_2 plants. The pedigree selection method is a long process from parents being hybridised to the release of a new variety (10–15 years). However, the most recalcitrant problem in the use of pedigree selection is the difficulty of identifying high-yielding genotypes in the early generations, where only a limited amount of seed is available and this is usually heterozygous, i.e. has both alleles. To overcome this problem breeders may delay selection until a sufficient number of plants is obtained. Once this number is achieved selection for the desired characteristic can begin until progenies are approaching homozygosity, that is, both alleles are the same and plants with the desired characteristic can be recognised. Two main approaches are used to achieve this: single-seed descent and doubled haploidy. Haploids are plants that contain a single chromosome which can be doubled to produce homozygous generations of plants in a single generation. Two methods are generally used to produce haploids. First, cultured gamete cells may be regenerated into haploid plants, and second, haploids can be induced by rare crosses between species in which the genome of one parent is eliminated after fertilisation. Doubled haploids (DH) result in an increase in selection efficiency due to the fact that homozygosity is established in a single generation. DH is now routinely used in major cereal and brassica breeding programmes and used for rapid inbred line production in maize. A new method for inducing haploids in *Arabidopsis*, a small weed long used and characterised by geneticists, has been reported based on the modifications of what is called a single centromere-specific histone protein CENH3 [5]. Both maternal and paternal haploids can be generated by crossing this *cen3* mutant having an altered CEN protein with wild-type plants. Potentially, this technique may be extended to a broad range of plants including crop species.

For outbreeding crop species, allele frequencies in populations need to be considered. In 1908 Hardy and Weinberg independently realised that the consequences of Mendelian inheritance in a large random mating population with no selection, mutation or migration was that there was no change in gene frequencies and genotype frequencies from generation to generation. In other words the mechanism of inheritance maintains variation. Furthermore, Hardy and Weinberg showed that there is a simple relationship between gene and genotype frequencies, which became known as the Hardy–Weinberg law.

In maize, and more recently in sorghum and pearl millet, indirect approaches are deployed based on inbreeding and cross-breeding.

Hybrid maize (*Zea mays*) is a model example of genetic theory being successfully applied to a food crop (Duvick, 2001). The classic work of inbreeding and cross-breeding was performed on maize, where the phenomena of inbreeding depression, or reduced performance, and heterosis, or hybrid vigour, were discovered. The poor vigour and seed production of the inbred strains was overcome for commercial exploitation by the use of two vigorous single crosses to be used as parents to produce double-cross hybrid seed combining the desired characteristic with the rigour of the single-cross parents. Maize yield doubled between the 1930s to 1960s with 60–80% of the gain being attributed to heterosis. Since the 1960s high-performing inbreds have been generated for use as single-cross hybrids. In addition to their use in maize, F_1 single-cross hybrids have been a significant driving force behind vegetable and tomato breeding creating uniformity, greater stability and sustainability for mechanical harvesting by using this technique of single-cross hybridisation.

Clonally propagated crops (vegetatively from parts of the plants), e.g. potato and cassava, are usually highly heterozygous and outbreeding. Breeding methods are based on intercrossing between highly heterozygous variants and selection among seed progenies. Poly-cross designs similar to that used in forage crops are also used to efficiently generate large quantities of seed. A major challenge for clonally propagated crops is the generation and maintenance of collections of genes for the crops which are free of disease. Nassar and Ortiz (2010) provide a comprehensive account of the breeding strategy for cassava that includes wide hybridisation, and introgression programmes, i.e. movement of a gene from one plant into another by the repeated backcrossing of an interspecific hybrid with one of its parent species. For example, hybridisation between cultivated cassava, *Manihot esculenta*, and *M. glaziovii* is deployed to produce disease-resistant clones, plants from parts of others, with elevated protein. For many perennial crops, breeding is challenging due to polyploidy, i.e. the plant having more than one set of paired chromosomes. This is illustrated by bananas (*Musa* spp.) that are made up of *M. acuminata* (A genome) and *M. balbisiana* (B genome) and are generally triploid (i.e. have three sets of chromosomes) and seedless (sterile). Such hybrids are widely distributed by vegetative propagation and are vulnerable to diseases such as Panama disease fungus (*Fusarium oxysporum* f. sp. *cutsense*). Recently D'Hont *et al.* (2012) published the draft genome sequence of the 523 megabase genome of a *Musa acuminata* doubled haploid genotype. This genome sequence is providing fundamental new knowledge and insight into plant immunity genes and soluble sugar balance in ripening banana.

Bananas, plantains, cassava, potato and sweet potato as well as other indigenous African root vegetables are key to solving Africa's food and income security challenges. These vegetatively propagated crops are an excellent source of cheap energy and are a key staple food in sub-Saharan Africa. A uniquely African Green Revolution requires urgent improvement in the supply of new and improved cultivated varieties of these vegetative crops. New genome sequencing approaches of the hereditary information, coupled with new computing techniques, will enable discovery of genes controlling disease resistance in clonally propagated crops and improve our understanding of the biological mechanisms controlling post-harvest physiology.

In this section we have focused on the concepts and practices that underpin classical plant breeding. We consider that a comprehensive understanding of plant breeding methods is of critical importance to analyse how the benefits of crop genomics can best be deployed to maximise the improvement of plants to further help feed the world in a sustainable and equitable manner.

OPPORTUNITIES FROM GENOMIC TECHNOLOGY

Genomics research and its application to crop breeding are at an interesting stage where two main drivers are beginning to converge. First, we have a democratisation of genomics technology where the speed (and accessibility) of sequencing is accelerating and the cost is dropping. On average we have a doubling of sequencing data output every 5 months and this has resulted in a free fall in cost per DNA base sequenced (Varshney and Dubey, 2009). In parallel, food and nutrition security is a global priority and agriculture is recognised as pivotal to generating a sustainable planet based on a viable green economy. The convergence of technological advances and the need to address global challenges has dramatically changed the perspective on the breeding of orphan crops where genomics-based approaches are no longer rate limiting (Fridman and Zamir, 2012). Future developments and applications to crop breeding are likely to be more dependent on population development, the ability to reproducibly phenotype important traits, access and utilisation of genetic variability together with the status, capacity and scale of the breeding programme. This coupled with the leadership and support of breeding programmes will be key determinants of success.

Genetic variation is the engine that drives advances in plant breeding (Morrell *et al.*, 2011). Advances in next-generation sequencing have resulted in the completion of reference genome sequences for many important crops and model plants (Ehrhardt and Frommer, 2012). For example, the draft genome sequence of pigeon pea (*Cajanus cajan*), an orphan legume crop, was published in 2011 (Varshney *et al.*, 2011), providing a platform for gene identification and for the large-scale discovery of genetic markers, particularly single-nucleotide polymorphisms (SNPs) where a single nucleotide is altered. However, further advances and application in crop breeding will require the development of genotyping by sequencing (Elshire *et al.*, 2011), resequencing of different varieties of crop plants coupled with the assembly of next-generation populations (Cavanagh *et al.*, 2008). Rapid developments in these areas are improving our understanding and knowledge of crop genomes and enabling direct translation of gene annotation to allele and haplotype (i.e. groups of genes) discovery in relation to trait and phenotype diversity. Resequencing studies and comparative population genomics of maize landraces and improved maize lines have identified regions of the maize genome influencing gene expression and the removal of *cis*-acting regulatory variation (i.e. a length of DNA which can regulate several genes and, conversely, one gene can have several such lengths) during maize domestication and improvement (Hufford *et al.*, 2012). A genome-wide analysis of 278 maize lines has also been undertaken to quantify changes occurring during the breeding process (Jiao *et al.*, 2012). This study revealed that rare alleles were accumulated during the breeding process suggesting that the relative fractions of rare alleles could be used as a selective index in breeding. The methods described for maize and rice (Xu *et al.*, 2012) illustrate the power of next-generation sequencing and genotyping by sequencing to characterise collections of germplasm as genetic resources and identify novel variant alleles for deployment in breeding. These approaches also overcome the limitation of sampling bias that has impeded the genetic evaluation of diverse germplasm collections. These studies emphasise that for many crops of relevance to Africa the bottlenecks will not be genomic resources but carefully chosen and designed populations that sample genetic variation and increase the detail of the genetic analysis of complex characters.

Community-based populations of crop plants provide great opportunities to discover, characterise and deploy alleles that have direct relevance to breeders (Hamblin *et al.*, 2011). Essentially these approaches are designed to overcome the limitations of two-parent populations and manipulate population structure to increase genetic resolution. Although there are many potential designs, all have two components in common, the use of multiple parents to increase allelic diversity and the advancement of populations through several generations to improve genetic resolutions. One such population design is the so-called 'nested association mapping' (NAM) population first developed in maize and is based on crossing diverse strains to a reference parent B73 (Yu *et al.*, 2008) and has already been used to isolate a number of genes controlling complex traits (Kump *et al.*, 2011). Other designs involve intercrossing multiple parents to form a single large population and are often referred to as multi-parent advanced generation intercross (MAGIC) populations (Cavanagh *et al.*, 2008). This method is being deployed in wheat, rice and sorghum.

It was assumed that the dramatic drop in the cost of DNA-based assays would result in an accelerated use of marker-assisted selection (MAS) in plant breeding, in which a biochemical marker is used for selecting plants with the desired characteristic. MAS was used to introduce tolerance to submergence in water (*Sub 1*) into rice varieties grown in 15 million hectares of rain-fed lowland rice in South and Southeast Asia (Septiningsih *et al.*, 2009). While MAS has been used successfully in recurrent backcrossing (including importantly programmes to introgress transgenes from another species), its more general application has been limited (Langridge and Fleury, 2010). The reason for this is twofold: first, many of the traits of importance to breeding programmes are complex and are controlled by large numbers of genomic regions, each with relatively small genetic effect; second, the method of estimating the effect of stretches of DNA containing or linked to the genes that underlie a quantitative trait are often based on bi-parental populations that are non-representative of breeding germplasm. To overcome these deficiencies, Meuwissen *et al.* (2001) proposed a different approach known as genomic selection. In this method genome-wide panels of genetic markers are used to predict performance across the genome and has the promise to deliver more accurate predictions (Jannink *et al.*, 2010). This approach represents a paradigm change based on the ability to score large numbers of markers at low cost. The simultaneous estimate of all marker effects can be used to generate an estimated breeding value (EBV). This requires a 'training population' of individuals that have been both genotyped and phenotyped to develop a model to produce EBV (Heffner *et al.*, 2009). Such studies have been deployed in dairy cattle and are now being applied to crop plants.

This is an exciting time for plant breeders. The basic knowledge of crop genomes has increased exponentially providing new opportunities for crop improvement. However, there is a gap that may indeed be widening between this information and plant breeding practice. Plant breeding is resource intensive and very dependent on the accurate collection of data to support decision-making. It is also dependent on the migration of new information from the faster-moving fields of human genetics and evolutionary theory. This, coupled with new discoveries on the control of genes determining complex traits, means that plant breeding strategies and methodologies have become more advanced and dependent on new knowledge. The ever-increasing power of computer simulation (Li *et al.*, 2012) can compare different breeding strategies, include different modes of gene action and can help bridge the gap between theory and practice, but inevitably any simulation will require validation in plant breeding programmes.

In summary, plant breeding is a key technological platform that is of critical importance to feed a future 9 billion people equitably, healthily and sustainably (Beddington, 2010). The introduction of improved crop varieties has been most successful for the major crops in favourable agro-ecological areas where a combination of genetic and agronomic (fertiliser, pesticide and irrigation) factors have made approximately equal contributions. More attention needs to be focused on breeding for less favoured 'orphan crops' (Brummer *et al.*, 2011). These are of particular importance to smallholder farmers in sub-Saharan Africa and include sorghum and millets, groundnut, cowpea, common bean, chickpea and pigeon pea, cassava, yam and sweet potato. Breeding for orphan crops has lagged behind major crops but genomic technologies offer significant opportunities to reverse this trend and make a significant contribution to the key staple crops in many developing countries. To realise this goal will require the judicious application of basic plant breeding principles, coupled with appropriate experimental design to quantify phenotypic variation in a reproducible manner, to design and effect phenotypic selection regimes. There is also a need to harmonise current plant breeding practices with ecological and management practices that sustain and improve the natural resource base. In broad terms this will require plant breeders to identify and select genotypes that are climate resilient and able to maintain productivity in suboptimal conditions. In addition, creating new ideotypes (i.e. specifically designed and adapted crops with particular shoot or root architectures) that can contribute to creating and supporting ecosystems services is required (Ehrhardt and Frommer, 2012). A key objective will be to target specific environments and identify adaptive gene complexes to maximise productivity in individual landscapes.

Breeding crops that are appropriate for local conditions is vital for their widespread adaptation, and schemes to promote greater participation through the involvement of local farmers in the breeding process (participatory plant breeding) are important components of a decentralised plant breeding strategy.

WHAT DOES THIS MEAN IN THE CONTEXT OF AFRICAN AGRICULTURE?

- This is a good time to be considering plant breeding in the context of African agriculture. A comprehensive, strategic perspective is required that will not only consider the technology but the biology of the species that are of critical importance to Africa.
- Priority should be given to clonally/vegetatively propagated crops that are well placed to benefit from developments in genomics. Identification of elite clones can form part of a participatory farming approach to facilitate uptake and dissemination of planting material. Higher-yielding, disease- and pest-tolerant new cultivars can form the basis of multiplication chains that are connected to a phytosanitary testing scheme to control plant diseases.
- Education, training and skills with a strong orientation towards species of relevance to Africa are essential. This needs to include cutting-edge research in genetics, population biology, analysis of complex traits and experimental design/statistical analysis. The goal should be to inspire a new generation of African plant breeders who can assimilate and integrate new modern concepts and approaches into practical and productive plant breeding programmes.
- New innovative partnerships between funders, breeders, policy-makers, urban planners, ecologists, educational institutions and professional communicators are needed to articulate that plant breeding is of critical importance in harmonising agricultural productivity and environmental sustainability in the developing world.
- The culture of plant breeding needs to be compatible with the goals of African agriculture to embrace species diversity and farming systems. This needs to consider the agro-ecology and socio-economic context. A bold vision for African plant breeding is needed with a focus on crops that promote not only

food security but income and productive employment. Such an holistic approach is needed to embrace the need for sustaining the natural resource base.

REFERENCES

Beddington, J. (2010). Food security: contributions from science to a new and greener revolution. *Philosophical Transactions of the Royal Society B* 365, 61–71.

Bernardo, R. (2002). *Breeding for Quantitative Traits in Plants.* Stemma Press, Woodbury, MN.

Brummer, C., Barber, W. T., Collier, S. M., et al. (2011). Plant breeding for harmony between agriculture and the environment. *Frontiers in Ecology and Environment* 9, 561–568.

Cavanagh, C., Morell, M., Mackay, I., and Powell, W. (2008). From mutations to MAGIC: resources for gene discovery, validation and delivery in crop plants. *Current Opinion in Plant Biology* 11, 215–221.

D'Hont, A. D., Denoeud, F., Aury, J.-M., et al. (2012). The banana (*Musa acuminata*) genome and the evolution of monocotyledonous plants. *Nature* 488, 213–217.

Duvick, D. N. (2001). Biotechnology in the 1930s: the development of hybrid maize. *Nature Reviews* 2, 69–74.

Ehrhardt, D. W., and Frommer, W. B. (2012). New technologies for 21st century plant science. *Plant Cell* 24, 374–394.

Elshire, R. J., Glaubitz, J. C., Sun, Q., et al. (2011). A robust, simple genotyping-by-sequencing (GBS) approach for high diversity species. *PLoS ONE* 6, e19379.

Fridman, E., and Zamir, D. (2012). Next-generation education in crop genetics. *Current Opinion in Plant Biology* 15, 218–223.

Hamblin, M. T., Buckler, E. S., and Jannink, J. L. (2011). Population genetics of genomics-based crop improvement methods. *Trends in Genetics* 27, 98–106.

Heffner, E. L., Sorrells, M. E., and Jannink, J. L. (2009). Genomic selection for crop improvement. *Crop Science* 49, 1–10.

Hufford, M. B., Xu, X., van Heerwaarden, J., et al. (2012). Comparative population genomics of maize domestication and improvement. *Nature Genetics* 44, 808–811.

Jannink, J. L., Lorenz, A. J., and Iwata, H. (2010). Genomic selection in plant breeding: from theory to practice. *Briefings in Functional Genomics* 9, 166–177.

Jiao, Y., Zhao, H., Ren, L., et al. (2012). Genome-wide genetic changes during modern breeding of maize. *Nature Genetics* 44, 812–817.

Kump, K. L., Bradbury, P. J., Wisser, R. J., et al. (2011). Genome-wide association study of quantitative resistance to southern leaf blight in the maize nested association mapping population. *Nature Genetics* 43, 163–169.

Langridge, P., and Fleury, D. (2010). Making the most of 'omics' for crop breeding. *Trends in Biotechnology* 29, 33–40.

Li, X., Zhu, C., Wang, J., and Yu, J. (2012). Computer simulation in plant breeding. *Advances in Agronomy* 116, 219–264.

Lusser, M., Parisi, C., Plan, D., and Rodríguez-Cerezo, E. (2012). Development of new biotechnologies in plant breeding. *Nature Biotechnology* 30, 231–239.

Mather, K., and Jinks, J. L. (1971). *Biometrical Genetics: The Study of Continuous Variation.* Chapman and Hall, London.

Meuwissen, T. H. E., Hayes, B. J., and Goddard, M. E. (2001). Predicition of total genetic value using genome wide dense marker maps. *Genetics* 157, 1819–1829.

Morrell, P. L., Buckler, E. S., and Ross-Ibarra, J. (2011). Crop genomics: advances and applications. *Nature Reviews Genetics* **13**, 85–96.

Nassar, N., and Ortiz, R. (2010). Breeding cassava to feed the poor. *Scientific American* **302**, 78–82, 84.

Naylor, R. L., Falcon, W. P., Goodman, R. M., *et al.* (2004). Biotechnology in the developing world: a case for increased investments in orphan crops. *Food Policy* **29**, 15–44.

Ravi, M., and Chan, S. W. L. (2010). Haploid plants produced by centromere-mediated genome elimination. *Nature* **464**, 615–618.

Septiningsih, E. M., Pamplona, A. M., Sanchez, D. L., *et al.* (2009). Development of submergence-tolerant rice cultivars: the *sub I* locus and beyond. *Annals of Botany* **103**, 151–160.

Varshney, R., and Dubey, A. (2009). Novel genomic tools and modern genetic and breeding approaches for crop improvement. *Plant Biochemistry and Biotechnology* **18**, 127–138.

Varshney, R., Chen, W., Li, Y., *et al.* (2011). Draft genome sequence of pigeonpea (*Cajanus cajan*), an orphan legume crop of resource-poor farmers. *Nature Biotechnology* **30**, 83–89.

Xu, X., Liu, X., Ge, S., *et al.* (2011). Resequencing 50 accessions of cultivated and wild rice yields markers for identifying agronomically important genes. *Nature Biotechnology* **30**, 105–114.

Yu, J., Holland, J. B., McMullen, M. D., and Buckler, E. S. (2008). Genetic design and statistical power of nested association mapping in maize. *Genetics* **178**, 539–551.

6

Using molecular breeding to improve orphan crops for emerging economies

INTRODUCTION

This chapter makes the case for more investment in the application of advanced molecular breeding techniques not only to improve the world's major food crops but also the numerous orphan crops that play an essential role in global agricultural production, particularly in the economies of developing countries. Orphan crops, also referred to as underutilised or neglected crops, have been domesticated through centuries of human selection for their food, feed, fibre, oil and medicinal properties. They are culturally valued and often adapted to harsh agro-ecological niches and marginal lands. Collectively, orphan crops play a major role in feeding the poor in developing countries of sub-Saharan Africa and South Asia. A thorough comparison of orphan food crops with the four major food crops was carried out by Naylor et al. (2004). At that time wheat, rice, maize and soya beans collectively covered 580 million hectares of land worldwide, with two-thirds of this total in developing countries. Twenty-seven orphan crops within developing countries collectively occupied 250 million hectares. However, due to the lack of genetic improvement, orphan crops typically produce inferior yields compared to the major food crops. In addition, other issues such as production of toxic substances and susceptibility to environmental factors including drought, soil salinity, pests, diseases and weeds mean that the massive potential of orphan crops to help meet the demands of the growing human population in the twenty-first century is not being realised.

The last decade has seen major advances in the molecular genetic understanding of how model plants such as *Arabidopsis* and major crops

Successful Agricultural Innovation in Emerging Economies: New Genetic Technologies for Global Food Production, eds. David J. Bennett and Richard C. Jennings. Published by Cambridge University Press. © Cambridge University Press 2013.

such as maize and rice grow, produce and interact with their environment. The remarkable advances made by next-generation sequencing technologies since 2005, resulting in orders of magnitude increases in output sequence per unit cost (Mardis, 2011), mean that acquisition of genomic sequence information is cheaper and faster than ever before. Consequently, we now have the ability to gain rapid low-cost access to comprehensive information on genes and DNA sequence variations (known as polymorphisms). It is these genes and DNA sequence variations that are the essential tools for molecular breeding. New high-throughput, low-cost Next Generation Sequencing (NGS) based applications for plant genetics and plant breeding are now being developed. The transformational nature of these technologies in terms of both throughput and cost makes it now realistic to think about advanced and affordable molecular breeding programmes in orphan crops. The challenge is to develop programmes that will rapidly apply the knowledge and technologies gained from work on model plants and major crops to produce incremental gains in performance of orphan crops.

MOLECULAR PLANT BREEDING

Molecular plant breeding is defined here as the use of advanced molecular biology techniques for crop improvement. The objective of plant breeding is to identify and cross the best parents and recover progeny with improved performance. Genetic variation among individuals in a population is essential for both natural selection and plant breeding which is a form of artificial selection. Figure 6.1 considers three types of variation: (i) natural variation which arises as a result of accumulation of random mutations or polymorphisms (from the Greek meaning 'having multiple forms') in a population over evolutionary time; (ii) induced variation, caused by exposure to ionising radiations or chemical agents that increase the number of mutations in DNA; (iii) engineered variation caused by what is commonly referred to as genetic modification (GM). Powerful DNA-based technologies have been developed over the last 20 years to exploit these different forms of variation in the major food crops. These technologies have the potential to deliver incremental increases in the performance of orphan crops in several ways.

NATURAL VARIATION AND MARKER-ASSISTED SELECTION

DNA polymorphisms are spread throughout the genome, which typically codes in the order of 30000 genes, not considering the number of sets of chromosomes in a cell. Some characteristics or 'traits' such as flower

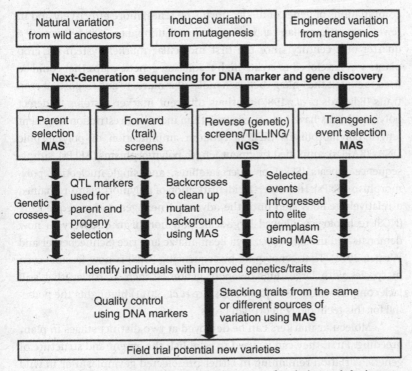

Figure 6.1 Molecular plant breeding: sources of variation and the impact of next-generation sequencing (NGS) and marker-assisted selection (MAS). The sources of variation and the use of NGS and MAS are discussed in the text.

colour can be controlled in a Mendelian fashion by a single gene. However, many other agriculturally important traits such as crop yield or disease resistance do not follow the rules of simple Mendelian genetics because they are influenced by multiple genes. These are known as 'quantitative traits' and the locations of the genes on the chromosomes that make up the plant's genome are known as quantitative trait loci or QTL.

Conventional plant breeding relies on selection based on visible or measurable traits referred to as the phenotype. While this approach obviously works, it is relatively slow in terms of number of generations required to get to new varieties. Furthermore, plants have to be grown to maturity before selection can take place, environmental effects can play a confounding role in selection of the best germplasm from the range of plants produced, and there is limited ability to select for multiple beneficial genes that have a quantitative effect on a single trait.

Marker-assisted selection (MAS) is a DNA-based technology that increases our ability to identify genetic variation and develop new combinations of genes or alleles (forms of genes) more effectively and in fewer generations than can be done by conventional breeding. It is now a quarter of a century since the first molecular marker system for crop plants was developed to create high-resolution genetic maps that could be used to exploit the linkage between molecular markers and important crop traits (Edwards *et al.*, 1987). Various different marker systems to detect polymorphisms have been used since then including restriction fragment length polymorphisms (RFLPs), random amplification of polymorphic DNAs (RAPDs), amplified fragment length polymorphisms (AFLPs), simple sequence repeats (SSRs) or microsatellites, and single-nucleotide polymorphism (SNPs). However, the identification of polymorphisms remained a relatively tedious step until the advent of next-generation sequencing (NGS) technologies allowed large-scale polymorphism discovery as now demonstrated in *Arabidopsis*, soya bean, maize and rice (Schneeberger and Weigel, 2011). That genotyping by sequencing (GBS) is now feasible even in species with large, high-diversity genomes such as barley that still lack complete genome sequence (Elshire *et al.*, 2011) highlights the potential for this technology in orphan crops.

Molecular markers can be deployed at two distinct stages in plant breeding. First, they can be used to establish the extent and structure of genetic variation remaining in either preselected germplasm or in wild relatives of a particular crop species. Information on genetic diversity gained from this type of analysis can be used to select parents with an increased probability of producing improved progeny. Second, molecular markers that show linkage to important crop traits can be used to track inheritance of these traits. This is particularly important for quantitative traits. Identification of molecular markers that are linked to QTL allows the rapid stacking of multiple positive QTL in a single individual plant. Automated high-throughput technologies have advanced to the stage where it is now possible to screen for thousands of molecular markers such as SNPs in thousands of individual progeny derived from just two parents. Thus in one generation it is possible to identify rare individuals that contain multiple beneficial QTLs affecting multiple traits.

MARKER-ASSISTED SELECTION IN ORPHAN CROPS AND THE IMPACT OF NGS

Marker-assisted selection works very well in those model and crop plants that have well-established molecular marker systems together with QTL maps that establish linkage between markers and traits. Next-generation sequencing technology means that it is now possible to rapidly establish the platform for similar approaches in orphan crops. In one of the first examples of this we performed NGS on the medicinal plant *Artemisia annua*, which produces the potent antimalarial compound artemisinin (Graham *et al.*, 2010). This compound is produced in microscopic structures called glandular trichomes that occur on the surface of the *A. annua* leaves. The set of gene sequences that are expressed by being transcribed into messenger RNA is known as the transcriptome. By sequencing the glandular trichome transcriptome from F_1 progeny derived from a cross between two genetically divergent parents we were able to rapidly discover a large number of genes associated with the production of artemisinin and other traits. This also allowed us to identify thousands of SNP and SSR markers associated with the expressed genes. Identifying molecular markers in this way has the advantage that they all reside in gene sequences that are expressed and thus could affect gene function. The use of a high-throughput genotyping platform together with F_1 mapping populations allowed genetic linkage maps for both parents to be rapidly established and QTL for various traits including artemisinin yield to be mapped and shown to be present in hybrids with increased yields. Thus, in a time-frame of less than 5 years it is possible to establish the toolkit for marker-assisted breeding in an orphan crop where previously little or no genetic information had been available and to begin to see incremental improvements in crop performance.

A second example of a non-food orphan crop that can now benefit from molecular breeding is *Jatropha curcas*, a perennial plant of the spurge family (Euphorbiaceae). It has received much attention as a potential source of vegetable oil to replace petroleum through the production of biodiesel (King *et al.*, 2009). However, despite the interest shown in the large-scale cultivation of this crop, until recently genetic resources have been poorly characterised and there has been very little plant breeding for improved traits. Varieties being used to establish plantations in Africa and Asia at present are inedible and the meal obtained after the extraction of oil is not fit for use as an animal feed. Naturally existing edible varieties occur in Mexico, which is also a rich source of genetic diversity for this species (He *et al.*, 2011). We have used NGS to characterise the developing seed transcriptome of *J. curcas* (King *et al.*, 2011) and we are now engaged in a molecular breeding programme similar to that performed for *A. annua* to identify new high-yielding varieties of *J. curcas* that can be used for animal feed as well as biodiesel production.

INDUCED VARIATION AND MUTATION BREEDING

Mutagenesis using ionising radiation or chemical agents to increase the amount of genetic variation available for selection was used in breeding many of the major crops during the second half of the twentieth century. While the approach has been successful it does have its limitations. For example, because induced mutations are mostly recessive and do not produce an effect when present with a dominant allele, resulting phenotypes, or characteristics, are only observed when the mutant allele is present in the homozygous state. Consequently, forward genetic or phenotypic screens work best in diploid crops with cells containing two sets of chromosomes rather than polyploid crops with cells containing many. In addition, we now know that many proteins are encoded for by small redundant gene families and recessive mutations are often silent even when in the homozygous state.

In the last decade an updated version of mutation breeding known as TILLING (targeting induced local lesions in genomes) has been developed for use in both plants and animals (Till *et al.*, 2006). This approach, which requires prior knowledge of the target gene sequence, can be used to identify allelic series of point mutations in specific genes as demonstrated for the granule-bound starch synthase I (*GBSSI*) gene in hexaploid and tetraploid wheat (Slade *et al.*, 2005). This reverse genetic method, so called because it goes from gene sequence to phenotype, can be used to test a hypothesis for the function of any candidate gene, since it can identify individual plants carrying induced mutations or naturally occurring polymorphisms in the target gene. As such it is an excellent tool for evaluating candidate genes in orphan crops that have been previously functionally characterised in model plants.

TILLING relies on a method mediated by an enzyme for the detection of heteroduplexes within DNA in which the two strands are derived from different individuals. Heteroduplexes arise wherever there is a mismatch in double-stranded DNA due to mutations or polymorphisms. EcoTILLING is a variation of the TILLING method that is used to identify natural allelic variation in germplasm. This is a low-cost, high-throughput method for the discovery and characterisation of SNPs and insertions/deletions ('indels') and has been demonstrated in a wide range of animal and plant species including the *Musa* genus that includes banana and plantain (Till *et al.*, 2010). By applying ecoTILLING to 14 gene targets and 80 *Musa* accessions, over 6000 polymorphisms representing 870 unique alleles were rapidly discovered.

Prior to the advent of NGS, a major obstacle to the use of TILLING in orphan crops was identification of candidate genes that are typically selected on the basis of similarity to genes of proven function in model plants. As demonstrated by our work in A. annua and J. curcas, NGS can now be used to rapidly identify numerous candidate genes associated with specific traits (Graham et al., 2010; King et al., 2011). As demonstrated through the work in wheat, the impressive range of alleles that can be developed by TILLING approaches provides a rich source of material for continued breeding efforts. Crosses can be done to combine individual mutations and progeny carrying multiple mutant alleles selected. This selective approach to combine beneficial mutant alleles is simply not possible using forward genetic screens to identify and select individuals with a phenotype of interest.

Each individual in a mutant population carries multiple induced mutations depending on the initial level of mutagenesis. In Arabidopsis a mutation frequency of one mutation per 170 kbp has been reported (Till et al., 2003) which equates to 771 mutations per individual plant given a genome size of 135 Mbp. To remove additional potentially deleterious mutations it is essential to perform multiple rounds of backcrossing to parents while constantly selecting for the beneficial mutant. The availability of molecular marker systems that allow selection of progeny that carry maximum amounts of the 'clean' genome can significantly decrease the number of backcrosses required to clean up a mutant line.

Reports of ongoing TILLING projects in orphan crops including cassava, banana and tef (Eragrostis tef) are now beginning to appear. Tef is a staple cereal in East Africa where it is grown on over 2.5 million hectares of land, mainly in Ethiopia. Desirable characteristics attributed to this crop include tolerance to drought stress and waterlogging, exceptional nutritional value, and relatively few postproduction problems with pests or diseases. However, annual tef yields in Ethiopia can be as low as 0.9 tonnes per hectare (t/ha) compared to wheat at 1.3 t/ha. The major trait that limits yield is lodging of the stem, causing it to fall to the ground under wind and rain, resulting in poorer quality and quantity of grain. Conventional breeding has so far not led to significant improvement of this species. The classic Green Revolution genes that boosted the yield of wheat and rice in the 1960s and 1970s produce dwarf varieties that are more resistant to lodging and produce higher grain yields. The two best-characterised dwarfing genes for production of semi-dwarf varieties are the sd1 (semi-dwarf) and rht1 (reduced height) genes from rice and wheat, respectively (Ashikari et al., 2002;

Ellis *et al.*, 2002). The *sd1* mutation in rice results in loss of function of the GA20 oxidase, which is involved in the biosynthesis of the growth-promoting plant hormone, gibberellic acid. The *rht1* mutation in wheat however is an example of a gain-of-function allele since it produces a more active form of a growth-repressing protein causing the dwarf phenotype. Gain-of-function alleles are much less frequent than loss of function in mutation breeding programmes and therefore the *sd1* locus would be the preferred TILLING target in this case. Next-generation sequencing of the tef genome is already under way (Plaza *et al.*, 2009) and a recent report indicates that the tef genome carries three *sd1* and two *rht1* genes (Smith *et al.*, 2012) and these can now be used as targets for TILLING with the aim being to produce semi-dwarf higher-yielding varieties of this important orphan crop. In this context it is interesting to note that the Syngenta Foundation for Sustainable Agriculture has funded the University of Bern to develop a TILLING platform for tef improvement (www.syngentafoundation.org/index. cfm?pageID=529).

APPLICATION OF NGS-BASED TECHNOLOGY IN ORPHAN CROPS

The applications of NGS in plant genetics and plant breeding are now extending well beyond marker and gene discovery. The identification of mutant loci responsible for interesting phenotypes identified through forward screens has typically taken years of work even in well-established models and crops. The examples given above for the *sd1* and *rht1* dwarfing genes from rice and wheat demonstrate this very well, with decades elapsing before the genes were identified. However, new NGS-based applications are now having a big impact on the time it takes to identify mutant loci even in model species such as *Arabidopsis*. For example, using the principle of bulk segregant analysis, it is possible in a single sequencing reaction to map a mutation responsible for a specific phenotype and identify the causal change in an approach known as 'SHOREmapping' (Schneeberger *et al.*, 2009). Since this method requires a highly accurate and complete reference genome sequence it is only applicable to those species where this is available. However, NGS technology may not be that far away from being able to deliver highly accurate reference genome sequences for orphan crops at an affordable price. Exciting new single-molecule sequencing technologies, such as that recently introduced by Oxford Nanoparticle Technologies, will facilitate fast and accurate genome assembly (Metzker, 2010; www. nanoporetech.com/). The availability of reference genome sequences for orphan crops will further facilitate the rapid deployment of NGS technologies for molecular breeding and allow transfer of knowledge from the established models for trait improvement.

Screening methods based on NGS for reverse genetic analysis are also being developed and these may well replace the TILLING and ecoTILLING methods described above which rely on using enzymes such as CEL1. One such method, termed KeyPoint™, is based on NGS sequencing of target genes amplified from mutant or natural populations (Rigola *et al.*, 2009). Multidimensional pooling of large numbers of individual DNA samples combined with sample barcoding allows identification of mutations in target genes from both mutant and natural populations. Reverse genetic methods based on NGS provide immediate information on the sequence of mutant loci allowing alleles to be selected that are predicted to impact on gene function.

Thus NGS-based technologies are now available for both characterisation of mutants and natural variants identified from classic forward screens and for the identification of induced mutations or natural allelic variants in target genes in reverse genetic screens. Together with the proven utility in marker and gene discovery, NGS is poised to have a massive impact on genetic characterisation and improvement in orphan crops.

ENGINEERED VARIATION THROUGH GENETIC MODIFICATION

Genetic modification or GM has been used to improve the yield, quality and environmental impact of some of the major crop plants including soya bean, maize, rice and oilseed rape; and GM foods have had a major presence in global markets since 2006. However much controversy remains about the use of GM foods and crops, particularly in Europe, despite a scientific consensus that those currently on the market do not pose a risk to either human health or the environment. Regulatory approaches to GM crops are very different in Europe compared to the USA where they have been widely consumed since the mid-1990s. Europe requires new and separate laws and institutions for the regulation of GM food and crops while in the USA the same laws and institutions are used to regulate both GM and non-GM foods and crops. Also, in Europe the precautionary principle can be followed whereby a GM product can be declined on the grounds of 'uncertainty' alone, which can be a major barrier to commercial release. To date, African countries have tended to adopt the European model in terms of regulatory approval of GM food and crops due, it has been argued, to a number of external influences (Paarlberg, 2010). Consequently, 15 years after GM crops were first planted commercially in the USA, only two governments in sub-Saharan Africa had allowed commercial release: the Republic of South Africa (for maize, soya bean and cotton) and Burkina Faso (for cotton). Not surprisingly given this context, there has been little investment in GM improvement of orphan crops for Africa despite the massive potential benefits that can be gained from the technology. The success of programmes such as BioCassava Plus, which involves a team of scientists from around the world, with the objective of reducing malnutrition in the 250 million people in sub-Saharan Africa who rely on cassava as their staple food, could go a long way to convincing African politicians of the benefits of GM technology. It has been reported that in the last 5 years this team has developed cassava plants that have 30 times as much beta-carotene, four times as much iron and four times as much protein as traditional cassava – levels that represent the minimum daily requirements for a child (www.danforthcenter.org/science/programs/international_programs/bcp/). This project has now reached the stage of field-testing and safety assessments in Nigeria and Kenya in preparation for regulatory approval and release.

The cost of compliance with each country's regulatory processes can itself represent a major barrier and deterrent to investment in development of GM crops, and this is particularly relevant to GM orphan crops. However, a recent analysis carried out in the Philippines on four GM products – *Bt* (insect-resistant) eggplant, *Bt* rice, ringspot-virus-resistant papaya and virus-resistant tomatoes – suggests that regulatory costs are significant but generally smaller than technology development costs (Bayer *et al.*, 2010). Furthermore, regulatory costs appear to decline within countries as they gain experience with more products. There is therefore cause for optimism that investment in the development of GM orphan crops that deliver incremental increases in performance for the benefit of society and the environment will gain both public and regulatory approval and at a cost that is not prohibitive in developing as well as developed countries.

CONCLUSIONS

This short perspective demonstrates the potential for molecular plant breeding to enable significant rapid improvement in orphan crops. It also highlights the pervasive nature of NGS technologies and the role that these are now playing across the various stages of crop improvement. It is now possible to rapidly establish molecular breeding platforms in orphan crops and target traits for fast-track improvement. Indeed, several of the technologies are now sufficiently 'plug and play' that the concept of rapid domestication of wild species can also be considered. With the decrease in costs and increase in throughput brought about by NGS and associated technologies we can also realistically plan for advanced molecular breeding projects on orphan crops to be carried out in research institutes, academic laboratories and small to medium-sized companies as well as large multinational organisations. This diversity will be essential if we are to realise the full potential that orphan crops offer society in the twenty-first century.

REFERENCES

Ashikari, M., Sasaki, A., Ueguchi-Tanaka, M., et al. (2002). Loss-of-function of a rice gibberellin biosynthetic gene, GA20 oxidase (*GA20ox-2*), led to the rice 'green revolution'. *Breeding Science* **52**, 143–150.

Bayer, J. C., Norton, G. W., and Falck-Zepeda, J. B. (2010). Cost of compliance with biotechnology regulation in the Philippines: implications for developing countries. *AgBioForum* **13**, 53–62.

Edwards, M. D., Stuber, C. W., and Wendel, J. F. (1987). Molecular marker-facilitated investigations of quantitative-trait loci in maize. I. Numbers, genomic distribution and types of gene action. *Genetics* **116**, 113–125.

Ellis, M. H., Spielmeyer, W., Gale, K. R., Rebetzke, G. J., and Richards, R. A. (2002). 'Perfect' marker for the *Rht-B1b* and *Rht-D1b* dwarfing genes in wheat. *Theoretical and Applied Genetics* **105**, 1038–1042.

Elshire, R. J., Glaubitz, J. C., Sun, Q., et al. (2011). A robust, simple genotyping-by-sequencing (GBS) approach for high diversity species. *PloS ONE* **6**, e19379.

Graham, I. A., Besser, K., Blumer, S., et al. (2010). The genetic map of *Artemisia annua* L. identifies loci affecting yield of the antimalarial drug artemisinin. *Science* **327**, 328–331.

He, W., King, A. J., Khan, A. M., et al. (2011). Analysis of phorbol-ester content, curcin content and genetic diversity, in edible and non-edible accessions of *Jatropha curcas* (L.) from Madagascar and Mexico. *Plant Physiology and Biochemistry* **49**, 1183–1190.

King, A. J., He, W., Cuevas, J. A., et al. (2009). Potential of *Jatropha curcas* as a source of renewable oil and animal feed. *Journal of Experimental Botany* **60**, 2897–2905.

King, A. J., Li, Y., and Graham, I. A. (2011). Profiling the developing *Jatropha curcas* L. seed transcriptome by pyrosequencing. *BioEnergy Research* **4**, 211–221.

Mardis, E. R. (2011). A decade's perspective on DNA sequencing technology. *Nature* **470**, 198–203.

Metzker, M. L. (2010). Sequencing technologies: the next generation. *Nature Reviews Genetics* **11**, 31–46.

Naylor, R. L., Falcon, W. P., Goodman, R. M., et al. (2004). Biotechnology in the developing world: a case for increased investments in orphan crops. *Food Policy* **29**, 15–44.

Paarlberg, R. (2010). GMO foods and crops: Africa's choice. *New Biotechnology* **27**, 609–613.

Plaza, S., Bossolini, E., and Tadele, Z. (2009). Significance of genome sequencing for African orphan crops: the case of tef. *African Technology Development Forum Journal* 6, 55–59.

Rigola, D., van Oeveren, J., Janssen, A., *et al.* (2009). High-throughput detection of induced mutations and natural variation using KeyPoint technology. *PLoS ONE* 4, e4761.

Schneeberger, K., and Weigel, D. (2011). Fast-forward genetics enabled by new sequencing technologies. *Trends in Plant Science* 16, 282–288.

Schneeberger, K., Ossowski, S., Lanz, C., *et al.* (2009). SHOREmap: simultaneous mapping and mutation identification by deep sequencing. *Nature Methods* 6, 550–551.

Slade, A. J., Fuerstenberg, S. I., Loeffler, D., Steine, M. N., and Facciotti, D. (2005). A reverse genetic, nontransgenic approach to wheat crop improvement by TILLING. *Nature Biotechnology.* 23, 75–81.

Smith, S. M., Yuan, Y., Doust, A. N., and Bennetzen, J. L. (2012). Haplotype analysis and linkage disequilibrium at five loci in *Eragrostis tef. G3 (Bethesda)* 2, 407–419.

Till, B. J., Reynolds, S. H., Greene, E. A., *et al.* (2003). Large-scale discovery of induced point mutations with high-throughput TILLING. *Genome Research* 13, 524–530.

Till, B. J., Zerr, T., Comai, L., and Henikoff, S. (2006). A protocol for TILLING and ecoTILLING in plants and animals. *Nature Protocols* 1, 2465–2477.

Till, B. J., Jankowicz-Cieslak, J., Sági, L., *et al.* (2010). Discovery of nucleotide polymorphisms in the *Musa* gene pool by Ecotilling. *Theoretical and Applied Genetics* 121, 1381–1389.

KEY FURTHER RESOURCES

1. *African Technology Development Forum (ATDF) Journal* 6 (3 & 4), 2009 – African Orphan Crops (www.atdforum.org/spip.php?article350).
2. The International Resource Centre is an online support tool for marker assisted breeding with protocols, tutorials, learning modules, literature and general resources, such as information on writing proposals. Funded by the Syngenta Foundation for Sustainable Agriculture (http://irc.igd.cornell.edu/).
3. The CNAP Artemisia Research Project is developing improved varieties of the medicinal plant *Artemisia annua*. This plant is the primary source of the leading antimalaria drug artemisinin (www.york.ac.uk/org/cnap/artemisiaproject/index.htm).
4. BioCassava Plus is an integrated team of scientists from Africa, Asia, Europe, Latin America and North America whose objective is to reduce malnutrition among the 250 million people in sub-Saharan Africa who rely on cassava as their staple food by delivering a more nutritious and marketable cassava (www.danforthcenter.org/science/programs/international_programs/bcp/)
5. RevGenUK is a BBSRC-funded technology platform for plant reverse genetics by TILLING (Targeting Induced Local Lesions IN Genomes) (http://revgenuk.jic.ac.uk/).

Part 2 New genetic crops across the emerging world

Introduction

CHRISTOPHER J. LEAVER

He who has bread may have troubles. He who lacks it has only one.

Old Byzantine proverb

During the last 50 years the world's population has doubled to more than 7 billion and the median UN prediction is that it will reach 9 billion by 2050. Eighty percent of this increase in population will live in the developing and transition countries of China, India, Africa, Southeast Asia and South America, with the majority (more than 70%) living in an urban environment in megacities. In Africa, for example, the population is predicted to double in the next four decades. Each hectare of arable land in 2050 will need to feed five people compared to just two in 1960. This must be achieved in the face of climate change, due in part to massive fossil fuel usage and the intensification of agriculture which results in increased greenhouse gases and carbon dioxide levels (contributing directly 10–12% of emissions). This will be coupled to decreased water availability because most of the agriculture is rain-fed, increased and changing biotic stress due to pests and pathogens, environmental pollution, loss of biodiversity, urbanisation and dietary upgrading.

Successful Agricultural Innovation in Emerging Economies: New Genetic Technologies for Global Food Production, eds. David J. Bennett and Richard C. Jennings. Published by Cambridge University Press. © Cambridge University Press 2013.

It has been estimated that existing food stocks are adequate and, with political and financial will, these could be sufficient to address current acute hunger problems associated with civil unrest and with adverse climatic effects on local food production and availability. However, they are not sufficient to solve the long-term chronic problems of food security and nutrition associated with the increase in population. To meet the needs of 9 billion people the United Nations Food and Agricultural Organization (FAO) estimates that food production will have to rise by around 70% in the next 40 years on essentially the same area of land. This will require 'sustainable intensification – growing more from less' with the aim of meeting the current needs while improving the ability of future generations to meet their own needs.

This global demand for food and increased consumption of meat in the developing world coupled with the increasing use of grain for biofuels is already having an impact on commodity prices which are both increasingly volatile and near an all-time high. In contrast, grain reserves are near an all-time low. Of course efforts are also needed to stabilise the global population, not least by economic development and as a result of the education and empowerment of women.

Global food production has kept pace with the more than doubling of the population since 1960. This dramatic increase in crop yields has primarily been due to five innovations: genetics and plant breeding, the use of NPK (nitrogen, phosphorus and potassium) fertilisers and nitrogen in particular (essentially by conversion of fossil fuels, in the form of natural gas, by the Haber–Bosch process, into food), pesticides/agrochemicals which enabled farmers to grow high-density crops year after year without severe loss to pests and weeds, irrigation (agriculture currently consumes about 70% of global water use) and mechanisation which decreased intensive manual labour and freed up perhaps 25% of extra land to grow human food instead of fodder for horses and oxen. The effect of these five innovations was to allow more and more food to be produced from less and less land. The developed world became complacent! Unfortunately the year-on-year increases in yield of the major world crops in the developed countries have plateaued in recent years while the yields in the developing countries have failed to improve significantly.

Despite this increase in productivity almost 1 billion do not have enough food to meet their daily calorific needs and go hungry each day; about 250 million of these are in Africa and more than 75% live in rural areas. A large proportion of the 1 billion undernourished people are either small landholder farmers or labourers on these farms who depend on agriculture for their income. More than 3 billion people are living in absolute poverty on less than 2 dollars a day and are generally deficient in at least one nutrient necessary for maintaining their health. About 30000 people, half of them children, die every day due to hunger and malnutrition. Increasing poverty, particularly in rural areas, also leads to forced economic migration to cities and neighbouring countries and as a consequence leads to an exacerbation of food insecurity.

The success of modern agriculture has also had a high environmental cost. We have lost about 20% of our topsoil due to erosion, desertification and salinity, about 20% of our agricultural land due to overgrazing and the generation of marginal land, and about 30% of our forests. While there is certainly a role for local, organic/diverse agroecological basic farming practices in many developing countries, these should complement rather than replace the fossil-fuel-driven industrialised farming on which we are dependent on our overpopulated planet.

It is obvious that the Earth's carrying capacity has been exceeded (to live sustainably as the world does now, we would need more than one-and-a-half Earths) and it is unrealistic to assume that the developing world can be fed adequately in a sustainable manner by continuing with 'business as usual'. By contrast, a significant proportion of the population of the developed world is obese (due to overconsumption of calories) and food waste is reaching epidemic proportions. The developed world (some would say 'overdeveloped') is living beyond its means and must 'shrink and share' while increasing food and energy security for the rest of the world using all available technologies and interventions.

To achieve this the scientific community should be engaged in transferring the capacity for research and development leading to advances in agricultural productivity, and should also endeavour to see that the benefits associated with such advances accrue to the benefit of the poor. Efforts should focus on access for poor farmers in the developing world to improved crop varieties adapted to their local conditions. Particular attention is required to local needs and crop varieties and to the capacity of each country to adapt its traditions, social heritage and administrative practices to achieve the successful introduction of such crops. In Africa in particular, agriculture has the added potential to become the engine for economic growth. Food security and the need for healthy staple food are no longer just the concerns of developing countries – they have become a global issue, not least in Europe which is a net food importer.

If we are to conserve biodiversity and preserve ecosystems and ecological resources while minimising the effects of climate change, the necessary increase in crop yield must be accomplished by sustainable intensification. We need to 'grow more from less' by using land and resources more efficiently. Ideally this should be addressed by meeting the needs of farmers at every level, from small plots to the prairie-style farming practised in the Americas.

The chapters in this section of the book address the ways in which the revolution in plant science – including our ability to modify plant traits by both marker-assisted plant breeding and genetic modification – is being applied to both staple and orphan crops to address the challenges of food security in the developing world. These improvements depend upon our increasingly detailed understanding of the information content of plant genomes and of their pests and pathogens to produce better crops capable of coping with climate change. Integration of all these so-called '-omics technologies' using bioinformatics and modelling coupled with the appropriate use of agrochemical inputs and suitable farming practices is the only way in which the global challenges outlined above can be addressed. A recurrent theme in this section is that all available technologies should be evaluated and, subject to appropriate and realistic biosafety regulation, those that are most effective need to be deployed.

Pamela Ronald in her Chapter 11 provides examples of the successes of both conventional plant breeding and first-generation genetically engineered traits (insect resistance and herbicide tolerance) which were designed to complement the use of agrichemicals. These crops provide better pest and weed control thus improving crop productivity in many of the major staple crops (such as maize, rice, cotton and soya bean) grown worldwide. She concludes by explaining that the future for sustainable agriculture lies in harnessing the best of all agricultural technologies, including the use of genetically engineered seed, within the framework of ecological farming.

In Chapter 8 Trigo and Cap describe the dramatic economic and environmental benefits (including 'no-till' farming) of the adoption and rapid transformation of the agricultural sector in Argentina by the introduction of genetically modified crops (primarily herbicide-tolerant soya beans). They emphasise that this success was only possible because Argentina had in place the institutions required to deal with technology transfer and the diffusion process together with the biosafety regulations and infrastructure necessary to assess the technologies. The existence of an active and efficient seed industry was also vital for the rapid introduction of new genes into commercial varieties that were already adapted to the wide range of agro-ecological conditions in the different growing regions.

Diran Makinde in Chapter 7 provides a comprehensive overview of the major challenges facing African agriculture and the significant role that crop biotechnology, in combination with conventional approaches, can have in enhancing agricultural productivity and eliminating poverty. Makinde describes the influence of attitudes in Europe to GM technology on the drafting and enacting of biosafety regulations – a prerequisite for the commercial planting of GM crops – in African countries. This influence is partly responsible for the fact that only three African countries have embraced GM crops: South Africa, Burkina Faso and Egypt. Future progress will require a step change in collaboration between African countries coupled with increased government investment, capacity building and research to deliver the range of improved genetic traits in relevant African crops. There is a need to build and sustain collaborative relationships for exchange of information, knowledge and expertise leading to acceptance of a common regulatory framework across the continent.

Adrian Dubock in Chapter 12 asserts that the primary goals of plant breeding to date and current farming practices have been to improve the harvestable macronutrients and calorific yield in the major grain crops such as wheat, maize and rice. In contrast, improvement of other important staple crops in Africa and elsewhere ('orphan crops') such as cassava and plantain (cooking banana) has been relatively neglected. One of the goals for the future must therefore be to improve all crops for both macronutrients – carbohydrate, proteins and fats (of the appropriate nutritional quality) – and, of equal importance, micronutrients vital for human health and development. Dubock also describes recent progress and tribulations in the biofortification of staple crops for essential vitamins (e.g. beta-carotene, a source of vitamin A in Golden Rice) and minerals (iron and zinc) by both conventional breeding and, in instances where that was not possible, by genetic engineering.

Murdock, Sithole-Niang and Higgins in Chapter 13 tell the story of cowpea improvement. Cowpea, also known as black-eyed pea, is the archetypal nitrogen-fixing, protein-rich, legume crop of Africa predominantly grown by women. This versatile but neglected orphan crop is the most nutritionally and economically important legume grown in Africa and thrives on poor soils and can survive extremes of rainfall. In recent years breeders in Africa have developed cowpea with high yield potential but, depending on the agro-ecological zones, high yields are rarely obtained because of insect infestation before and after harvest which dramatically lowers yields. Insecticides can be used to combat the insect pests but are considered an expensive and poor solution for field-grown cowpeas. The authors describe progress in the successful development of insect-resistant transgenic Bt cowpea and the challenges facing its deployment for the benefit of smallholder farmers in sub-Saharan Africa.

In Chapter 10 Chavali Kameswara Rao addresses the food security needs facing India with its widespread poverty, malnutrition and predicted increase in population from 1.21 to 1.6 billion by 2050 coupled with a failure to increase productivity compared to the developed world. He explains that following the impact of the Green Revolution in the 1960s, which turned India into a net food-exporting country, it has failed to adopt new agricultural technologies, including genetically engineered food crops, to develop seed with traits relevant to regional needs and to increase crop yield. The success of commercialisation of Bt cotton is contrasted with the current impasse facing the adoption of a range of genetically engineered food crops, including Bt brinjal (aubergine) and Golden Rice, due to widespread activism against GM crops. The government of India has failed to resolve biosecurity issues from the tangled mix of political, economic, societal and ethical issues which have been hijacked by the activists. As a consequence, no improved crops developed by genetic engineering are being approved for commercialisation and the Indian poor are the net losers.

Lu Bao-rong in Chapter 9 brings a Chinese perspective to the mounting tension between the decreasing land available for agriculture (China has about 7% of the world's arable land) due to industrialisation, urbanisation and economic development and an increasing demand for food by the Chinese population which at 1.35 billion is currently 20% of the world total and set to increase to about 1.5 billion by 2050. Lu describes the success of the development of new germplasm and hybrid breeding technologies since 1978, coupled with high agrichemical inputs, on increasing the productivity of rice, maize and wheat which resulted in lifting many millions out of poverty. In the belief that far greater effort is required to increase crop productivity and ensure food security and stability the Chinese government has over the last 30 years invested more than US$3.5 billion in transgenic research in more than 52 species of crop plants and some animals. Except for the successful commercialisation of Bt cotton (which is described in detail) little is known of the deployment of other transgenic crops but the expectation must be that they will be a feature in addressing the challenge of improving agricultural production and long-term food security.

Jonathan Gressel in Chapter 14 provides a thought-provoking look at the future where tropical developing countries may be able to

improve their economies by growing transgenic marine algae, engineered with value-added traits and modified for maximum photosynthetic efficiency and cultured on seawater with waste carbon dioxide.
The vision being that transgenic marine microalgae could replace plant
protein from grain legumes as well as from fishmeal, as a protein and
lipid feedstock for aquaculture and animal fishmeal, from sea-caught
fish which is a resource required for aquaculture, and is being rapidly
depleted from the oceans, so such a replacement is imperative for more
efficient aquaculture and for export.

While science can provide technological solutions, these have to
be implemented in a responsible and fair way for them to have impact.
And this is not the job of scientists but of politicians. Now, and in the
future, making sure everyone has enough to eat is about both politics
and science ...

7

Status of crop biotechnology and biosafety in Africa

INTRODUCTION

The application of recombinant deoxyribonucleic acid (rDNA) techniques to artificially transfer useful genes from one organism to another in order to enhance the expression of desirable traits or to suppress undesirable traits has resulted in the development of transgenic or genetically engineered (GE) crops. Such crops first became commercially available in 1996 and in countries where they have been officially approved, growers have rapidly adopted GE crop traits of herbicide resistance (HR), insect resistance (IR) or a combination of both.

Current literature and data support the overall positive impacts of agricultural biotechnology globally. Savings in terms of increased gross margins (114%), reduced pesticide costs (62–96%), beneficial human and environmental effects and improved yields (18–29%) over conventional crops in the presence of pest pressure have been documented for small-scale African farmers growing commercial GE crops (IFPRI, 2011).

The yield losses and reduced product quality caused by insect pests have long been a major agricultural concern. Each year, insect pests destroy about 25% of food crops worldwide and this can be as high as 90% of some African commodities. Plant diseases caused by a wide variety of bacteria, viruses and fungi as well as other organisms such as plant-eating nematodes and mites also reduce plant growth and yield.

Adoption of IR and HR crops has generated significant economic benefits. For IR crops, this has come with higher yields due to greater insect control, reduced insecticide costs or both (Brookes and Barfoot, 2010). Other studies have considered benefits such as simplicity,

Successful Agricultural Innovation in Emerging Economies: New Genetic Technologies for Global Food Production, eds. David J. Bennett and Richard C. Jennings. Published by Cambridge University Press. © Cambridge University Press 2013.

convenience, flexibility, safety for human health and the environment and compatibility with conservation tillage. Genetically engineered crops also provide environmental benefits such as the return of a diversity of pollinating insects to fields that were hitherto treated with agrochemicals. Adoption of IR cotton has led to shifts to insecticides with lower toxicity and fewer harmful environmental effects (Brookes and Barfoot, 2010). Interestingly, the adoption of Bt maize (with a gene inserted from the microorganism *Bacillus thuringiensis* to provide resistance to insects) has had a more limited impact on insecticide use (Qaim, 2009). This is thought to be because the primary target pest, the maize stem-borer, is not effectively controlled with insecticides. Thus the main benefits to maize growers has been yield gains rather than reduced insecticide applications.

Others argue that GE crops in the USA were not designed to increase yields, as pesticides and herbicides already maximised yields, but rather to reduce farm expenditures on labour and inputs (Ledermann and Novy, 2012). However, it was acknowledged that if Bt crops are compared to conventional crops that are grown outside of an integrated pest management system or with no application of insecticide, the Bt crop will outperform the comparator crop, in terms of yield (Heinemann, 2009). This is relevant to Africa, as most African smallholder farmers operate a traditionally low-input agriculture (Concern, 2008).

NEEDS

African agriculture has been underperforming for over three decades now despite providing for the livelihoods of about 60% of the continent's active labour force, contributing 17% of Africa's total gross domestic product and accounting for 40% of the foreign currency earnings. African farmers' yields have essentially stagnated for decades but the total output has been rising steadily because of the extension of land area under cultivation. However, this growth rate has not kept pace with the increasing population. The reasons for this stagnation are multiple. They include:

- Continuing dependence on uncertain rainfall
- Africa's degraded soil health (nutrient mining)
- Small and dispersed domestic markets
- The instability and decline of world prices for agricultural exports
- The size of most small farm holdings (less than 3 acres (1.2 hectares) on average)
- Farmers' lack of organisation

- The lack of rural roads and bridges
- The low technology input into farming
- Neglect of the particular needs of women farmers (who produce most of the continent's food)
- The pandemic of HIV/AIDS.

It has been suggested that Africa's food consumption patterns will change over the coming decades as a result of the rising urban population and growing per capita incomes. These will bring about a rapid change in African food systems in such a way that a doubling of the marketed volumes of foodstuffs and increased demand for high-value foods (dairy, meat, fresh fruits and vegetables) and processed foods will occur. This could encourage the training of more scientific and technical experts in policy and regulatory issues, among others.

Within the context of sustainable development, Africa needs to focus on the provision of sufficient and affordable nutrition, access to improved, affordable and effective healthcare, protection of the African environment, development of new and sustainable energy sources, creation of jobs and reduction of poverty. Biotechnology can play a significant role in all of these areas. Agricultural biotechnology, therefore, is needs-based in Africa.

Having recognised the importance of harnessing science and technology for Africa's development, the African Union (AU)-New Partnership for Africa's Development (NEPAD) Planning and Coordinating Agency (NPCA) has adopted a common approach to address issues pertaining to agricultural biotechnology and biosafety by endorsing the AU's decision EX.CL/Dec.26 (III) which calls for a common African position on biotechnology. The AU has thereafter put in place the African Model Law on Biosafety to guide Member States in drafting their national legislation. Other initiatives include the African Strategy on Biosafety with the objective of providing Member States with a framework for regional, subregional and national initiatives on biosafety; guiding and promoting regional coordination and harmonisation of biosafety on the continent; and enhancing regional capacity in biosafety. These complement some of the recommendations contained in the recent AU-NEPAD publication, *Freedom to Innovate* (Juma and Serageldin, 2007) which emphasises, amongst others, the following two recommendations:

> *The AU's Regional Economic Communities (RECs) need to be staffed with appropriately trained experts who can advise states on regional and international agreements, guidelines and conventions on all aspects of biotechnology. The AU Secretariat and NEPAD need to build further capacity in biotechnology regulation.*

They could also provide assistance to states on multilateral mechanisms and agreements. (AU-NEPAD Recommendation 14)

AU Member States should consider adopting a consistent Africa-wide position on food and environmental standards, commensurate with international obligations. Taking such a step will help to ease inter-AU trade, among other activities. (AU-NEPAD Recommendation 15)

Other recommendations include the need to outline priority areas in biotechnology that are of relevance to Africa's development. Along this line, southern Africa is to concentrate on health biotechnology, North Africa on biopharmaceuticals, West Africa on crop biotechnology, East Africa on animal biotechnology and Central Africa on forest biotechnology. Other areas of focus are identifying critical capabilities for the development and safe use of biotechnology and the need to establish appropriate regulatory measures that can advance research, commercialisation, trade and consumer protection, and setting strategic options for creating and building regional biotechnology innovation communities and local innovation areas in Africa (Juma and Serageldin, 2007).

In addition to the above recommendations, the Conference of African Union Ministers of Agriculture held in Libreville, Gabon on 27 November to 1 December 2006 took an African position on genetically modified organisms (GMOs) in agriculture and made the following recommendations under capacity building for scientific and regulatory institutions:

- *encourage Member States to develop policies that enhance training in biosafety and biotechnology;*
- *develop policies for on-the-job training on the safe management of biotechnology;*
- *establish regional GMO testing laboratories by developing norms and standards;*
- *promote policies to enhance and encourage public-private partnerships in biotechnology; and*
- *encourage the development of policies that enhance Member States regulatory capacity on issues of biosafety and biotechnology.* (AU position document on GMOs, 2007)

In some African economies, agricultural biotechnology is being adopted for harnessing innovation and efforts are ongoing for integrating biotechnology in the national development agenda. A few countries have made excellent progress but many others continue to face challenges, despite the well-intentioned accession to the Cartagena Protocol on Biosafety, particularly for the wider application and resultant commercialisation of biotechnology. Despite the fact that GE crops confer agronomic, environmental, nutritional and health benefits, only

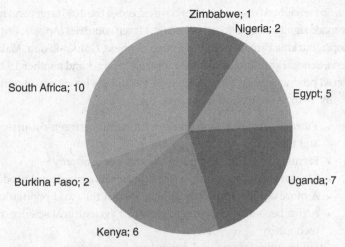

Figure 7.1 Number of confined field trials (CFTs) in African countries.
(NEPAD Agency ABNE website, www.nepadbiosafety.net)

three African countries have commercialised GE crops, namely South Africa, Burkina Faso and Egypt. However, a good number of African countries are conducting field trials for GE crops, namely South Africa, Egypt, Burkina Faso, Kenya, Uganda, Nigeria and Zimbabwe (Figure 7.1).

African leaders have repositioned agriculture in their development agenda by establishing the Comprehensive Africa Agriculture Development Program (CAADP) and in adopting the co-evolution of biotechnology and biosafety. Implementation of the vision for agricultural research and development in Africa is derived from the AU-NEPAD Agency CAADP process. CAADP consists of four pillars:

- Pillar 1: Sustainable land and water management
- Pillar 2: Rural infrastructure and trade-related market access
- Pillar 3: Food security and hunger reduction
- Pillar 4: Agricultural research, technology dissemination and adoption

with each pillar being led by Africa-based institutions. This arrangement brings together key players at the continental, regional and national levels to improve coordination, share knowledge and to promote joint and separate efforts to achieve CAADP goals. This compelling vision lays down a dramatic change in direction and approach in preparing for a new tomorrow of Africa's agricultural research for development. African leaders agreed to increase public investment in agriculture by a minimum of 10% of their national budgets and to raise agricultural productivity by at least 6%.

So far, eight African countries have exceeded the 10% target and most have made significant progress towards it. Eleven countries (Angola, Eritrea, Ethiopia, Burkina Faso, Republic of Congo, Gambia, Guinea-Bissau, Malawi, Nigeria, Senegal and Tanzania) have met the 6% target and another 19 have achieved productivity growth of between 3% and 6% (www.nepad.org).

By 2015, African leaders expect:

- Vibrant agricultural markets within and between countries and regions in Africa
- Farmers as active players in the market economy
- The continent becoming a net exporter of agricultural products
- A more equitable distribution of wealth for rural populations
- Africa becoming a strategic player in agricultural science and technology
- Environmentally sound agricultural production
- A culture of sustainable management of the natural resources in Africa.

In implementing activities leading to the realisation of this vision, it is imperative to note that whereas there is a possibility for alternatives to fossil fuels, there is no alternative to food.

Achieving the CAADP vision also entails a substantial productivity shift in smallholder farming practices – a revolution which must be led by robust producer and consumer-centric research approach. Africa's agricultural research, therefore, needs to respond to the socio-economic, environmental and policy constraints impeding development in Africa. Access to technology and to markets must be created for producers.

BENEFITS ACCRUING FROM GENETIC ENGINEERING TECHNOLOGY IN AFRICA

Information from South Africa and Burkina Faso serves to highlight the demonstrable benefits of GE crops in Africa. Egypt has a number of crops under confined field trials (CFTs), but only maize has been commercialised.

South Africa

South Africa is the leading country in the cultivation, research and development of GE crops in Africa. Research is carried out on IR and HT cotton, virus-resistant and drought-tolerant maize, fungal-resistant and virus-resistant grapevine, starch-enriched cassava and sugar cane with higher yields and enhanced sugar content. These new GE crops are being developed by national and international seed companies, South African research institutions, academic institutions and industry. Thirteen GE crops, consisting of eight cotton lines, four maize lines and one soya bean line have been approved for commercial release. The benefits that accrued to South Africa with the cultivation of GE crops included 31% increased yield for hybrid maize over the corresponding conventional hybrid and 134% over conventional open-pollinated varieties planted by some small-scale farmers. In addition, smallholders who cultivated GE cotton experienced yield increases of 11% and average benefits of US$35/ha; on irrigated land, average benefits were US$117/ha (South Africa Department of Agriculture, 2010).

Burkina Faso

Prior to 2003, Burkina Faso was spending over €30 million (US$40 million) on pesticides as a result of high pest pressure that annually caused 50–70% production losses in cotton. The annual seed cotton yield decreased from 1200 to 900 kg/ha. In adopting Bt cotton, the country focused on the benefits that the technology could offer. The first CFT for Bt cotton started in 2003 with American cultivars to evaluate the efficacy of the Bollguard II (BGII) gene under climatic and insect pest conditions in the country. Multi-location trials were carried out in 2004–5. The first test of BGII technology with local germplasm varieties started in 2006 and was extended to 20 demonstration tests on farms with 100m isolation distance from conventional cotton the following year. Thereafter the commercialisation of Bt cotton started in 2009 and currently 80% of cotton production in Burkina Faso is Bt cotton and on about 350,000 ha. In addition to this, approval for CFT HT (glyphosate-resistant gene) Bt cotton started in 2010; Maruca-resistant cowpea in 2011; biofortified sorghum is under review; and marker-assisted selection of cowpea resistant to Striga started in 2009.

The benefits that accrued to Bt cotton farmers included increased yields by an average of 18.2% over the conventional varieties; and a profit of between US$39/ha to US$61.88/ha over the conventional cotton (James, 2010). Farmers also observed an increase in honey production in the areas of Bt cotton cultivation. Other benefits derived from this technology in Burkina Faso were the co-ownership of the GE varieties by the government and the technology developer; accrued royalties go into research infrastructure; two-thirds of the profits go to the cotton producers; and one-third is divided between the technology owner and the local seed companies that do seed multiplication. Estimated annual economic benefits from *Bt* cotton is over US$100 million arising from yield increases of close to 30%, plus at least a 50% reduction in insecticides sprays, from a total of eight sprays required for conventional cotton, to only two to four sprays for *Bt* cotton (Vitale *et al.*, 2011).

CHALLENGES

The potential of modern biotechnology to contribute to addressing Africa's food security has been recognised by the AU Commission, NPCA and by Member States themselves. After 15 years of safe application of modern biotechnology to crop production with novel traits, a period that witnessed a phenomenal global growth in adoption of the technology from 1.7 million ha in 1996 to 148 million ha in 2010, the pace of adoption in Africa has been relatively slow.

The limitations to biotechnology growth in Africa agriculture include:

- Absence of biosafety regulations in most African countries
- Lack of/inadequate political will to implement biotechnology/ biosafety policies
- Poor critical mass of expertise in biotechnology/biosafety (compounded by the brain drain of experts) and inadequate infrastructure
- Lack of access to accurate information
- Lack of viable seed industries
- Nascent public–private sector partnerships and weak linkages between industry and R&D institutions
- Lack of funding and investment in the technology including the 2003 Maputo Declaration of allocating 10% of the national budget to agriculture each year. To date, Burkina Faso, Ethiopia, Ghana, Guinea, Malawi, Mali, Niger and Senegal have exceeded this target and most countries have made significant progress towards achieving this goal.

In a recent study report entitled the *Status of Biotechnology and Biosafety in Sub-Saharan Africa* by the Forum for Agricultural Research in Africa (FARA) and the Syngenta Foundation for Sustainable Agriculture (FARA Study Report, 2011), it was revealed that even where several national research centres had the potential to apply biotechnology, levels of scientific expertise were generally low and there were insufficient infrastructural facilities for GE research. It was suggested that research capacity needs to be strengthened in such areas as molecular biology,

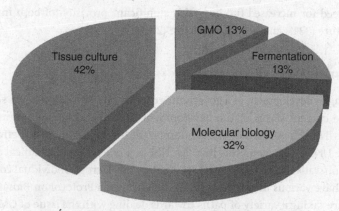

Figure 7.2 Most commonly available biotechnologies in laboratories in African countries. (FARA 2009 Study Report)

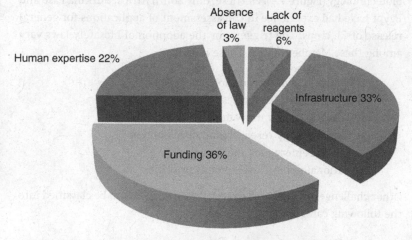

Figure 7.3 Constraints encountered by institutions conducting biotechnological research. (FARA 2009 Study Report)

biochemistry, genomics, plant breeding, bioinformatics and policy-making for the effective application of home-grown GE to African agriculture. Of the laboratories studied, 42% were using molecular biology tools to assist plant breeding, 32% applied tissue culture and 13% of the laboratories were involved in research on GE crops (Figure 7.2). In the same study, 50% of the respondent institutions had never applied for permission to conduct GE research. The reason given was that the countries in which such institutions were based had not adopted biosafety laws or had no biosafety regulatory procedures in place to facilitate GE research. About 36% of institutions studied (Figure 7.3) indicated the need for increased funding with significant provision of both infrastructure (33%) and human expertise (22%).

BIOSAFETY REGULATORY REGIME IN AFRICA

The lag in development of a governance capacity for biotechnology is seen in the current status of the development of National Biosafety Frameworks (NBFs) in Africa (Figure 7.4), despite the huge investment and efforts of the UNEP–GEF along this line. An NBF is a policy, legal, technical and administrative instrument to address safety of GE crops. Individual countries have various interpretations of the Cartagena Protocol on Biosafety and are taking a variety of paths towards dealing with the issue of GMOs.

Out of the 54 Member States of the African Union, only 18 countries have laws, regulations, guidelines or policies related to modern biotechnology (Figure 7.4). Of these, only South Africa, Burkina Faso and Egypt have had experience in the assessment of applications for general release of GE crops. The triggers for the adoption of biosafety laws vary among these Member States. These include:

(1) Obligations in international agreements
(2) Precautionary measures
(3) Trans-boundary movement of GMOs
(4) Biotechnology research and development
(5) Specific emergency measures
(6) National policy development priorities.

Other challenges with regard to regulatory regimes can be classified into the following categories:

(a) Policy that:
- Is heavily focused on risks and does not reflect the benefits
- Uses non-science elements to assess the technology, e.g. socio-economic factors,
- Includes 'strict liability clauses' that are a disincentive to public researchers, private investors and technology developers,
- Imposes risk assessment requirements inconsistent with the product development cycle,
- Presents regulations that are typically unaffordable (costs) and unenforceable,
- In some cases promotes national and regional conflicts on biosafety harmonisation.

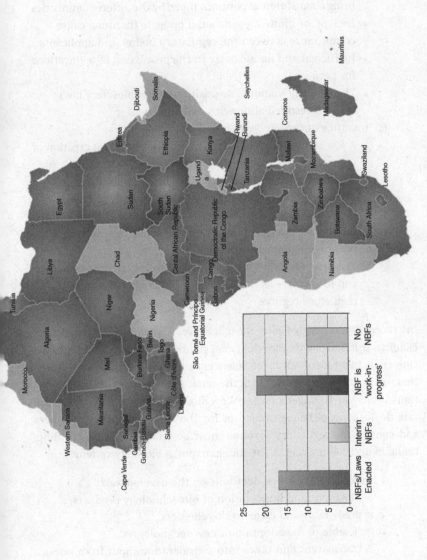

Figure 7.4 Status of development of National Biosafety Frameworks (NBFs), February 2012. (NEPAD Agency ABNE website, www.nepadbiosafety.net)

(b) Lack of cooperation and coordination among government departments that includes:
- Non-harmonisation of roles and responsibilities among government ministries involved with biosafety which often bring about different positions taken by the different ministries
- Lack of, or limited, operational budgets that may cause compromise between the regulatory bodies and applicants
- Inefficient and undue delay in the processing of applications for permits
- The frequent stand-alone position of the biosafety laws without reconciliation with existing laws.

(c) Insufficient capacity which includes:
- A low critical mass of experts with the attendant creation of opportunities for activists
- A low critical mass of experts which has permitted the creation of many biosafety service providers in Africa with little or no synergies among them and much duplication of efforts
- Biosafety service providers that are donor-dependent and possibly hold opposing viewpoints
- The need for long-term biosafety training in environmental and food safety as it relates to specific products and regulatory regimes.

In a recent report entitled *Internationally Funded Training in Biotechnology and Biosafety: Is It Bridging the Biotech Divide?* published by the United Nations University Institute of Advanced Sciences (Johnston *et al.*, 2008) it is stated that many countries in Africa lack the capacity to implement their obligations under the Cartagena Protocol on Biosafety. In order to ensure the safe deployment of transgenic crops for the benefit of African farmers and consumers, regulatory systems must be established that are functional in decision-making. A functional national biosafety system:

- makes science-based decisions on the development, deployment and importation of biotechnology products,
- is predictable and clear to stakeholders,
- is flexible to the adoption of new technologies,
- is transparent and takes into consideration input from public and other stakeholders,
- has biosafety policies and implementing regulations in place that are workable and utilised.

However, such decision-making systems are largely absent in Africa.

Recently, it was pointed out that there are increasing calls for the benefits of GE crops to be considered as well as the risks, and for a risk-benefit analysis to form an integral part of GMO regulatory frameworks (Johnston *et al.*, 2008; Morris, 2010). This trend represents a shift away from the strict emphasis on risks, which is encapsulated in the Precautionary Principle that forms the basis for the Cartagena Protocol on Biosafety, and which is reflected in the national legislation of many countries. The governments of countries that have been at the forefront of the adoption of the technology have inherently recognised the benefits although they may not have explicitly captured this aspect in their legislation (Laursen, 2010). In contrast, countries that have been slow to approve commercial plantings of GE crops have tended to focus on risks rather than benefits (Morris, 2010). The combination of inadequate policies and legal frameworks requires urgent revision that should be led primarily by Africans if credibility is to be achieved in the eyes of African governments, African civil society and African people (Johnston *et al.*, 2008).

Information is an integral component of functional biosafety systems. Access to up-to-date biosafety information is an important element in empowering African regulators. Information must be relevant to Africa and scientifically validated so that regulators can make informed and confident decisions on biotechnology products based on needs. In a recent publication (Maredia *et al.*, 2011), it was suggested that capacity building in biosafety must be viewed as a part of overall biotechnology capacity building and must fit national policies and priorities for agricultural research and development. In addition, regional and global cooperation and collaboration should be continually fostered to encourage the exchange of information and experiences acquired in different parts of the world.

GENETICALLY ENGINEERED CROPS IN THE PIPELINE IN AFRICA

The goal of plant breeding and genetic engineering is to make changes in the genetic make-up of the plant to improve characteristics considered desirable. Traits of interest (Figure 7.5) are based on the following:

- Farmer-oriented traits (pest resistance – insects, diseases, weeds)
- Stress resistance (frost, drought, salinity) and growth/performance (earliness, yield)

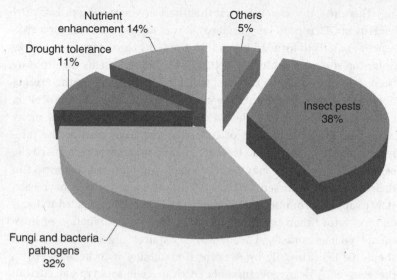

Figure 7.5 Traits focused on by biotechnology laboratories.
(FARA 2009 Study Report)

- Processing traits (altered oil, carbohydrates, protein)
- Consumer-oriented traits (flavour, nutritional quality – vitamin A, vitamin E and protein and post-harvest storage).

Initially, African scientists lacked the requisite infrastructure for the technology. However with the assistance of development partners and technology developers significant progress has been made in building the capacity of African scientists in biosciences engaged in developing transgenic African indigenous crops. Many of them nevertheless still face the challenges of infrastructure and financial resources. With the creation of the African Agricultural Technology Foundation (AATF), which is generously supported by development partners and technology developers, a number of crops are being field-tested in many institutions in Africa using modern biotechnology (Table 7.1). These include: drought-tolerant crops (water-efficient maize for Africa, WEMA), nutrient-enriched crops (biofortified sorghum, sweet potato and banana, biocassava plus), and crop protection against pests and diseases (cassava mosaic disease, *Maruca*-resistant cowpea, banana brown streak, etc.).

CONCLUSION

The transformation of Africa's agriculture system will require new approaches, new methodologies, new efficiencies and the accompanying political focus needed to effect change. Since most of the analysis

Table 7.1 *Details of some biotech R&D activities under way in Africa*

Country	Crop	Trait	Institutions/companies involved
South Africa	Maize	Drought tolerance, herbicide tolerance, insect resistance, insect resistance/herbicide tolerance	Monsanto, Syngenta, Pioneer
	Cassava	Starch enhancement	Agricultural Research Council–Institute for Industrial Crops
	Cotton	Insect resistance/herbicide tolerance, herbicide tolerance	Bayer
	Potato	Insect resistance	Agricultural Research Council-Onderstepoort Veterinary Institute
	Sugar cane	Alternative sugar	South African Sugarcane Research Institute
Kenya, Tanzania, South Africa, Mozambique	Maize	Drought tolerance	African Agriculture Technology Foundation, National Agricultural Research Institutes, CIMMYT (International Maize and Wheat Improvement Center), Monsanto, Bill and Melinda Gates Foundation, Howard G. Buffet Foundation
Kenya	Maize	Insect resistance	Kenya Agricultural Research Institute, CIMMYT, Monsanto, University of Ottawa, Syngenta, Rockefeller Foundation
	Cotton	Insect resistance	KARI/Monsanto
	Cassava	CMVD resistance	KARI, Danforth Plant Science Center
	Sweet potato	Viral disease resistance	KARI/Monsanto
Uganda	Cotton	Insect resistance/herbicide tolerance	National Agricultural Research Organisation/Monsanto, Agricultural Biotechnology Support Project II, USAID, Cornell University

Table 7.1 (cont.)

Country	Crop	Trait	Institutions/companies involved
	Banana	Black sigatoka	NARO-Uganda, University of Leuven, International Institute of Tropical Agriculture, USAID
	Cassava	Cassava mosaic virus disease, resistance to Cassava Mosaic and Cassava brown streak disease	National Crops Resources Research Institute, CIP (International Potato Center), Danforth Plant Science Center
Nigeria, Burkina Faso, Ghana	Cowpea	*Maruca* resistance	Institute for Agricultural Research, Zaria/INERA (Institute de l'environment et de recherches agricoles)/SARI (Savanna Agricultural Research Institute)
South Africa, Burkina Faso, Kenya	Sorghum	Nutrient enhancement	Consortium of nine institutions led by Africa Harvest Biotechnology Foundation International and funded by Bill and Melinda Gates Foundation
Nigeria	Cassava	Nutrient enhancement	National Root Crops Research Institute, Umudike, Danforth Plant Science Center, International Institute of Tropical Agriculture, USAID
Egypt	Maize	Insect resistance	Monsanto, Pioneer
	Cotton	Salt tolerance	Agricultural Genetic Engineering Research Institute
	Wheat	Drought tolerance, fungal resistance, salt tolerance	Agricultural Genetic Engineering Research Institute
	Potato, Banana, Cucumber, Melon, Squash, Tomato	Viral resistance	Agricultural Genetic Engineering Research Institute

Source: ISAAA AfriCenter (2009).

demonstrates weak human capacity and weak regulatory structures, efforts to upgrade and strengthen capacity and support science-based, cost-effective regulatory systems should be seen as a high priority.

Realising Africa's vision of achieving a 6% growth in agricultural productivity will need increased and sustained investments in agricultural research and development. The vision is not about fixing yesterday's mistakes; it is about preparing for a new tomorrow.

Biotechnologies could help Africa if governments would invest in human and infrastructural capacities and not leave things entirely to development partners. Policy reforms that encourage access to biotechnology to produce more food and feed through increased yield and reduced losses, including those caused by climate change, are urgently needed. There is also the necessity to enhance nutritional and industrial quality of agricultural products.

The science and policy issues surrounding biotechnology are new to much of the developing world, especially African countries. Fear and scepticism have made biotechnology a sensitive trade issue. A biotechnology information gap has contributed to a global misunderstanding that influences public perception and acceptance of this technology. The NEPAD Agency African Biosafety Network of Expertise, as with other biosafety initiatives, is ever ready to assist Member States with information, training, technical support, networking and linkages on biotechnologies to make informed decisions.

It is clear that there is insufficient integration between regional biotechnology research developments and the biosafety approaches of the individual countries. For most countries with little or no capacity in biotechnology and biosafety, there is insufficient consideration of possibilities to pool expertise within the subregion, or to consider mutual acceptance of data or regional risk assessment and decision-making.

Despite the existence of overarching policy statements at the level of the AU and NEPAD Agency, a number of African countries are taking their own paths in the development of safe biotechnology. No one country can ensure biosafety on its own without the engagement of neighbouring countries in a coordinated approach. In addition, information sharing in Africa has been one major challenge in the implementation of the Cartagena Protocol on Biosafety. For instance, the Biosafety Clearing House mechanism is seldom used. If effective higher-level regional policies are to be implemented in the future, a binding Directive (in the manner of the Cartagena Protocol on Regional Trade and Technology Development) might be necessary to ensure that

African countries implement their national frameworks according to a common agreement and create mechanisms that will lower the barriers to cross-border trade.

The Sub-Regional Research Organisations show promise in generating a critical mass of scientific expertise, reducing cost through the sharing of facilities and equipment and shortening the time for technology development. However, the cost and complexity of taking GE crops through the regulatory processes will be prohibitive to public-sector progress if approval has to be obtained separately for each country in Africa.

REFERENCES

Brookes, G., and Barfoot, P. (2010). *GM Crops: Global Socio-Economic and Environmental Impacts, 1996–2008*. PG Economics Ltd, Dorchester, UK. www.pgeconomics.co.uk.

Concern (2008). *Responding to the Needs of Marginal Farmers: A Review of Selected District Agriculture Development Plans in Tanzania*. www.concern.net/sites/concern.net/files/resource/2009/03/3515-tanzania-research.

FARA (Forum for Agricultural Research in Africa) (2011). *Status of Biotechnology and Biosafety in Sub-Saharan Africa: A FARA 2009 Study Report*. FARA Secretariat, Accra, Ghana.

IFPRI (International Food Policy Research Institute) (2011). *Draft Report: A 'State of Affairs' Assessment of Agricultural Biotechnology for Africa for the African Development Bank*. IFPRI, Washington, DC.

Heinemann, J. (2009). *Hope not Hype: The Future of Agriculture Guided by the International Assessment of Agricultural Knowledge, Science and Technology for Development*. Third World Network. http://twnside.org.sg/title2/books.

James, C. (2010). *Global Status of Commercialized Biotech/GM Crops: 2010*, ISAAA Brief No. 42. International Service for the Acquisition of Agri-biotech Applications, Ithaca, NY.

Johnston, S., Monagle, C., Green, J., and Mackenzie, R. (2008). *Internationally Funded Training in Biotechnology and Biosafety: Is It Bridging the Biotech Divide?* United Nations University Institute of Advanced Studies, Yokohama, Japan. www.ias.unu.edu/sub_page.aspx?catID=111&ddlID=673.

Juma, C., and Serageldin, I. (2007). *Freedom to Innovate: Biotechnology in Africa's Development*. African Union and New Partnership for Africa's Development. http://nepadst.org/doclibrary/pdfs/biotech_africa_2007.pdf.

Laursen, L. (2010). How green biotech turned white and blue. *Nature Biotechnology* **28**, 393–395.

Ledermann, S. T., and Novy, A. (2012). *GMOs and Bt Cotton in Tanzania: The Smallholder Perspective*, ANSAF Report. Agricultural Non-State Actors Forum, Dar es Salaam, Tanzania.

Maredia, K. M., Weebadde, C., Komen, J., and Ghosh, K. (2011). Capacity building in biosafety. In R. Grumet, J. F. Hancock, K. M. Maredia, and C. Weebadde (eds.) *Environmental Safety of Genetically Engineered Crops*, pp. 189–207. Michigan State University Press, East Lansing, MI.

Morris, E. J. (2008). The Cartagena Protocol: implications for regional trade and technology development in Africa. *Development Policy Review* **26**, 29–57.

Morris, E. J. (2010). A semi-quantitative approach to GMO risk-benefit analysis. *Transgenic Research* 20, 1055–1071.

National Research Council (2010). *Impact of Genetically Engineered Crops on Farm Sustainability in the United States.* National Academy Press, Washington, DC.

Qaim, M. (2009). The economics of genetically modified crops. *Annual Review of Resource Economics* 1, 665–693.

South Africa Department of Agriculture (2010). www.nda.agric.za/docs/Fact Sheet/field crops

Vitale, J., Ouattarra, M., and Vognan, G. (2011). Enhancing sustainability of cotton production systems in West Africa: a summary of empirical evidence from Burkina Faso. *Sustainability* 3, 1136–1169.

KEY FURTHER RESOURCES

1. International Service for the Acquisition of Agri-Biotech Applications (ISAAA). www.isaaa.org/resources/publications/briefs/42/executivesummary/default.asp
2. Cartagena Protocol on Biosafety. http://bch.cbd.int/protocol/cpb_faq.shtml
3. South African Association for Science and Technology Advancement (SAASTA). *Public Understanding of Biotechnology.* www.pub.ac.za/resources/docs/survey_pub_feb2005.pdf
4. AfricaBio (Biotechnology Stakeholders Association). www.africabio.com
5. African Agricultural Technology Foundation (AATF). www.aatf-africa.org
6. Food and Agriculture Organization of the United Nations (FAO). www.fao.org/documents
7. International Food Policy Research Institute (IFPRI) *Program for Biosafety Systems (PBS).* www.ifpri.org/themes/pbs/pbs.htm
8. Thurow, R., and Kilman, S. (2009). *Enough: Why the World's Poorest Starve in an Age of Plenty.* PublicAffairs, New York.
9. Juma, C. (2011). *The New Harvest: Agricultural Innovation in Africa.* Oxford University Press, New York.
10. NEPAD Planning and Coordinating Agency. www.nepad.org
11. NEPAD Agency African Biosafety Network of Expertise. www.nepadbiosafety.net

8

Transforming agriculture in Argentina: the role of genetically modified (GM) crops

INTRODUCTION

Science, innovation and technology have always been strongly correlated with progress in agriculture. In many countries that have a long-standing tradition as agricultural commodities exporters, productivity gains have been higher in agriculture than in any other sector for a good part of the twentieth century. That has been the case, for instance, in the United States, up until the dawning of the digital era (computers, communications and information).

Nowadays even more is expected from science and technology applied to agriculture: it has become, especially in the less developed nations, one of the key factors determining the rate of progress of agricultural development, food security, the improvement of farmers' incomes and poverty alleviation. Its relevance derives from the strong dependence of the supply function of agricultural commodities on two very specific features: a fixed factor, i.e. land, and the dynamics of highly competitive agricultural commodity markets. At the level of individual farms, which are price-takers, income is tied directly to the productivity level of the production factors under their control. It is in this context that the role of genetically modified (GM) crops in reshaping world agriculture should be raised, and Argentina's experience offers concrete evidence of the full expression of its potential and of the key role played in the entire process by a friendly and effective institutional environment. This chapter is based on a recent research report (Trigo, 2011) and looks into the impacts of the availability and the adoption of GM crops in that country, emphasising both the benefits to

Successful Agricultural Innovation in Emerging Economies: New Genetic Technologies for Global Food Production, eds. David J. Bennett and Richard C. Jennings. Published by Cambridge University Press. © Cambridge University Press 2013.

the domestic economy as well as those accruing to global consumers. In addition it analyses some of the non-economic benefits arising from the adoption by farmers in Argentina of GM technology and briefly discusses some of the underlying drivers behind this process.

ONE HUNDRED YEARS OF AGRICULTURE AT A GLANCE

The history of agriculture in Argentina over the last century is one that shows a positive long-term trend in terms of area planted with grains and oilseeds, with ups and downs in the middle. As can be seen in Figure 8.1, between 1900 and 2008 the increase is more than fivefold (from 5 million to almost 28 million hectares). Strong growth took place in the first three decades, driven mostly by mechanisation, until both the Great Depression and restrictive domestic policies induced a reduction of the investment levels in the sector, all of which led to a severe reduction in area planted (from 18 million hectares in 1940 it went down to about 10 million in 1950). In the 1960s, the availability of both the Green Revolution's improved dwarf wheat varieties and the highly productive hybrid maize turned out to be two major technological milestones that ushered in a renewed innovation-based growth cycle that has continued unabated ever since. In 1991 no-till farming (NTF[1]) started to be massively adopted and the availability of the first GM varieties (herbicide-tolerant soya beans) in 1996 induced a synergy with NTF of such a magnitude that surpassed all expectations, vastly outperforming even its counterparts in the United States, the centre of origin of both technologies.

As further evidence of the importance of innovation and technology in the history of agriculture in Argentina, Figure 8.2 shows how both labour and land productivity have evolved between 1908 and 2007. In 1908 Argentina was already a major player in international agricultural commodity markets and for each farmworker, 7.5 tons of grains and oilseeds were produced. Back then it took 1 hectare of land to produce 1 ton of those commodities. A century later, each farmworker produced 58.2 tons of grains and oilseeds and 1 hectare of land yielded 3.14 tons.

A recent study on sources of growth in Argentina's farm sector (Reca *et al.*, 2010) has confirmed the overwhelming pre-eminence of

---

[1] No-till consists basically of the placement of the seed in the soil at the required depth with a minimum disturbance of the soil structure. This is achieved through the use of machinery specially designed to eliminate the need for plowing and other tillage practices that were previously required to plant the crop.

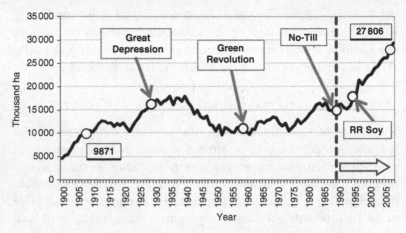

Figure 8.1 Argentina: the evolution of the area planted with grains and oilseeds, and milestones (1900–2008). (Cap, 2012)

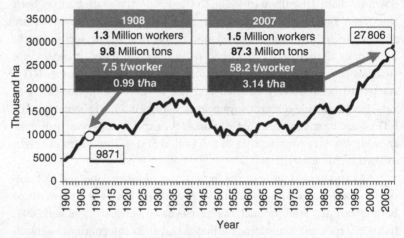

Figure 8.2 Technological change in Argentina: evolution of productivity of labour and land, tons of grains + oilseeds per worker and per hectare (1908–2007). (Cap, 2012)

technical change among them (Figure 8.3). Over two-thirds of growth recorded for the agricultural sector from 1963 (approximately the time when the second wave of sustained expansion got started) and 2009 is explained by increases in total factor productivity (TFP),[2] a figure rarely

---

[2] It is a ratio of index numbers, with quantity of output in the numerator and quantities of inputs in the denominator.

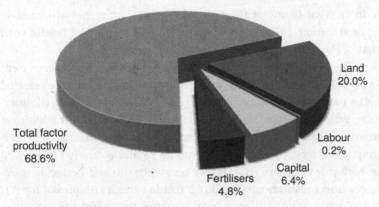

Figure 8.3 Argentina: sources of growth in grains and oilseeds (1968–2008). (Lema, 2010)

seen in farm sectors the size of Argentina's. One caveat that should be taken into account: the reported contribution of land to this process (20%) is most likely an overestimate, since expansion of land use cannot always be regarded as linearly independent of technical change (an example that lends support to this hypothesis will be presented in the next section). In other words, at least a fraction of that 20% should be credited to innovation (and thus TFP's contribution would end up being even higher than 68.6%). Current econometric tools used to estimate TFP, however, do not allow for special situations like the one described to be built into the formulae used in the calculations. Theoretical econometricians should probably look into how to remove methodological restrictions like this in order to improve the existing tools.

GENETICALLY MODIFIED CROPS IN ARGENTINA

The story of genetically modified (GM) crops in Argentina is still unfolding and it is one that the world has been paying close attention to over the last 15 years. A number of variables that have played a role in this story were not present in any other country. Unforeseen synergies on a scale never seen before have occurred. The sheer magnitude of the figures involved has shaken the foundations of traditional econometric analytical tools. Agricultural economics is at a loss when trying to model the process using those same tools. Prices have not been driving this unprecedented shift in the supply function of grains and oilseeds. Technological change has, and it continues to do so. That is one of the reasons why this story has attracted so much attention. In the next sections we will try to deal briefly with most of the issues raised in the preceding paragraph.

The first GM varieties introduced in Argentine agriculture were herbicide(glyphosate)-tolerant soya bean materials which were released by the national regulatory authority and subsequently made commercially available in the 1996/7 crop season. Since then, 20 additional varieties have been approved for commercialisation, planting and consumption as food, feed or fibre, including 15 maize varieties (improved for herbicide tolerance (HT), insect resistance (IR) and herbicide tolerance + insect resistance (HT + IR)), 3 cotton varieties (improved for HT, IR and HT + IR) and 2 soya bean varieties (improved for resistance to herbicides other than glyphosate).[3] In the 2010/11 crop season, these technologies were used in nearly 22.9 million hectares, of which 19 million were grown with HT soya beans, 3.5 with GM maize (1.6 million IR, 300000 HT, 1.6 million with both traits combined) and 614000 hectares with GM cotton (56000 HT, 8000 IR and 550000 with both traits combined) (ArgenBio, 2011). These figures represent approximately 100%, 86% and 99% respectively of the total area planted with each of these species (Figure 8.4).

These numbers place Argentina third, behind the USA and Brazil, in GM crop area at the world level; India and Canada being ranked immediately following (James, 2010). This adoption dynamic represents an almost unprecedented process in the history of world agriculture only comparable to what happened with hybrid maize in the state of Iowa (USA) in the 1930s, but much faster than the adoption rates of GM technologies reported in other American 'Corn-Belt' states and, later on, in other parts of the world with the so-called 'Green Revolution' varieties. Even within the Argentine experience, the evolution of the adoption of GM technologies by farmers compares very positively with

[3] Since the Argentine government created, in 1991, the National Advisory Commission on Agricultural Biotechnology (CONABIA) 1721 applications for field trials have been granted. Maize, soya beans, cotton and sunflower are the crops with the greatest number of implemented field trials, followed by wheat, rice, potato and forage crops (alfalfa), among others. In terms of technologies (traits) there has been a significant evolution from single traits (herbicide tolerance, insect resistance) to combined or stacked traits, which clearly prevail towards the end of such periods following the trend observed by GM technologies elsewhere around the world (James, 2010). The vast majority of technologies subjected to field trial were of foreign origin (www.minagri.gob.ar/site/agricultura/biotecnologia/20-CONABIA/index.php).

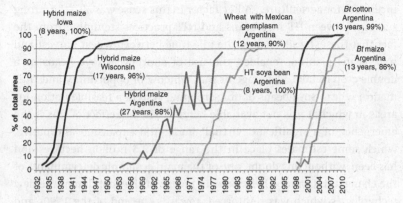

Figure 8.4 Rate of adoption for different GM technologies compared to other technological milestones. (Trigo, 2011)

other previous cases such as hybrid maize and wheat with Mexican germplasm (Figure 8.4). It took hybrid maize 27 years to reach the percentage adoption recorded for GM maize after only 13 years, and it took Mexican wheat 12 years to be as widely adopted as soya beans in only four crop seasons (90% of planted area). It is worth noting, however, that none of the major technologies introduced have evolved from local R&D. All the GM technologies that have received approval for commercial use have been generated outside the country by multinational seed companies and introduced into the local genetic pool. The predominance of externally generated technologies has remained unchanged since the first HT soya bean varieties were introduced for field-testing during the early 1990s as no applications for field-test permits have been filed for locally developed innovations in the major crops. In spite of this, there is general agreement that both the strength of local breeding programmes and the existence of a consolidated seed industry have played a key role in the rapid adoption and diffusion of these new technologies.

THE ECONOMIC IMPACTS OF GM TECHNOLOGIES

The economic impacts of the above-mentioned introduction of GM technologies – HT soya beans in particular – in Argentina have been highly significant, not only due to the reduction in costs of production that they made possible, but also because they provided renewed thrust to an agricultural growth process that had started a few years earlier as the consequence of a change in the overall framework of economic incentives for agriculture.[4] A key factor in this sense was also the strong synergy between HT soya beans and NTF practices stemming from the fact that these, by shortening the idle time between wheat harvest and soya beans planting, enabled the use of short-cycle soya beans as a double crop to take advantage of that window of opportunity and thus made a wheat–soya beans double-cropping system a feasible option for areas in which it was not available before.[5] The net effect of this is the emergence of a significant 'virtual' horizontal expansion alternative, which some estimates place in the range of 3.5 million hectares and has been, without any doubt, one of the main economic determinants of the changes in farmers' behaviour in favour of the adoption of the new technologies which was reinforced, towards the end of the 1990s and the beginning of the 2000s, by the free fall in the price of glyphosate (it went from US$10/litre at the end of the 1990s to less than US$3/litre in 2000). At the same time, the new technologies made soya beans strongly competitive relative to other crops (maize and sunflower) and livestock production (beef and milk) which in turn induced a process of technological intensification that, in the end, resulted in productivity increases in these activities even though the area devoted to them was reduced in favour of soya beans. It has been estimated that, during the period 1996–2005, the area under pasture suffered a reduction of more that 5 million hectares without a decrease in output of beef while strong recovery in the production of milk and similar trends is observed for maize and sunflower (Trigo and Cap, 2006).[6]

In this context the cumulative gross benefits to Argentina resulting from the use of GM crops during the period 1996/7–2010/11 have been estimated to amount to US$72.65 billion. Out of that total figure, US$65.44 billion comes from HT soya beans (US$3.52 billion from the reduction of production costs, and US$61.92 billion from the expansion of planted area), US$5.38 billion from the use of (Bt) IR and HT maize (single and combined events) and US$1.83 billion stem from

---

[4] Incentives included the elimination of export taxes and the reduction or elimination of import duties for farm machinery.

[5] This is a real-world example that lends support to the hypothesis put forward in the previous section which casts doubt on the validity of the implicit assumption of linear independence between the contribution to output growth of the expansion of area planted on the one hand and changes in factor productivity on the other.

[6] These increases in productivity have not been recorded by the statistics due to the fact that the yield indicators commonly used, that is, extraction rate (slaughtered heads per year/stock), in the case of beef and volume of milk for dairy, are computed without reference to the area on which that output is produced.

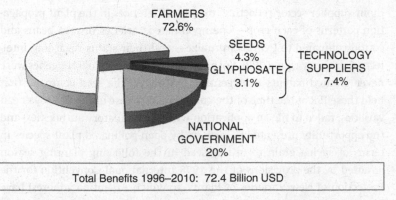

Total Benefits 1996–2010: 72.4 Billion USD

Figure 8.5 Argentina: distribution of cumulative benefits generated by GM crops (soya beans, maize and cotton) during the period 1996–2010. (Trigo, 2011)

the use of IR and HT cotton (single and combined events).[7] The bulk of these benefits (72.67%) have gone to farmers, while the agricultural inputs industry has received about 7.38% of the total and the national government – through export duties – the remaining 19.95% (Figure 8.5). At the crop level, the distribution of benefits follows a similar pattern when it comes to the fraction accruing to farmers, with soya beans and maize farmers capturing 72.4% and 68.2% of the grand total respectively, but there are significant differences in regard to benefits accruing to the

[7] The analytical tool used to estimate the economic impacts of GM events availability in Argentina's agricultural sector is a dynamic simulation model (SIGMA), developed by INTA (National Institute of Agricultural Technology). The model replicates, through simulations, the situations that occur in the field in countries like Argentina, which show a great diversity of technological and productive realities that cannot be attributed to agro-ecological differences but to socio-economic factors. The key component of the model is the replication of the farmers' adoption process of technological innovations that introduce changes in the production function, inducing a more efficient use of resources, which in turn leads to an increase in crop yields and/or to a reduction in unit costs and/or to an improvement in the quality of the product and/or to an expansion of the area potentially suitable for its commercial production. The model may be used for ex-ante and ex-post studies, and the final result is an estimate of the effects of alternative technology generation and adoption scenarios (regional or national) on aggregate production. This means that SIGMA calculates social benefits (rather than private benefits). That is to say, how much more could be produced (in volume and value) compared with a defined baseline owing to the adoption (through pathways that vary according to farmer's profile) of technologies already available on the market or to be generated in the future by the R&D system (for further details, see Trigo (2011), Appendix I).

input supplier sector reflecting, mostly, differences in the plant propagation patterns of each crop – open-pollinated varieties in soya beans and cotton compared to hybrids in maize – and their status regarding intellectual property rights (IPR). Both the fact that the original HT genes were never granted patents in Argentina (ASGROW, the seed company that held those IPR at the time of the commercial release of the GM soya bean varieties, failed to file an application with the regulatory authorities) and the opportunity presented to farmers by open-pollinated plant species in terms of saving grain to use as seed in the following planting season (granted by the provisions of the 1978 International Convention for the Protection of New Varieties of Plants, to which Argentina adheres) have been powerful incentives for the development of an illegal seed market, both for soya beans and cotton. This has not happened with maize, though. The explanation is linked to the logic behind the process of producing hybrid seed which results from crossing genetically homogenous parent lines, not available outside the seed company itself and thus precludes that option.[8] As a consequence of this differential status, seed companies reaped 19% of total benefits in the case of maize, while capturing only 3.2% in the case of soya beans and 3% in cotton (Trigo, 2011).

In addition to the direct economic benefits reported above, the introduction of GM crops into Argentine agriculture has also had, through its multiplier effects, a significant impact on the economy as a whole, particularly in terms of job creation. According to Trigo (2011), during the 1996–2010 period 1.8 million jobs were created as an indirect positive impact of the introduction and adoption by farmers of agricultural GM technologies in Argentina.[9] This constitutes an impressive figure in itself considering the relatively small size of the Argentine economy (with a workforce estimated at 17 million for 2010).

---

[8] For an extensive discussion of the situation regarding intellectual property rights see Trigo et al. (2002).

[9] Job creation estimates were made based on the assumption that for each additional dollar in goods generated by the adoption of GM materials (valued at border price, that is to say, FOB prices at Argentine ports), another dollar is generated in the services sector (transportation, storage, etc.), and following a procedure based on the actual 'cost' of adding one job to the economy for each year during the period – estimated in terms of GDP, assuming a baseline stock of 10 million jobs in 1996 with annual cumulative increases or subtractions to account for the evolution of the growth or contraction of the economy during the different stages of the cycle of the economy over the period (for further details of the procedure and the complete estimation see Trigo, 2011).

Its relevance is further emphasised when one considers that the period under analysis includes the crises years of 2001–3 when, following the abandonment of the fixed peso–dollar exchange rate (pegged at a value of 1) the economy contracted in 2002 by 10.9% and the unemployment rate skyrocketed to 21.5% (INDEC, 2003).

Environmental benefits

The expansion of GM varieties in Argentina has moved along, as mentioned above, *pari passu* with a dramatic increase in the area under no-till farming. This is particularly meaningful from the viewpoint of its environmental impact since, on the one hand, it has made it possible to partially reverse the negative consequences that conventional tilling and ploughing practices, prevailing until the beginning of the 1980s, have had on the physical structure of Pampean soils (Viglizzo *et al.*, 2010) and, on the other hand, to significantly increase the efficiency standards in the energy balance of agricultural production (Pincen *et al.*, 2010).

No-till farming began to gain significance in Argentine agriculture by the end of the 1980s, due to the fact that in many of the most important areas of the Pampas region the cumulative effects of water and wind soil erosion, resulting from 'agriculturalisation' based on traditional farming practices,[10] had already began to manifest itself in the bottom-ranking farms (SAGyP–CFA, 1995). Such effect on yields and, therefore, on the economic viability of agriculture, together with a greater availability of proprietary farm machinery (as a result of deregulation and opening of the economy) and the reduction in direct costs (due to the elimination of tillage practices), were the optimal platform for the spread of NTF and, partly at least, the recovery of lost productivity. But it was not until the introduction of HT soya beans that the process picked up speed and consolidated as the predominant agronomic strategy in the country's farming systems (Figure 8.6). This is clearly reflected in the evolution of the area under NTF which went from about 300000 hectares in 1990/1 to nearly 25 million hectares at present (for an in-depth discussion on this process, see Trigo *et al.*, 2010). The combination NTF + HT soya bean integrates two technological concepts; one of them consisting of new mechanical technologies that modify the soil–crop interaction, and the other is based on the use of a

---

[10] 'Agriculturalisation' is understood as the substitution of crop–pastures rotation systems for pure crop patterns, which had been the prevailing productive strategy in Argentina until the mid-1970s.

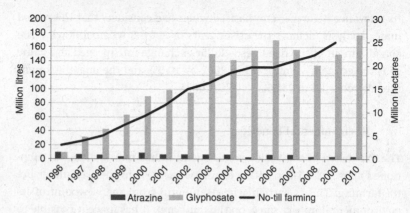

Figure 8.6 Evolution of planted area with NTF and type of herbicide used. (Based on data from AAPRESID (www.aapresid.org.ar) and CASAFE (www. casafe.org))

total herbicide (glyphosate, which causes a lesser environmental impact than other herbicides) that is highly effective to control every kind of weed and has virtually no residual effect.[11] The use of mechanical technologies and total herbicides imply an intensified use of inputs, which is generally described as 'hard' (agricultural) intensification. However, as can be seen in Figure 8.6, this 'hard' intensification is, at the same time, environmentally friendly because it has resulted in a parallel reduction in the use of other herbicides with higher residual effect such as atrazine. It is difficult to quantify the benefits of this synergy between HT soya beans and NTF but one cannot ignore the potential positive effects on soil fertility and thus on present and future land productivity as well as some other positive consequences such as its contribution to mitigating the so-called 'greenhouse effect'. From the standpoint of soil organic matter content recovery, Casas (2005) indicates that in NTF systems with crop rotations including wheat, maize or sorghum, the annual soil losses are under 2 t/ha, much lower than the tolerable maximum (10 t/ha), and the average amounts observed under other types of management practices.

Apart from the benefits derived from increased sustainability of farming systems due to the recovery of soil fertility, the GM soya beans + NTF package shows another type of environmental benefit in terms of

[11] According to Pincen et al. (2010), the persistence of glyphosate in the soil ranges between 12 and 60 days and it carries a low risk of polluting underground waters. Its toxic effects on animals are mild. It does not bioaccumulate in animal tissues.

fuel consumption, emissions reduction and carbon sequestration, all of which are worth mentioning. According to Brookes and Barfoot (2011), between 1996 and 2009 total fuel consumption in soya bean farming increased by 201.3 million litres (95.1%), going from 211.6 to 412.9 million litres annually, but the average consumption per hectare dropped by 38%, falling from 35.8 to 22.2 litres/ha, which facilitated a reduction of 5.19 million tons of carbon dioxide emissions compared to the ones that would have occurred if the 2009 output levels had been based on conventional farming practices. In annual terms this means the use of 13.5 million litres of fuel less than the volume required under conventional farming. The same authors report similar effects regarding the carbon sequestration impact resulting from the application of reduced or zero tilling practices. Although it is admitted that data on Argentina is not always accurate, by using conservative indicators such as 100 kg carbon/ha per year for such farming practices, the authors estimate that the resulting cumulative amount of carbon may reach 13.82 million tons, which is equivalent to retaining in the soil 50.71 million tons of carbon dioxide. Impacts of this same nature have been described for maize and cotton but they are of a lesser magnitude as the area planted and the time elapsed since these technologies were first adopted differ significantly from the case of soya beans.

Sustainability issues associated with the expansion of soya beans

Underscoring the above-mentioned synergies and benefits does not necessarily mean ignoring the potential risks implied in the massive transformation of farming systems that appears to have been triggered by the introduction of GM crops in Argentina during the mid-1990s. Of particular importance here is the significant soil nutrient losses resulting from the increasing predominance of monoculture practices and the relatively low fertilisation levels recorded in Argentina and the potential impairment of the more fragile ecosystems of the country's new emerging agricultural frontier in the northeastern and northwestern subhumid areas which have gradually become suitable for growing soya beans. The issue of the environmental effects of changes in land use is a relevant one, but it is beyond this discussion, and it is worth pointing out that, even though soya beans represent a core component of present-day 'agriculturalisation', this process started long before soya beans made their appearance in the farming scene of Argentina and most of the areas where they are planted today had already been allocated to other crops.[12] Regardless of these facts, which should be the subject of further analysis and discussion, the issue of sustainable productive strategies is highly relevant, given the magnitude of the figures involved. The issue at stake here is the long-term effect of the continued 'exports' of soil nutrients, particularly phosphorus, since replacement is either non-existent or insufficient.[13] Trigo (2011) has estimated the cumulative amount of phosphorus exported between 1996 and 2010, denominated in tons of triple super phosphate (TSP), as more than 14 million tons. Its restocking cost, for the 15-year period under study, comes to US$7.95 billion. Although this is obviously a high figure, it only accounts for 8.41% of total cumulative benefits for the period mentioned above.

IMPACT ON WORLD CONSUMERS

Argentina is one of the key players in the international soya beans market – it is the third largest producer, exporting almost 100% of its production – so in addition to the impact on the country's own economic and social dimensions, significant impact can also be found at the world level. In particular these benefits accrue to consumers, through reduction in the price of the commodity with respect to what it would have been if Argentina had not adopted the new technologies and if its production trajectory continued along a production pattern based on conventional practices. Trigo and Cap (2006) and Trigo (2011) have estimated these benefits – expressed in terms of savings in consumer expenditure – as US$89.0 billion for the period 1996–2010. This is a result of the additional output of soya beans attributable to the adoption of GM technology by Argentine farmers. This has been estimated at 216.1 million tons, which would account for 22.53% of the total world soya beans production expansion during that period.[14]

---

[12] Changes in rainfall patterns, enabling crop production in areas where it was not possible before, has been identified as one of the probable drivers of this process (Grau *et al.*, 2005).

[13] More recently hard evidence has become available showing that soya beans positively respond to phosphorus fertilisation (between 500 and 730 additional kg of soya bean per hectare) (FERTILIZAR, 2011).

[14] These figures were estimated on the basis of soya beans supply–price flexibility, which measures the response of prices to changes in output, using as a starting point an estimation of the supply–price elasticity of soya beans for the United States, the world's biggest producer, of 0.80, which implies (given that certain assumptions hold) a price flexibility value of 1.25. For a more complete discussion of the methodology and the calculation process, see Trigo and Cap (2006).

It remains to be seen whether these savings were effectively passed on to consumers by the other stakeholders in the value chains in which this commodity is involved as an input or if, on the contrary, it was captured as rent by them. In any case it is a good indicator of the power of the technologies in question when assessed from the perspective of their potential social impact.

CONCLUDING COMMENTS

The story of GM crops in Argentina is, undoubtedly, one of success, and highlights a number of issues that deserve to be raised. The story clearly reveals the sheer magnitude of the economic benefits resulting from these technologies, but it also reveals some of the necessary conditions that need to be in place for a country to be able to benefit from these technologies which may have little to do with their own R&D capacities. A key feature in this story is the fact that the country was an early adopter of the new technologies and this was made possible because when they were made available Argentina already had in place the institutions required to deal with the technology transfer and diffusion process. By the early and mid-1990s it had already in place the biosafety regulations and the infrastructure to assess the technologies together with a very active and efficient seed industry that enabled rapid intro-duction of the new genes into commercial varieties that were already well adapted to the multiple agro-ecological conditions in the different growing areas, a necessary condition to ensure a rapid diffusion of the new GM materials. Should this particular feature not have been present the story would, most likely, have evolved in a different way. The story of the Argentine case highlights that, at least at the present state of GM R&D technologies, having biosafety institutions and a well-functioning seed market in place appears to be more important than a functioning local R&D apparatus to generate the innovations. Not a single one of the GM technologies used today in Argentina has been developed in the country, and the same holds true for many of the GM crops in other parts of the world as well. GM technologies seem to 'travel well' and for a country to extract maximum benefits from them it is essential to have the right tools in place at the time of their availability.

Another relevant issue deals with the non-economic implications of the introduction of these technologies. The synergies between GM technologies and no-till practices point in a direction of a 'win–win' outcome, bringing together productivity and output increases with the positive conditions associated with no-till practices in terms of both micro- and macroenvironmental impacts. Recognising this does not mean to overlook – as it has been stressed above – that alongside this virtuous cycle there are other issues that need to be looked at closely. It would have been surprising if that were not the case, given the magnitude of the changes that have been described. The dramatic expansion of soya beans and the increase in output of grains and oilseeds as a whole has resulted in significant benefits, but also in a shift in the allocation of land to competing crops with a significant increase in the share of soya bean and, in turn, concerns about the long-term sustainability of the whole process, given its detrimental effects on soil nutrients and the potentially negative impacts of its expansion into more fragile ecosystems. These concerns are legitimate but, nevertheless, they do not detract from the clearly positive net balance of the first 15 years of GM crops in Argentina. However, they do highlight the need for appropriate policy responses aimed at optimising the management of this particular kind of innovation. GM technologies are path-breaking events and for a successful implementation at the farm level they require biosafety and intellectual property frameworks to properly contain technology transfer and diffusion. Additionally, there is also the need for policy action in other areas – such as land use, input prices, market regulations, etc. – so as to minimise its potentially negative effects.

REFERENCES

ArgenBio (Argentine Council for Information and Development of Biotechnology) (2011). www.argenbio.org.

Brookes, G., and Barfoot, P. (2011). *GM Crops: Global Socio-Economic and Environmental Impacts 1996–2009*. PG Economics Ltd, Dorchester, UK. www.pgeconomics.co.uk

Cap, E. (2012). Productivity increases + sustainability: the case of no-till in Argentina. *Humboldt Forum for Food and Agriculture: Sustainable World Agriculture – How to Measure Progress*, Berlin, January.

Casas, R. (2005). Keynote speech. *National Academy of Agronomy and Veterinary Medicine (Disertación ante la Academia Nacional de Agronomía y Veterinaria)*. Buenos Aires, Argentina.

Cruzate, G., and Casas, R. (2003). Balance de nutrientes: Revista fertilizar INTA. '*Sostenibilidad*', 7–13.

FERTILIZAR (2011). La Argentina pierde rendimentos y empobrece su sueto por no fertilizar. www.fertilizar.org.ar/vertext.php?id_nota=651

Grau, H. R., Gasparri, N. I., and Aide, T. M (2005). Agriculture expansion and deforestation in seasonally dry forests of north-west Argentina. *Environmental Conservation* 32, 140–148.

INDEC (Instituto Nacional de Estadistica y Censos de la República Argentina) (2003). www.indec.mecon.ar.

James, C. (2010). *Global Status of Commercialized Biotech/GM Crops: 2010, ISAAA Brief No. 42*. International Service for the Acquisition of Agri-biotech Applications, Ithaca, NY.

Paruelo, J., and Oesterheld, M. (2004). *Patrones espaciales y temporales de la expansión de la soja en la Argentina: Relación con factores socioeconómicos y ambientales*. Facultad de Agronomía-UBA, Buenos Aires.

Pincen, D., Viglizzo, E. F., Carreño, L. V., and Frank, F. C. (2010). La relación soja-ecología-ambiente: Entre el mito y la realidad. In E. F. Viglizzo and E. G. Jobbágy (eds.) *Expansión de la Frontera Agropecuaria en Argentina y su Impacto Ecológico-Ambiental*, pp. 53–63. Ediciones INTA, Buenos Aires.

Qaim, M., and Traxler, G. (2002). *Roundup Ready Soybeans in Argentina: farm level, environmental and welfare effects*. 6th ICABR Conference on Agricultural Biotechnologies: New Avenues for Production, Consumption and Technology Transfer. Ravello, Italy.

Reca, L., Lema, D., and Flood, C. (2010). *El crecimiento de la agricultura argentina: medio siglo de logros y desafíos*. Facultad de Agronomía, Universidad de Buenos Aires. Buenos Aires.

SAGyP (Secretaría de Agricultura, Ganadería y Pesca) – CFA (Consejo Federal Agropecuario) (1995). *El deterioro de las tierras en la República Argentina*. Buenos Aires.

Trigo, E. (2011). *Fifteen Years of Genetically Modified Crops in Argentine Agriculture*. ArgenBio, Buenos Aires. www.argenbio.org.

Trigo, E., and Cap, E. (2004). The impact of the introduction of transgenic crops in Argentinean agriculture. *AgBioForum* 6, 87–94.

Trigo, E., and Cap, E. (2006). *Ten Years of Genetically Modified Crops in Argentine Agriculture*. ArgenBio, Buenos Aires. www.argenbio.org.

Trigo, E., Chudnovsky, D., Cap, E., and López, A. (2002). *Los Transgénicos en la Agricultura Argentina: Una historia con final abierto*. Libros del Zorzal, Buenos Aires.

Viglizzo, E. F., Carreño, L. V., Pereyra, H., *et al.* (2010). Dinámica de la frontera agropecuaria y cambio tecnológico. In E. F. Viglizzo and E. G. Jobbágy (eds.) *Expansión de la Frontera Agropecuaria en Argentina y su Impacto Ecológico-Ambiental*, pp. 9–16. Ediciones INTA, Buenos Aires.

9

China: earlier experiences and future prospects

SETTING THE SCENE: THE CHALLENGE OF FOOD SECURITY IN CHINA

There is an old Chinese saying: 'Food is heaven for the people.' This dictum well mirrors the great importance of sufficient food supply for Chinese people, and that is probably also true for people in other countries. In Chinese history, food supply usually determines the stability and prosperity of the country. Therefore, food security is always the most important core issue for the sustainable development and peace of China.

China has a huge population (currently over 1.3 billion), which is expected to increase to about 1.46 billion by the year 2030 (Niu, 2010). The continued increases in population and demands for sufficient food supply have placed an extremely heavy pressure on agricultural production in China. On the other hand, China has a comparatively small proportion of farming lands with about 122 million hectares in total area (MLRC, 2006). And unavoidably, the area of farming land may be further reduced (Figure 9.1) due to the rapid development of industrialisation, urbanisation and infrastructure construction. Such a contrasting situation makes the feeding of the huge population in China extremely challenging! The current fact is that China feeds more than 20% of the world's population using less than 7% of the world's arable land. All these together indicate the severe challenge of continued demands for food and agriculture production in this country.

China is an agriculture-based country that has a long history of traditional agriculture. Before the 1980s, the major proportion (<85%)

Successful Agricultural Innovation in Emerging Economies: New Genetic Technologies for Global Food Production, eds. David J. Bennett and Richard C. Jennings. Published by Cambridge University Press. © Cambridge University Press 2013.

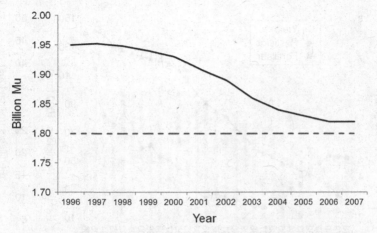

Figure 9.1 The trend of rapid decrease in farming lands in China since 1996. The broken line at 1.8 billion Mu indicates the bottom level that the Chinese government resolves not to touch (1 Mu = 0.067 hectare).

of the Chinese population was involved in the agriculture sector. However, within the last three decades rapid industrialisation, urbanisation and economic development have caused the relocation of the country population to the cities, which has resulted in the evident losses of agricultural labourers. Currently, there is less than 50% of the Chinese population working in agriculture. This indicates the rapid change of population structure in terms of the assignment of social responsibility (Ru et al., 2012).

As in many developing countries, the progressive shortage of agricultural lands and resources (e.g. water) typifies modern agriculture in China. In addition, the application of pesticides and chemical fertilisers has quickly increased during the past 30 years although the increase in crop production was not closely associated (Figure 9.2). At the same time, the overuse of pesticides and chemical fertilisers has caused severe degradation of agricultural ecosystems. All these harsh circumstances together with global climate change have placed Chinese agricultural production and food security in a worse situation. It is estimated that by the year 2020, there will be a shortage of 70 billion m^3 of water for agriculture and 8 million hectares of farming land (NDRC, 2008). By 2030, there may be a shortage of 120–170 million tons of food supplies. Therefore, China will be faced with the most severe challenge of food production and food security in the future under the traditional mode of agricultural management.

Another problem in the pursuit of high agricultural production in China is the 'high-input and high-output' mode that has not only caused

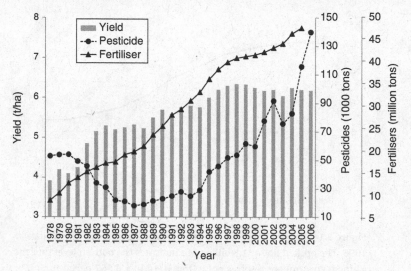

Figure 9.2 The change of crop yield against the rapid increase in the application of chemical fertilisers and pesticides in the past 30 years. The increased use of chemical fertilisers and pesticides has not significantly affected the crop yield increase since 1995.

rapid exhaustion of agricultural resources, but also the serious degradation of the agricultural environments. Conventional ways of agricultural production cannot be sustainable for long-term food security in China. Therefore, how to enhance crop productivity per unit area significantly in an environmental friendly and resource cost-effective manner is the question that needs to be addressed immediately.

The appropriate utilisation of genetic resources (germplasm) combined with adoption of new technologies (including biotechnology) may provide greater opportunities to considerably increase crop productivity. Such an efficient means of agricultural production will help to meet the increased demands of food supply and maintain a more sustainable and healthy agricultural production system in China.

UTILISATION OF NEW GERMPLASM AND BREEDING
TECHNOLOGY: IMPACT ON RICE PRODUCTION

Rice is one of the most important staple foods in China, accounting for the third largest world crop in terms of its total production (FAO, 2004). It is the second largest food crop in China. Understanding the role of rice improvement in fighting the difficulties of attaining food security will help to appreciate the importance of utilising genetic resources and adoption of new breeding technologies for crop production in China. It is well known that the production level of rice was relatively low for a long time under the traditional style of rice farming. Up to the end of 1950s, the yield of rice was less than 2.7 t/ha per harvest (NBSC, 2006).

The first milestone in the development of rice production was the promotion of semi-dwarf rice varieties bred by the utilisation of semi-dwarf germplasm (e.g. from the traditional rice varieties Dijiaowujian and Ainante) hybridised with high-yielding rice varieties. The combination of the elite germplasm and genetic breeding technology resulted in the production of a number of new rice varieties with short plant height, stronger tillering (grain-bearing shoot) ability and high grain yield. Such high-yielding rice varieties could not only resist lodging (lying down) before harvest under more fertiliser application but also enhance the harvest index (the proportion of panicle (spikelet that develops into grain) production against that of total plants) to 0.55 from 0.38 of the traditional rice varieties (Cheng et al., 2007a). A few famous semi-dwarf rice varieties, such as Guang-chang-ai, Zheng-zhu-ai and Er-jiu-ai (where 'ai' stands for dwarf in Chinese) had been developed and widely cultivated in China since the early 1960s. This took place slightly before the Green Revolution in which many semi-dwarf crop varieties were produced and cultivated in different parts of the world.

The application of the genetic breeding technology coupled with the utilisation of elite germplasm had brought an additional rice grain yield of 0.75~1.5 t/ha. Consequently, the average rice yield exceeded 4 t/ha by 1978 from the level of about 2.7 t/ha as mentioned previously (NBSC, 2006). The change of rice breeding technology had significantly helped to solve the food security problems in the 1960s–1970s in China.

The second milestone in the development of rice production was associated with the utilisation of heterosis (hybrid vigour) and the improvement of breeding technology (three-line system), where hybrid rice with high-yielding and early-ripening characters was developed at the end of the 1970s (Cheng et al., 2007b). The three-line system (including a sterility line and its restorer, and maintainer) of hybrid rice breeding technology was developed, with the introduction of a male-sterile alien gene from wild rice (Oryza rufipogon) germplasm through interspecific hybridisation and selection. Consequently, many semi-dwarf hybrid rice varieties such as Guichao-2, Teqing-2, and Shanyou-63 with strong hybrid vigour were developed. These hybrid rice varieties brought an additional grain yield of up to 20%, compared to the conventional semi-dwarf rice varieties (pure lines). The cultivation area for hybrid rice varieties expanded quickly from 5 million hectares in 1978

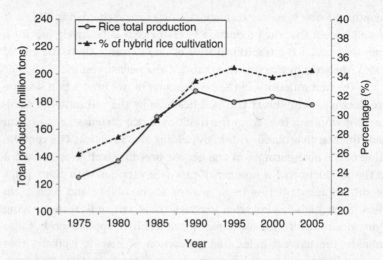

Figure 9.3 The trend of rapid increase in total rice grain yield with the increase in the percentage of hybrid rice cultivation during the 30 years 1975–2005. Data sources: CAAS and HNAAS (1991) and Mao *et al.* (2006).

to about 16 million hectares in 1990, which accounted for more than 60% of the total rice cultivation area in China (CAAS and HNAAS, 1991; Mao *et al.*, 2006; Cheng *et al.*, 2007a). Such a rapid growth indicates the popularity of the new technology products at that time. Although the total cultivated acreage of rice in China (including conventional and hybrid) was gradually reduced during the period 1975–2005, the total rice grain yield was significantly increased with the increase in proportion of hybrid rice cultivation (Figure 9.3).

The extensive promotion of hybrid rice varieties by the Chinese government and research institutions had largely increased rice production per unit area. Rice grain yield was significantly increased to about 5.8 t/ha in 1997 from about 4 t/ha in 1978. As a result, the total rice production increased to 186.2 million tons in 1992.

The example of rice development and improvement demonstrates the power and benefit of the utilisation of elite genetic resources and new technologies in agricultural production. Importantly, similar trends in development are also seen in other crops such as maize (semi-dwarf varieties and hybrid vigour), wheat (semi-dwarf varieties) and oilseed rape (hybrid vigour). Such changes have in general led to a great increase in crop productivity, which is essential for solving the problem of food security. However, the extensive adoption of genetically improved semi-dwarf pure-line and hybrid crop varieties has had significant impacts on traditional farming styles, and has promoted a 'high-input and high-output' model of rice production (also with other crops), with much increased use of chemical fertiliser and pesticides, in addition to great losses of genetic resources in traditional rice varieties. The change to 'high-input and high-output' agricultural management has given rise to considerable degradation of agricultural systems, which has affected the long-term productivity of the land. How to maintain the continued high productivity of agriculture and at the same time to promote environment-friendly agricultural ecosystems under more efficient management is a crucial question that needs to be addressed for the development of modern agriculture.

TRANSGENIC BIOTECHNOLOGY AND ITS APPLICATION IN CHINA: THE CASE OF COTTON

Because of the unique situation in China having a huge human population and limited agricultural resources, the pursuit of more efficient means of food production through the application of new technologies is the unremitting endeavour of the country. The Chinese government always encourages the application of biotechnology in agriculture to increase the crop productivity (Huang and Wang, 2002).

As early as the mid-1980s, the State Science and Technology Commission (now Ministry of Science and Technology) launched the first National High Technology Research and Development Programme (known as the '863' Programme), following the rapid development of new technologies worldwide. The Programme identified research and application of transgenic biotechnology as an important strategy in the national development programme. Research and application of transgenic biotechnology in the agricultural and pharmaceutical sectors was especially targeted. Together with other national research and development programmes such as the National Key Basic Research and Development Programme (known as the '973' Programme), China launched national research projects specifically focused on the research and development of transgenic crops and their commercial application.

In the 30 years since the first launch of transgenic biotechnology programmes, China has established a comprehensive platform for biotechnology research, development and application, with strong capacity in gene cloning, gene transformation, transgenic breeding, biosafety assessment and management, and commercialisation of transgenic products. This comprehensive ability makes China a leading country within the developing world in terms of biotechnology and transgenic breeding. To date, more than 52 plant species have been explored for transgenic research and development in China, including the top 10 species: cotton, rice, maize, potato, tomato, wheat, oilseed rape, tobacco, poplar tree and soya bean. More than 100 types of transgenes have been used in genetic transformation. Scientists in China have produced a large number of transgenic lines of different crop species with diverse traits, such as insect resistance, herbicide tolerance, high grain quality and drought tolerance, all with Chinese intellectual property right. By the end of March 2011, approximately 1560 biosafety certificates had been issued for various crop species (many for transgenic cotton), and many more transgenic lines had entered biosafety assessment, including transgenic cotton, rice, maize, wheat, soya bean, oilseed rape, potato and tree species (Table 9.1). Up to 2011, transgenic varieties involving seven plant species are in commercial production, and two insect-resistant transgenic (Bt) rice lines and one phytase transgenic maize line (which can reduce a usable form of inorganic phosphorus) have been granted biosafety certificates.

As reported in *Science* magazine, the Chinese government is expected to budget a total of US$3.5 billion to promote the intensive research and development of GM organisms, including five plant species and three animal species used in livestock farming (Stone, 2008). This is the first time in Chinese history that such a huge amount of funds has been invested especially to support the technological development of only a few crop species. The objective of such a massive research pro-gramme is to further enhance the general capacity of transgenic breed-ing for producing varieties with novel traits that are targeted for commercial application. Chinese transgenic cotton is the most success-ful transgenic crop in use in agricultural production. Therefore, the rationale and procedure of transgenic cotton research, development and commercial application will now be reviewed to illustrate the wider issue of transgenic biotechnology and its application in China.

China is one of the largest cotton-producing countries in the world, where insect-resistant transgenic cotton is the major biotechnol-ogy crop under commercial production so far (James, 2011). In the 1990s, China usually lost hundreds of million Yuan per year in the cotton industry, due to severe attacks by lepidopteron pests (mostly cotton bollworm). Heavy applications of insecticides not only caused severe environmental damage to agricultural ecosystems but also caused rapid development of resistance in the insect pests, making control of the pests by insecticides ineffective. The introduction of insect-resistant transgenic cotton has significantly changed the situ-ation, and has saved the Chinese cotton industry, in addition to its positive ecological impacts (Pray *et al.*, 2002; Wu *et al.*, 2008). The great

Table 9.1 Genetically modified plant species that have been commercialised or have received a biosafety certificate in China

Species		Year of commercialisation or biosafety certificate	Trait	Status of application
Cotton	Gossypium hirsutum L.	1997	Insect resistance (Bt, Bt/CpTI, Bt/API)	3.9 million hectares
Petunia	Petunia hybrida Vil.	1997	Flower colour change (CHS)	Small-scale cultivation
Tomato	Lycopersicon esculentum Mill.	1998 2000	Virus resistance (CMV CP) Storage endurance (Antisense EFE)	Small-scale cultivation
Sweet pepper	Capsicum annuum L.	1998	Virus resistance (CMV CP)	Small-scale cultivation
Poplar tree	Populus tremula L.	2005	Insect resistance (Bt)	Small-scale cultivation
Papaya	Carica papaya L.	2006	Virus resistance (PRV CP)	About 10000 hectares
Rice	Oryza sativa L.	2009	Insect resistance (Bt)	Biosafety certification
Maize	Zea mays L.	2009	Phytase (phyA)	Biosafety certification

success of the commercial application of transgenic cotton has greatly inspired further research and development of transgenic biotechnology, and has supported the decision by the Chinese government to utilise biotechnology more widely in agricultural production.

During the course of the development and application of transgenic cotton, some important events occurred that can help us to understand the rapid progresses of transgenic biotechnology in China.

1991: The launch of the '863' research programme 'Research and Development of Transgenic Insect-Resistant Cotton'

1992: The successful synthesis of artificial GFM *cryIA(b)* and GFM *cry1A (c) Bt* transgenes by the Biotechnology Research Center (BRC) of the Chinese Academy of Agricultural Sciences (CAAS)

1993: These synthesised *Bt* transgenes were transferred to cotton by *Agrobacterium*-mediated transformation under the cooperative research of the Chinese Academy of Agricultural Sciences

1994: *Bt* insect-resistant transgenic cotton lines were produced by transgenic breeding

1995: Insect-resistant transgenic cotton and other crops were recognised as the important breakthrough of the key technologies in China by the '863' programme, and these crops were under preparation for application

1996: The transgenic *Bt* insect-resistant cotton varieties were produced and patents were applied for to protect the relevant technologies

1997: The relevant technologies of *Bt* insect-resistant transgenic cotton received patents and at nearly the same time the insect-resistant transgenic cotton received biosafety permission for commercial production in China

1998: China launched the commercialisation of locally produced *Bt* insect-resistant transgenic cotton varieties with Chinese intellectual property; the cultivation area was about 5% of the total GM cotton plantings in China, and in the following years the commercialisation of insect-resistant cotton varieties spread rapidly to the major cotton planting areas in China

2003: The cultivation of Chinese locally produced insect-resistant transgenic cotton varieties in China reached about 50% of the local crop

2011: China has issued a total of approximately 1550 biosafety
certificates since 1997, of which more than 95% were
for Chinese locally produced GM cotton varieties; the
total cultivation area reached 3.9 million hectares,
accounting for 75% of the total cotton
cultivation area.

According to the most recent statistics, 7 million small farmers
benefit from the cultivation of transgenic Bt cotton in China (James,
2011). The accumulative net income from the cultivation of the Bt
cotton varieties is in excess of 44 billion Yuan (approximately US$7
billion). In addition, due to the extensive commercial cultivation of
the Bt insect-resistant cotton, the application of chemical insecticides
was significantly reduced to less than 30% of that in the conventional
cotton fields, which has largely improved the agricultural environment,
including reducing poisoning by spraying of farmers and local populations (Wu et al., 2008).

The great success of transgenic Bt insect-resistant cotton has promoted the research, development and application of other transgenic
crops such as poplar trees, papaya, oilseed rape, wheat, potato, tomato
and many vegetable species in China (Table 9.2). Having experienced the
changes brought about by the commercial application of transgenic
cotton, biotechnology researchers and government are convinced of
the correctness of their decision to invest in developing biotechnology
for efficient agriculture production in China.

It is evident from the cotton case history that the promotion of
transgenic biotechnology and its products in China is largely due to the
needs of farmers for a more efficient way of crop production. It is also
important that there is great support from the Chinese government in
terms of policy and finance. The Chinese government has provided
tremendous financial investment for the research and development of
transgenic biotechnology and transgenic breeding through different
channels and this is the crucial factor in the success of the technology
and its application. The financial support is mainly from the four
sources:

(1) Direct funding from central government through national
research programmes such as the '863' and '973' programmes
funded through the Ministry of Sciences and Technology, as
well as the Agricultural 'Leap Programmes' funded through
the Ministry of Agriculture
(2) Direct fund allocation by provincial governments

Table 9.2 *Transgenic plant species with various transgenic traits that have entered biosafety field tests in China since 1997*

Plant species	Trait and transgene (in parenthesis)
Cotton	Insect-resistant (*Bt, Bt/CpTI, Bt/API, API, GST*); abiotic-stress-resistant (*RFdd205, BcDh2, CBF1, NPTII*)
Tomato	Insect-resistant (*Bt*); virus-resistant (*CMV CP*); quality improvement (*Antisense EFE*)
Papaya	Virus-resistant (*PRV CP*)
Soya bean	Insect-resistant (*Bt, GNA, Bt/GNA, Bt/CpTI, Vip3H*); herbicide-resistant (*EPSRS, Bar, PsbA135, EPSPS, G2-aroA, HTG7-aroA*); virus-resistant (*Chi, Rip, nptII, SMV CP, pBch*); quality improvement (*Antisense PEP, Antisense ACC, PhyA, GmDGAT, GmDofC, GmDofA*); abiotic-stress-resistant (*ABP9, GmPHD2, BADH, GmHSFAI, GmSARK/rlpk2*)
Poplar tree	Insect-resistant (*Bt*); abiotic-stress-resistant (*Tacap, MnSOD, TaLEA, eIF1A*)
Maize	Insect-resistant (*Bt, Bt/CpTI*); environment protection (*phyA*)
Rice	Insect-resistant (*Bt, CpTI, Bt/CpTI*); herbicide-resistant (*Bar, Bar/GUS, EPSPS*); bacterium-resistant (*Xa21*)
Oilseed rape	Insect-resistant (*Bt*); virus-resistant (*OXO, GIII, GO, CH5B*); quality improvement (*Antisense PEP, Antisense Fad2*)
Wheat	Herbicide-resistant (*HMW/PPT*)
Sweet pepper	Virus-resistant (*CMV CP*)
Petunia	Flower colour change (*CHS*)
Tobacco	Insect-resistant (*Bt, CpTI*); virus-resistant (*TMV CP, TMV CP/CMV CP*)
Potato	Virus-resistant (*Cecropin B*); change growth rate (*ipt*)
Beet	Insect-resistant (*Bt*)
Broccoli	Insect-resistant (*Bt*)
Cabbage	Insect-resistant (*Bt*)
Eggplant	Insect-resistant (*Bt*)
Apple	Insect-resistant (*Bt*)
Sunflower	Insect-resistant (*Bt*)

(3) Locally allocated funds by research institutions and universities
(4) Funds from local companies and non-governmental organisations.

This continuing financial support provides the guarantee of the sustainable development and application of transgenic biotechnology in China.

BIOSAFETY ISSUES: THE DILEMMA FOR TRANSGENIC BIOTECHNOLOGY AND GM PRODUCTS

The rapid development of biotechnology and the extensive application of genetically modified (GM) crop varieties with transgenes conferring diverse agronomic traits have opened up a new dimension for meeting the great demands for food by enhancing the efficiency of crop production with relatively less input. However, the extensive commercial cultivation of transgenic crops has aroused tremendous biosafety concerns and debates worldwide (Stewart et al., 2000; Ellstrand, 2001, 2003; Pretty, 2001). Biosafety issues have already become a crucial factor that can significantly influence the development of transgenic biotechnology and its wider application in agriculture. Therefore it is not possible to circumvent biosafety issues when discussing the development and application of transgenic crops in the world. Legally in all countries of the world, it is a 'MUST' requirement to receive a biosafety certificate or deregulation permit for any type of GM crop after appropriate biosafety assessment prior to its commercial cultivation. The following case indicates how essential it is to deal with the biosafety issue in a proper manner.

As early as 1992, China already cultivated transgenic tobacco resistant to tobacco mosaic virus (TMV) and cucumber mosaic virus (CMV) in Yunnan province (Yan, 2003). However, at that time, due to the lack of a well-established national biosafety regulation and assessment system in China, the regulatory procedure for transgenic tobacco cultivation was not properly carried out. The commercial cultivation of transgenic tobacco affected the export of Chinese tobacco to the international markets. Consequently, China stopped the cultivation of the first transgenic tobacco. This reflects the importance of the biosafety regulatory system in the development and application of transgenic products and biotechnology.

Biosafety is related to the safety aspects of genetically modified organisms (GMOs) produced via transgenic biotechnology. Hypothetically, GMOs may pose unwanted threats to human health and the environment during the entire procedures of research, exploration, production and application. There are quite a number of biosafety-related concerns being raised in general, but the most relevant ones can be summarised as follows:

- Food, feed and health biosafety risks potentially caused by the consumption of GM products
- Environmental and ecological biosafety risks caused by the extensive cultivation of GM crops

- Labelling systems of commercial GM products and the detection of possible transgene (or derived protein) presence in agricultural products if not labelled
- Socio-economic and ethical concerns raised by the application of GM products and biotechnology
- Regulatory procedures and acts for GM-related products
- General public perception or acceptance of GM products and biotechnology
- Risk assessment systems in relation to the commercial release and cultivation of GM products.

These biosafety issues are seriously discussed and debated by people around the world, and have considerably affected their attitude and decision-making on accepting or rejecting the GM products derived from transgenic biotechnology. However, the technology-derived biosafety issues are novel, technical and strongly influenced by the socio-economic, political and ethical background. This makes the prediction and understanding of long-term impacts caused by the application of GMOs very complicated, particularly the environmental impacts. The biosafety issues include the effects of the *Bt* transgene on target and non-target organisms, interactions and influences of transgenes and GM plants on biodiversity and ecosystem functions, and long-term ecological impacts from transgene flow to crop landraces and wild relative species. The resolution of these biosafety issues is challenging but extremely important for the promotion of transgenic biotechnology.

For example, the potential ecological impacts caused by transgene flow from a GM crop variety to its non-GM crop counterparts (particularly landraces or traditional varieties), or to its wild relatives (including weedy types of interbreeding related species) has aroused tremendous debate worldwide (Lu and Snow, 2005; Lu and Yang, 2009). Gene flow has been identified as a major environmental biosafety issue for GM crop cultivation. Therefore, relevant questions concerning environmental impacts caused by transgene flow still need to be addressed. Could transgene flow from a GM crop appreciably influence the conservation and safe use of crop biodiversity? Could transgene flow to wild relatives increase the weediness of the wild or cause the extinction of local wild populations? The uncertainty of these questions has largely affected the authority in terms of decision-making of the commercial application of GM crops. It also puzzles the public and consumers in deciding whether or not they should accept the GM products. Scientists and researchers have difficulty in determining how much more scientific research data should be provided to prove whether or not GM crops are safe to use as food and to be released to the environment. China has also struggled very hard to address the various biosafety issues during the process of development and application of transgenic biotechnology, including the establishment of its biosafety regulatory system.

Since 1993, China has gradually established the biosafety regulatory system to effectively assess and manage transgenic crops that are aimed for commercial production. On 24 December 1993, the State Science and Technology Commission (now Ministry of Science and Technology) issued the *Safety Administration Regulation on Genetic Engineering* that was the first biotechnology regulation in China at the highest level of administration. This regulation was aimed at promoting research and development of biotechnology and managing matters related to genetic engineering. On 23 May 2001, the State Council of China published the *Regulation on the Safety Administration of Agricultural GMOs* which was endorsed by Prime Minister Zhu Rongji, and is so far the most important legislation instrument of China for matters concerning biosafety. This legal document serves as a framework to provide policy guidance, with broad coverage from research and experimentation, commercialisation, labelling, to import and export. Subsequently, the Ministry of Agriculture issued four regulatory documents as implementing tools to deal with biosafety assessment, import control and labelling requirement of agricultural GMOs, respectively. China has established the National Biosafety Committee which is the highest authority, consisting of a panel of experts assisting the biosafety assessment and monitoring of GMOs.

In China, all proposals for GMOs, including plants, animals and microorganisms intended for commercial application, are obliged to proceed to biosafety assessment under the administration of the Ministry of Agriculture. For example, any GM crop line *must* receive a biosafety certificate before its developers can apply for a commercial production. To receive a biosafety certificate, a GM crop line should have completed a five-step biosafety assessment after a certain number of years as follows:

(1) Laboratory biosafety investigation (confined condition only)
(2) Restricted field biosafety test (<0.26 hectares)
(3) Enlarged field biosafety test (0.26–2 hectares)
(4) Field production biosafety test (>2 hectares)
(5) Application for a safety certificate.

Although biosafety issues are still under serious discussion and debate, it is necessary to face the challenge of biosafety issues raised by the application of GMOs and to try to understand what the potential biosafety issues are, and how the potential biosafety problems can be resolved. The sustainable and safe application of biotechnology products is determined by their relevant biosafety assessment and management. Proper biosafety assessment and management relies largely on solid data from sound scientific research. However, it is still challenging to deal with the biosafety-related issues raised by the new technology.

FUTURE PERSPECTIVES: PROMOTION OF NEW TECHNOLOGIES FOR FOOD SECURITY

As emphasised repeatedly in this chapter, China is a large country with a huge population, and food security is always the Number One important issue that has to be addressed. Food security is a key determinant for the stability of the country and the peace of the world, but food security is always challenging in China. It is very important but difficult to guarantee a sufficient food supply under the current situation of serious shortage of agricultural resources due to rapid industrialisation, urbanisation and economic development. Global climate change has added even more uncertainties to the situation of food production. To significantly increase the efficiency of crop production per unit area but at the same time to maintain environmentally friendly and sustainable agricultural ecological systems is the only choice to feed the increasing population of the country in a sustainable way. To achieve this goal, the effective application of new technologies, particularly transgenic biotechnologies, may provide a powerful tool to facilitate efficient crop breeding.

The world needs new biotechnologies to increase the efficiency of food production and to solve the food security problems, which can well be reflected in the progressively increasing global cultivation of GM crops by the increasing number of farmers, particularly in the developing countries during the past 15 years (James, 2011). Learning from the successful examples worldwide, China also needs biotechnology for food production. The Chinese government is very supportive of the development and application of new biotechnologies, including transgenic biotechnology, in fighting the challenge of long-term food security. Given the great success of the commercialisation of transgenic cotton, which has rescued the cotton industry in China, and other GM crops, the government has launched many projects particularly for research and development of transgenic biotechnology and its products

(Wu *et al.*, 2008). These projects not only focus on the development and application of GM organisms, but also on the biosafety-related knowledge and techniques for the sustainable and safe use of the GM products, including socio-economic impacts, technical training and public education and public awareness of biotechnology.

The development and application of new technologies in agriculture always brings new breakthroughs for crop production, such as the utilisation of semi-dwarf genes and genetic breeding, male-sterility genes and hybrid vigour mentioned earlier. Genetic engineering and biotechnology have significantly changed the style of agriculture characterised by largely reducing the input of resources (e.g. the application of insect-resistance transgenes) and labour (e.g. the application of herbicide-resistance transgenes), in addition to yield increases. This has made agricultural production more efficient, and eventually the improvement in agricultural production may meet the challenge for long-term food security.

ACKNOWLEDGEMENTS

This work was funded by the Natural Science Foundation of China (30730066), and the National Program of Development of Transgenic New Species of China (20011ZX08011–006).

REFERENCES

CAAS (Chinese Academy of Agriculture Science) and HNAAS (Hunan Academy of Agriculture Science) (1991). *The Development of Hybrid Rice in China*. China Agriculture Publishing Company, Beijing. (In Chinese)

Cheng, S. H, Cao, L. Y, Zhuang, J. Y., *et al.* (2007a). Super hybrid rice breeding in China: achievements and prospects. *Journal of Integrative Plant Biology* **49**, 805–810.

Cheng, S. H, Zhuang, J. Y, Fan, Y. Y., *et al.* (2007b). Progress in research and development on hybrid rice: a super-domesticate in china. *Annals of Botany* **100**, 959–966.

Ellstrand, N. C. (2001). When transgenes wander, should we worry? *Plant Physiology* **125**, 1543–1545.

Ellstrand, N. C. (2003). Current knowledge of gene flow in plants: implications for transgene flow. *Philosophical Transactions of the Royal Society B* **358**, 1163–1170.

FAO (2004). *The State of Food and Agriculture 2003–2004: Agricultural Biotechnology – Meeting the Needs of the Poor?* Food and Agriculture Organization of the United Nations, Rome.

Huang, J., and Wang, Q. (2002). Agricultural biotechnology development and policy in China. *AgBioForum* **5**, 122–135.

James, C. (2011). *Global Status of Commercialized Biotech/GM Crops: 2011*, ISAAA Brief No. 43. International Service for the Acquisition of Agri-Biotech Applications, Ithaca, NY.

Lu, B. R., and Snow, A. A. (2005). Gene flow from genetically modified rice and its environmental consequences. *BioScience* **55**, 669–678.

Lu, B.-R., and Yang, C. (2009). Gene flow from genetically modified rice to its wild relatives: assessing potential ecological consequences. *Biotechnology Advances* **27**, 1083–1091.

Mao, C. X., Wan, Y. Z., Ma, G. H., *et al.* (2006). Current status analysis of hybrid rice development in China. *Hybrid Rice* **21**, 1–5. (In Chinese)

MLRC (Ministry of Land and Resources of China) (2006). www.mlr.gov.cn/zt/dqr/38/xgxw/200711/t20071120_92688.html. (In Chinese)

NBSC (National Bureau of Statistics of China) (2006). *The Statistic Data of Agriculture from 1949–2004 in China*. Agricultural Scientific & Technical Publishing Corporation, Beijing. (In Chinese)

NDRC (National Development and Reform Commission) (2008). *Medium- and Long-Term Planning Framework for National Food Security (2008–2020)*. NDRC, Beijing.

Niu, W. Y. (2010). *China Scientific Development Report 2010*. Chinese Academy of Sciences, Beijing.

Pray, C. E., Huang, J., Hu, R., *et al.* (2002). Five years of Bt cotton in China: the benefits continue. *Plant Journal* **31**, 423–430.

Pretty, J. (2001). The rapid emergence of genetic modification in world agriculture: contested risks and benefits. *Environmental Conservation* **28**, 248–262.

Ru, X., Lu, X. Y., and Li, P. L. (2012). *Society of China Analysis and Forecast (2012)*. Beijing.

Stewart, C. N., Richards, H. A., and Halfhill, M. D. (2000). Transgenic plants and biosafety: science, misconceptions and public perceptions. *Biotechniques* **29**, 832–836, 838–843.

Stone, R. (2008). China plans $3.5 billion GM crops initiative. *Science* **321**, 1279.

Wu, K. M., Lu, Y. H., Feng, H. Q., *et al.* (2008). Suppression of cotton bollworm in multiple crops in china in areas with Bt toxin-containing cotton. *Science* **321**, 1676–1678.

Yan, X. F. (2003). *Transgenic Plants*. Science Press, Beijing. (In Chinese)

10

Genetically engineered crops would ensure food security in India

INTRODUCTION

Agriculture has been a way of life for most of the 73% of Indians living in 821000 villages (MIB, 2011). India with a population of 1.21 billion faces some of the worst challenges in meeting its future food security needs on account of: (a) its fast-growing population, projected to be 1.6 billion by 2050, (b) increasing incomes and increasing per capita food consumption which widen the yield–demand gap, and (c) shrinking area of cultivable land and inadequate and unexpandable irrigation facilities, all restricting prospects of increasing food production. India has the lowest productivity across developed and developing markets, leaving massive room for improvement. India's position in the 2011 Global Hunger Index was 67 of 81 countries. The Government of India (GoI) mooted an ambitious Food Security Bill in 2011, which will require over 65 million tons of additional annual food grain procurement.

While the Green Revolution has immensely helped India to become a food-exporting country, the advances in agriculture reached a plateau over a decade ago and the sizeable food surpluses of a few years are now fast dwindling. India's self-sufficiency and self-reliance in meeting the current food needs and catering for the demands of the increasing population in the coming decades is now a matter of serious concern. India urgently needs improved technologies to develop quality seed with traits relevant to regional needs and to increase crop yields by a wider adoption of such technologies as crop genetic engineering, to face future challenges. Technologies are needed to cope with

Successful Agricultural Innovation in Emerging Economies: New Genetic Technologies for Global Food Production, eds. David J. Bennett and Richard C. Jennings. Published by Cambridge University Press. © Cambridge University Press 2013.

biotic and abiotic stress factors, enhance nutritional quality of foods, enhance production without increasing arable land and/or irrigation facilities, and reduce cultivation costs through an efficient utilisation of inputs. The efficacy, biosafety, environmental safety and socio-economic benefits of biotech crops such as soya bean, maize, cotton, canola, etc., in cultivation in 19 developing and 11 developed countries since 1996, have been convincingly demonstrated. *Bt* cotton has achieved phenomenal success in India, reflecting benefits and farmer acceptance.

The other biotech crops in development in India hold a similar promise of enhancing production. However the current technological competence, infrastructural facilities and political and administrative atmosphere are not conducive to such a development. The GoI should ensure appropriate infrastructure and encourage the public and the private sectors to make concerted collaborative efforts to deploy modern technologies. The government should first overcome political compulsions from vested interest that vehemently opposes new technologies for concerns other than efficacy, safety and benefits to farmers, consumers and the country. Equally important is the removal of political and administrative hurdles in the path of agricultural development.

AGRICULTURE IN INDIA

Population growth

The 15th Indian National Census of the GoI indicates that the population was 1.21 billion in March 2011, with a growth rate of 17.6% during the decade of 2000–10, adding 181 million (GoI, 2011). The projection is that the Indian population will exceed 1.6 billion by 2050 and the rate of population increase is unlikely to stabilise before 2070. The growth of population, vastly out of tune with food production, is an important causative factor in the looming threat to India's food security.

Indian agriculture

Agriculture is the single most important means of livelihood of the Indian masses, accounting for 58% of direct or indirect employment (MoF, 2012). Rural incomes can grow to profitable levels only when the percentage of population dependent upon agriculture is no more than 5%, as in the developed countries. Such a shift requires consolidation of individual farm holdings to 15–20 hectares, and modernisation and

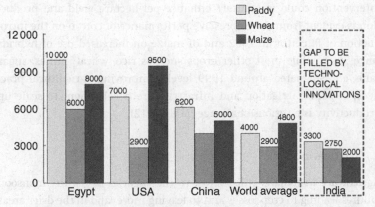

Figure 10.1 Comparison of Indian agricultural productivity with that of other countries. (Courtesy of Dr Gyan Shukla, Monsanto, Mumbai)

mechanisation of farm operations. There is hardly any scope and hope for such a development in the current political climate in India.

The per capita availability of agricultural land is expected to go down from the present 0.29 hectares to 0.23 hectares by 2025 and 0.19 hectares by 2050, due to the increase in population. This situation seriously affects agricultural production and food security, and could be partly compensated by (a) reclaiming 12 million hectares of degraded land, and (b) bringing under cultivation most of the 13 million hectares of cultivatable wasteland, and 29 million hectares left as fallow, both by enhancing soil health and irrigation, and by deploying modern technologies. Efforts should also be made to use 17 million hectares of presently uncultivatable barren land, through technological innovation.

Yield, production and productivity

Food grain production in India rose from 52 million metric tonnes in 1951–2 to 244.78 million metric tonnes in 2010–11. While the volume of food production has shown upward trends, the yields have been low. Indian crop yields have been nowhere near those in the USA, China and Egypt or the world averages (Figure 10.1). During 2009–10, Indian production of important food crops was behind that of China except for pulses (FAOSTAT, 2012).

India has not realised its full potential for productivity which can be made good only by technological innovation, since (a) the current technologies have reached a plateau, and (b) cultivatable land and irrigation facilities cannot be expanded appreciably. That technological intervention could immensely enhance per-hectare yield and production is evident from the impressive performance of cotton on the introduction of Bt cotton in 2002 and of maize on increased use of hybrids since 2003, while most other crops such as rice, wheat, pulses, sugar cane, etc. stagnated around 1999 levels. Improving agricultural practices, research, irrigation and infrastructure development to scale up productivity is a serious challenge (MoF, 2012).

Irrigation

Agriculture in India is still subject to the vagaries of nature, monsoon failures leading to crop losses and to leaving more land in the drier areas as fallow. The net irrigated area increased from 31.1% in 1970–1 to 45.3% in 2008–9, which is largely inadequate, even for the present needs.

Excessive and ill-advised use of bore-wells has pulled down the groundwater table to alarming levels, more particularly in central and northern India, as shown by the recent satellite images of NASA. It is estimated that 22% of the geographic area and 17% of the Indian population will be facing water scarcity by 2050.

Augmenting irrigation potential is the key to sustained growth in agriculture (MoF, 2012). Cultivation of oilseeds, pulses and millets, grown in the drier regions of the country, would greatly increase with expanded irrigation facilities. There is an ambitious and expensive mega-project in consideration for over half a century of linking the major rivers of the country to facilitate a wider and equitable distribution of water for agriculture.

Farm mechanisation

Farm mechanisation has immense potential to improve productivity, but this is a grossly neglected area of Indian agriculture. The small landholdings impose serious constraints on mechanisation. The GoI have proposed franchising rural entrepreneurs for establishing custom hiring and establishment of farm service centres to promote farm mechanisation that would include the small and marginal farmers (MoF, 2012).

Agricultural extension programmes

Agricultural extension programmes to guide farmers regarding choice of seed, fertilisers, insecticides, irrigation and alternate cropping patterns based on soil analysis and other factors have been largely ineffective except in a few states (MoA, 2012). Some areas are unsuitable for cultivating certain crops, as for example, cotton on red soils, particularly as a rain-fed crop. The farmers lured by higher financial returns often ignore advice and blame the dealers, government and the technology for crop failures. One of the most urgent needs is to establish and intensify grassroot guidance programmes.

The Green Revolution in India

The Green Revolution (GR), launched in the 1960s in India, is an intensive agricultural movement rooted in (a) establishment of scientific and agricultural research institutions to develop improved technologies and crop varieties, (b) chemicalisation of agriculture in terms of consistently improved fertilisers, pesticides and herbicides, (c) enhanced irrigation, and (d) effective management (Hazell, 2009).

The Green Revolution demonstrated that various barriers to food production could be overcome through technological intervention and handsomely transformed India from 'the ship to the lip' dependence on imports to the position of a food exporter. GR had a positive impact on food production in the developing countries and in such areas as reducing fluctuations in food production, increase in incomes and employment, reduction of poverty, enhanced nutrition, improved agricultural sustainability and enhanced availability of public goods (Hazell, 2009).

The benefits generated by GR were lost in the course of time due to lack of appropriate management practices and extension work to educate the farmer and to sustain the benefits. A number of warning signals were largely ignored by the Indian agricultural managers with the following serious consequences:

(a) An overzealous application of chemical inputs rendered large tracts of agricultural lands uncultivatable

(b) Excessive use of bore-wells for irrigation stressed power supply and also led to an alarming depletion of the water table

(c) In spite of numerous projects of different dimensions to expand irrigation facilities, their stretched use meant that no more additional lands could be brought under irrigated cultivation

(d) The benefits from improved crop varieties and development of new improved varieties reached a plateau over a decade ago

(e) The clamour of the farmers for cash crops severely brought down the production of millets, pulses and oilseeds, affecting dry land agriculture and the poor farmers who depend upon it

(f) Financial and power shortages have become a serious limitation even to sustain the gains, let alone to effect any improvements.

The managers of Indian agriculture failed to foresee that the factors contributing to gains from GR themselves would become limiting factors in time leading to failure of their continued expansion.

Demand and supply of food crops

Mittal (2008) reviewed the demand and supply position of major food crops from the base year 1999–2000, and projections for 2011, 2021 and 2026. In general, food production is expected to increase but the demand too would increase due to increase in population and increased per capita consumption. It is expected that the demand–supply gap will zero down in 2021 but will become significant by 2026, even when the total food production would be about 260 million metric tonnes, up from 190 million metric tonnes in 2005. Overall, the demand–supply position for rice and wheat would be satisfactory but the gap for pulses, edible oilseeds and sugar would be of concern by 2021 and alarming by 2026, if no effective correction measures are in place well in advance. Clearly, the need is for deployment of new technologies and management practices.

FOOD SECURITY, GLOBAL HUNGER INDEX AND ORGANIC FARMING

Food security

The Rome Declaration of the World Food Summit of 1996 defined food security as existing 'when all people at all times have access to sufficient, safe, nutritious food to maintain a healthy and active life'. The main thrust of the Rome Declaration, reiterated by the World Summit on Food Security in November 2009 (FAO, 2009), is to reduce the number of the world's 1 billion hungry and chronically undernourished people by half, by 2015. In this context, the need of the poor countries for development of economic and policy tools required to boost their agricultural production and productivity (FAO, 2009) becomes paramount.

Prerequisites of food security

The basic criteria in ensuring food security are:

- Availability of sufficient quantities of food on a consistent basis. The per capita availability of rice in 2010 in India was 185 g/day and of wheat 168 g/day. This is 26 g/day of rice and 1.1 g/day of wheat more than 1951 levels, but is lower than the recommended 300 g/day of rice and 200 g/day of wheat. Worse is the situation with pulses: the 31.6 g/day availability in 2010 is 29 g/day less than in 1951.
- Accessibility is having appropriate diverse foods for a nutritious diet within reach in sufficient quantities. In India, an impressive range of food items (grain, pulses, edible oil, vegetables, fruits and processed foods), a considerable percentage of them imported from different countries, are available. Unfortunately, these are largely inaccessible to the majority of Indians, particularly the semi-urban and rural population.
- Affordability is having sufficient resources to purchase what is accessible. During 2006–8, 19% of the total Indian population was undernourished, as against the global level of 13% (FAOSTAT, 2012). Four hundred million Indians now live on less than the government-defined poverty line of Rs.32/day (US$0.60). Another 200 million are classified as above the poverty line, but cannot afford to buy adequate quantities of appropriate food on account of high inflation and runaway food prices.
- Knowledge of basic nutrition and care. Indian tradition was rich in healthcare wisdom, but most contemporary Indians have lost track of the body of knowledge of balanced diets. The tragedy is that even educated people who can afford diverse nutritious foods suffer from nutritional deficiency disorders, while the encapsulated nutritional supplements they avidly consume largely fail them.

India's Food Security Bill, 2011

The recent Global Hunger Index placed India at the 67th position among 81 countries (Klaus et al., 2011), an alarming situation that requires urgent remedial measures. The GoI tabled the Food Security Bill, 2011 in the Parliament to facilitate supply of food grains at heavily subsidised costs to poor families, aimed to benefit 46–75% of the rural population and 28–50% of the urban population. This move, aimed at reducing malnutrition and hunger-related mortality, requires enormous financial inputs (Rs.1000–1500 million) and concerted technological and management efforts to produce over 65 million tons of additional food grains annually to realise the objectives of the Bill. There is no way to achieve this under the conventional agricultural practices, and much less through the highly touted organic farming. New technologies are the only option.

GENETICALLY ENGINEERED CROPS

Genetically engineered crops in India

There are about 60 agricultural and conventional universities, 10 autonomous institutes and 65 companies involved in R&D of genetically engineered (GE) crops in India, developing about 80 traits in 30 crops, among which pest tolerance is the most predominant trait. Only about 20 traits are being developed in the private sector. IGMORIS (2011) provides the details of the 22 traits in nine crops that are now under field trials.

Indian biosecurity regulatory system

The current Indian biosecurity regulatory system, drawing power from the Environment Protection Act 1986, its amendments and rules, is more elaborate and rigid than that of most countries and functions under the advice of four competent authorities constituted of representatives from government departments, scientific institutions, scientists, social scientists and other interest groups. There is elaborate documentation on the guidelines and standard operating procedures for the evaluation of product efficacy, biosafety and environmental safety. The R&D institutions of the Indian Council of Agricultural Research, the agricultural universities in different States and a dozen recognised public and private-sector research institutions are involved in gathering agronomic and biosafety data, and in recommending biotech crops for field trials, commercial release and their pre- and post-release monitoring. Basing the inputs and recommendations through this evaluation process and of specially constituted event-specific committees, the Genetic Engineering Appraisal/Approval Committee (GEAC), the apex body, permits multi-location field trials at two levels of biosafety research (BRLI and II) and approves successful events for commercial release. The GEAC is located in the Ministry of Environment and Forests, GoI, as in principle the GEAC's concern is only environmental safety. However, the GEAC has often dismayed the scientific community with irrational scientific decisions, such as those against the use of the β-glucuronidase reporter system (GUS) and antibiotic resistance marker genes, globally considered safe, in transgenic technology.

The GEAC came under heavy criticism from activists when it accorded proactive approvals of different GE events at different stages and levels, the worst being when the GEAC approved commercial release of Bt cotton in 2002 and Bt brinjal in October 2009.

Cognisant of the concerns about the current regulatory system, the GoI has now tabled the Biotechnology Regulatory Authority of India (BRAI) Bill in the Indian Parliament, providing for a three-stream (agricultural, medical and environmental) regulation of modern bio-technology, in a smooth single window clearance for each. The BRAI contains the judiciously balanced best of global regulatory procedures and the draft bill went through a lengthy process of inter-ministerial, inter-departmental and public consultation. Yet the activists are as vehemently against the BRAI as they have been with the present system. It would take four or five years to operationalise the BRAI and in the meantime the GoI should ensure that the present system functions without political interference or activist pressure, as happened with processing Bt brinjal.

Bt cotton in India

India, with only Bt cotton under commercialisation, occupied the fourth global position in area under GE crops in 2011–12. In the decade 2002–12, Bt cotton cultivation in India grew phenomenally from 0.5 million hectares to 10.6 million hectares (James, 2011), amounting to 88% of 12.1 million hectares under cotton in the country. The number of Bt cotton-farmers rose from 50000 in 2002 to 7 million in 2011.

The average yield of cotton in India increased from 309 kg/ha in 2001 to 499 kg/ha in 2010 and cotton production increased from 13.6 million bales (170 kg/bale) to 35.5 million bales during 2002–2012. The total area under cotton, the area under Bt cotton and the percentage of adoption from 2002 to 2011 in India are shown in Figure 10.2.

India has transformed from a cotton-importing country to the second largest exporter (Figure 10.3). During 2011–12, India exported 1.258 million metric tonnes of cotton, the top most exported agricultural

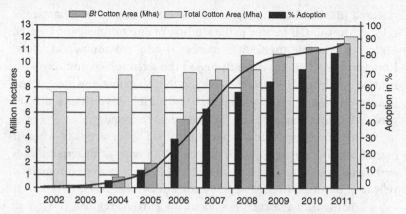

Figure 10.2 Total area under cotton, area under *Bt* cotton and percentage of adoption of *Bt* cotton in India (2002–2011). (Courtesy of ISAAA, Ithaca, NY)

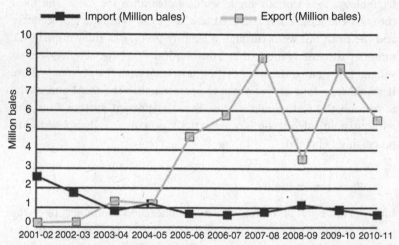

Figure 10.3 Import and export trends of cotton in India (2001–2011). (Courtesy of ISAAA, Ithaca, NY)

commodity, that increased by 36.11% over 2010–11 (MoA, 2012). *Bt* cotton helped India in enhancing farm income by US$9.4 billion in the period 2002–10 and US$2.5 billion in 2010 alone (James, 2011; Brookes and Barfoot, 2012), benefiting 7 million small resource-poor farmers in 2011.

The Indian Parliament was informed in December 2011 by the Minister of State for Agriculture, GoI, that the biggest gain from *Bt* technology was reduction of insecticide usage from 46% of total insecticide use in 2001 in the country to 21% during 2009–10. India has used 38.6 million kg less insecticide since 2002 (Brookes and Barfoot, 2012).

More than 25 farmer surveys in Indian cotton-producing states have recognised the overall positive impact of the *Bt* technology on cotton yields over time (Gruere and Yan, 2012). Several studies were cited by Gruere and Yan (2012) on the reduction of insecticide use and its consequent positive health benefits, increase of women's labour opportunities and contribution to poverty reduction, on adoption of *Bt* cotton and on the finding that farmer suicides are not related to *Bt* cotton (Gruere *et al.*, 2008; Gruere and Sengupta, 2011).

Bt brinjal in India

Bt brinjal (EE1) contains the *Cry1Ac* gene from the universally occurring soil bacterium *Bacillus thuringiensis*, the same gene as in *Bt* cotton. *Cry1Ac* controls the shoot- and fruit-borer (SFB) pests of brinjal. Several food crops, such as maize and potato, containing *Bt* genes have been consumed since 1996 and shown to be safe to consumers and the environment.

Bt brinjal has been in development since 2002 with all required official permissions from the Indian regulatory regime (Kameswara Rao, 2010). The efficacy, field performance and biosafety were evaluated by a dozen public and private-sector institutions involving some 200 scientists and experts, overseen by the Review Committee for Genetic Manipulation (RCGM). Two expert committees constituted by the GEAC have examined the entire dossier and approved it. The GEAC accepted their recommendation for commercial release in October 2009.

Being a statutory body, the GEAC need not seek the approval of the Minister of Environment and Forests (MoEF), GoI, but probably for the lack of courage to take responsibility on an issue in the midst of raging activism put the matter up for the approval of the MoEF, who chose to impose an indefinite moratorium on the commercialisation of *Bt* brinjal on 9 February 2010. That this moratorium was not based on any safety or environmental concerns was evident from the 535-page documentation released by the MoEF along with the moratorium order, as was discussed in detail (Kameswara Rao, 2010). That the MoEF had rejected the scientific regulatory process and the recommendation of the GEAC on political considerations is evident from his statements: (a) that '*it is for the political system to decide whether to introduce* Bt *brinjal* (*The Hindu*, 10 January 2011), (b) '*if I said yes to* Bt *brinjal, the civil society would have jumped on me*', and (c) '*I had a personal bias*' (b and c: Malhotra, 2011). He also had a baseless apprehension that '*90% of the GM seed is going to be controlled by one company*' (Malhotra, 2011), meaning Monsanto, which has nothing to do with *Bt* cotton and *Bt* brinjal, other than providing the *Cry1Ac* gene incorporated in them. Clearly, this is no basis for taking decisions on issues that seriously affect the prospects of the country's agriculture.

The moratorium on *Bt* brinjal had a much more serious and wider effect on the whole of GE crop R&D in India. The usually slow public sector became much slower and both Indian and foreign private investors became hesitant. In 2½ years no one in authority made a move to ease the impasse caused by the moratorium.

Golden Rice in India

Golden Rice comes to the developing countries free of technology cost burden, through the international charitable organisation, the Golden Rice Humanitarian Board, and developed in a public–private partnership as discussed by Adrian Dubock in Chapter 12 of this book. In India Golden Rice is being developed at the Indian Agricultural Research Institute, New Delhi, Directorate of Rice Research, Hyderabad and the Tamil Nadu Agricultural University, Coimbatore, all in the public sector. Activists have been opposing Golden Rice in India and elsewhere for the same reasons as for the other GE crops. Though India entered into an agreement in 2002 to receive the technology cost-free, there has been no progress in commercialising Golden Rice, due to political compulsions. Golden Rice is likely to go for open field trials in India some time in 2012, while the Philippines and Bangladesh are expected to commercialise Golden Rice in a year or two.

Public and private partnership in agriculture

The private sector plays a very crucial role in Indian agriculture. The Founder of the Maharashtra Hybrid Seed Company was awarded the World Food Prize in 1998 and regarded as the father of the Indian seed industry. Hybrids of *Bt* cotton and *Bt* brinjal are the contribution of the private sector. It is a common hope that a public–private partnership (PPP) would immensely help Indian agriculture. If past experience is an indicator, there is hardly any scope for benefits from PPP. The private sector, particularly the multinational companies (MNCs), has the state-of-the-art technological infrastructure, a very competent scientific force, financial strength, sense of time and a determination to produce, while the public sector largely lacks most of these important prerequisites for success. Besides, the private sector has a fiscal responsibility. Varieties of *Bt* brinjal and Golden Rice, the products of PPPs, are languishing due to the apathy of the government and public-sector establishments. Over a decade of effort and enormous expense in the public sector have not resulted in even a single marketed GE product. For PPPs to be productive and benefit the private-sector component as well and not just draw from it, a lot of committed effort is needed on the part of the government, managers of Indian agriculture and the public-sector scientific institutions.

Activism against GE crops in India

Development of GE crops in India has suffered from over a decade of persistent and vehement activism. Medical and environmental biotech products are hardly an issue for the activists, who singled out GE crops. The activists opposed commercialisation of *Bt* cotton, could not prevent it, but succeeded in condemning *Bt* brinjal to an indefinite moratorium. Opposing GE technology is a matter of livelihood and not a calling for the vast majority of the activist groups who pursue someone else's agenda. All those who support technology are branded as 'toadies of the industry'. The bureaucracy, scientific community and the product developers are all afraid of the aggressive stance of the activists who have no holds barred. Anti-tech activism which opposes new technologies is the most serious threat to India's future food security.

The pesticide industry and dealers, conventional seed developers and dealers and the organic lobby, who feel threatened by large-scale adoption of GE crops, support activism. Single-issue politics of all parties, ideological imperialism or mere prejudice against the Western countries and their MNCs have all fuelled the agitation. Activists use every trick of the trade, including petitions in the Supreme Court of India and vandalising R&D workplaces and crops in officially approved field trials, to mould public opinion against GE technology and to pressurise the GoI to ban it altogether.

The activists, through the media, spread misinformation and disinformation on issues that have an emotional influence on the public, though not supported by any scientific evidence. For example, it was argued that the introduction of *Bt* brinjal would jeopardise the use of brinjal in indigenous medicine, when brinjal is not used in any traditional medical system (Kameswara Rao, 2011). Consequences of deficient management and near absence of extension programmes to educate the farmers are projected as technology failures.

In spite of the voluminous peer-reviewed literature on the safety and benefits of GE crops (OECD, 2007; Chelsea *et al.*, 2012), their biosafety is the focus of activism. Among several other charges against GE crops that mostly are a ventriloquist's echo of Western activism, the often repeated issues are:

(a) *'The terminator gene in the GE crops seriously harms farmers' interests'*, when no crop anywhere in the world contains the so-called terminator gene

(b) *'GE crops negatively impact ecology and biodiversity'*, when no scientific evidence has emerged from over 10 years of experimental field research and commercial cultivation that the presently commercialised GE crops caused any environmental harm (Sanvido *et al.*, 2007),

(c) *'All the indigenous varieties and hybrids are lost because of crop GE'*, when the hybrids and varieties into which the *Bt* gene is incorporated are actually saved, as otherwise they would have been lost to the pests

(d) *'Failure of* Bt *cotton is responsible for innumerable farmer suicides in India'*, a charge that was not upheld by international (Gruere *et al.*, 2008; Gruere and Sengupta, 2011) or Indian reports accepted by the MoEF in his *Bt* brinjal moratorium order (Kameswara Rao, 2010).

There are important instances, which affect even the public sector, where activists scored:

(a) In 2002, the Indira Gandhi Agricultural University, Raipur, Chattisgarh State, and Syngenta International, the Swiss MNC involved in unravelling rice genome and co-developer of Golden Rice, proposed collaborative research on the university's rice germplasm to identity rice varieties and genes to develop new hybrids that meet with specific requirements of the farmers. There was instantaneous opposition from the activists since Syngenta is an MNC and some others who joined for their own reasons. Threatened by a violent agitation, Syngenta withdrew from the arrangement.

(b) The Indo-US Knowledge Initiative on Agriculture (KIA) was announced in 2005 and finalised in 2006, along with the Indo-US nuclear deal, by the US President and the Indian Prime Minister. The KIA is aimed at boosting agricultural cooperation between India and the USA, in sectors such as food processing and marketing, biotechnology, water management, and capacity-building at universities. Mainly because the

USA was the other partner, the activists and the opposition parties vehemently opposed both the deals. However, the GoI pressed for the approval of the nuclear deal by the Parliament risking even the survival of the government, but relegated the KIA to low priority, and it has not been heard of since.

(c) The MoEF, at the insistence of the Chief Minister of the Bihar State, issued a directive that the developers of GE crops should obtain the permission of the State governments even before they apply to the GEAC for permission to conduct field trials in the States, as agriculture is widely considered as a domain of the States. As most State governments side with the activists, this entirely political move has now become a serious impediment to the development of GE crops, blocking the process at the first step itself.

(d) The National Biodiversity Authority (NBA) is being pressurised by the activists to prosecute the developers of Bt brinjal, including the two public-sector agricultural universities, for using native germplasm in developing a GE crop without prior permission of the NBA. The hybrids and varieties used by the three Bt brinjal developers were developed by them a long time ago and have been marketed/distributed by them long before the NBA became functional. Since all germplasm of all crops grown in India can be interpreted as native, the NBA being under the influence of the activists, no product developer is likely to get permission to use Indian germplasm for any purpose.

Anti-GE activism in India is sustained by liberal foreign funding particularly from the EU countries (Kameswara Rao, 2010). The Prime Minster of India, while emphasising that *'biotechnology has enormous potential and we must make use of genetic engineering technologies to increase the productivity of Indian agriculture'*, observed that the *'NGOs often funded from the US and Scandinavian countries'* are in the way of development in India (PM's interview, 2012). What the Prime Minister missed was not lost on the Deputy Chairman, Planning Commission of India, who observed that *'many of the NGOs in India protesting against Bt brinjal were associated with and funded by European NGOs'* (India Today, 3 September 2011).

Untangling biosecurity issues, from the mix-up of political, economic, societal and ethical issues, projected as biosecurity issues by the activists, is the most urgent and important step the GoI needs to take, but it is not an easy task.

The government should respect the combined global and national scientific wisdom in evaluating GE products, and the decisions on their acceptance or rejection should not be allowed to be hijacked by the vested interest that uses junk science pursuing inept politics.

Today any GE crop can be released for cultivation in India, provided the developers do not say that it is GE, and in the process can also save enormous amounts of time and money by bypassing the regulatory regime, benefiting farmers and consumers. There are several unconfirmed reports of illegal cultivation of pest- and herbicide-tolerant gene stacked cotton, virus-tolerant papaya and *Bt* brinjal in some parts of the country.

Public awareness

The Indian public is not averse to technology. Acceptance of technology depends upon convincing the public on the safety, farmer profitability and consumer benefits. Robust public education programmes are essential to counter vehement anti-tech campaigns.

The scientists, product developers and governments have all neglected the very important aspect of educating the public on the benefits of GE crops in most parts of the world. The public now gets the activist propaganda through the media who are not concerned about the veracity of what is being fed to them by the activists, and project it as public opinion, which is only the publicised opinion of the activists, chronicling misery divorced from facts. Public education programmes are the immediate need and all those concerned should be alive to it.

Democratic governments need to respect public opinion, but should not allow ignorance and prejudice to dictate policy. The scientific establishments should educate and lead public opinion and the governments should not pander to vested interest with an eye on single-issue voters. Mere statements from the Prime Minister eulogising the potential of GE crops (PM's interview, 2012) are not enough.

REFERENCES

Brookes, G., and Barfoot, P. (2012). *GM Crops: Global Socio-Economic and Environmental Impacts 1996–2010*. PG Economics Ltd, Dorchester, UK. www.pgeconomics.co.uk/page/33/global-impact-2012

Chelsea, S., Bernheim, A., Bergé, J-B., *et al.* (2012). Assessment of the health impact of GM plant diets in long-term and multigenerational animal feeding trials: a literature review. *Food and Chemical Toxicology* **50** 1134–1148.

FAO (2009). www.fao.org/wsfs/world-summit/en/

FAOSTAT (2012). http://faostat.fao.org/site/339/default.aspx

GoI (2011). *Census of India 2011*. Government of India. http://censusindia.gov.in

Gruere, G. P., and Sengupta, D. (2011). Bt cotton and farmer suicides: an evidence-based assessment. *Journal of Development Studies* **47**, 316–337.

Gruere, G. P., and Yan, S. (2012). *Measuring the Contribution of* Bt *Cotton Adoption to India's Cotton Yields Leap*, Discussion Paper No. 01170. International Food Policy Research Institute, Washington, DC. www.ifpri.org/sites/default/files/publications/ifpridp01170.pdf

Gruere, G. P., Mehta-Bhatt, P., and Sengupta, D. (2008). Bt *Cotton and Farmer Suicides: Reviewing the Evidence*, Discussion Paper No. 00808. International Food Policy Research Institute, Washington, DC. www.ifpri.org/publication/bt-cotton-and-farmer-suicides-india

Hazell, P. B. R. (2009). *The Asian Green Revolution*, Discussion Paper No. 00911. International Food Policy Research Institute, Washington, DC. www.ifpri.org/sites/default/files/publications/ifpridp00911.pdf

IGMORIS (2011). *Indian GMO Research Information System*. Government of India, Department of Biotechnology, New Delhi. http://igmoris.nic.in/field_trials.asp

James, C. (2011). *Global Status of Commercialized Biotech/GM Crops: 2011*, ISAAA Brief No. 43. International Service for the Acquisition of Agri-biotech Applications, Ithaca, NY. www.isaaa.org/resources/publications/briefs/43/executivesummary/default.asp

Kameswara Rao, C. (2010). *Moratorium on* Bt *Brinjal: A Review of the Order of the Minister of Environment and Forests, Government of India*. Foundation for Biotechnology Awareness and Education, Bangalore, India. www.whybiotech.com/resources/tps/Moratorium_on_Bt_Brinjal.pdf

Kameswara Rao, C. (2011). *Use of Brinjal (Solanum melongena L.) in Alternative Systems of Medicine in India*. Foundation for Biotechnology Awareness and Education, Bangalore, India. http://plantbiotechnology.org.in/Brinjal%20alt%20syst%20med%203.pdf

Klaus, V. G., Maimo, T., Tolulope, O., *et al.* (2011). *2011 Global Hunger Index*. International Food Policy Research Institute, Washington, DC. www.ifpri.org/sites/default/files/publications/ghi11.pdf

Malhotra, R. (2011). Jairam Ramesh – in conversation. *Current Science* **100**, 1610–1612. http://cs-test.ias.ac.in/cs/Volumes/100/11/1610.pdf

MIB (2011). Agriculture. In *India 2011*, pp. 66–115. Publications Division, Ministry of Information and Broadcasting, Government of India, New Delhi.

MoA (2012). *State of Indian Agriculture 2011–12*. Report submitted to the Rajya Sabha. Ministry of Agriculture, Government of India, New Delhi. http://agricoop.nic.in/SIA111213312.pdf

MoF (2012). Agriculture. In *Economic Survey, India Budget, 2012–2013*, ch 8, pp. 179–201. Ministry of Finance, Government of India, New Delhi. www.indiabudget.nic.in/es2011–12/echap-08.pdf

Mittal, S. (2008). *Demand–Supply Trends and Projections of Food in India*, Working Paper No. 209. Indian Council for Research on International Economic Relations, New Delhi. www.icrier.org/pdf/Working%20Paper%20209.pdf

OECD (2007). *Safety Information on Transgenic Plants Expressing* Bacillus thuringiensis-*Derived Insect Control Protein*, Consensus Document No. 42. Organization for Economic Cooperation and Development, Paris. www.ENV/JM/MONO(2007)14.

PM's interview (2012). *Science* **335**, 907–908. www.sciencemag.org/content/335/6071/907.full

Sanvido, O., Romeis, J. A., and Franz, B. (2007). Ecological impacts of genetically modified crops: ten years of field research and commercial cultivation. *Advances in Biochemical Engineering/Biotechnology* **107**, 235–278.

11

Plant genetic improvement and sustainable agriculture

INTRODUCTION

One billion people go to sleep hungry each night, and that number grows daily. To accommodate an ever-increasing demand for food, world agricultural production must rise 50% by the year 2030. Simply increasing the area of cultivated land will not meet these needs, because arable land and fresh water supplies are limited. The predicted effects of climate change will compound these production challenges. Forecasts for the coming decades predict worsening droughts, increasing floods and new pest and disease outbreaks (Coakley et al., 1999). Regional temperature increases will also impact yield, with each additional degree of night-time temperature dropping yields of rice, maize and wheat 5–10% (Schlenker et al., 2009; Lobell et al., 2011).

Satisfying human food needs while adapting to the changing climate and reducing our environmental footprint presents a unique challenge to the agricultural community. Historically, advances in plant genetics provided the knowledge and technologies needed to address these challenges. Plant genetics remains a key component of global food security, peace and prosperity for the foreseeable future. For example, we will increasingly rely on the concerted efforts of plant geneticists to predict combinations of genes that can be used to produce crop varieties that will thrive in extreme environments. Millions of lives depend upon the extent to which crop genetic improvement can keep pace with the growing global population, changing climate and shrinking environmental resources.

Successful Agricultural Innovation in Emerging Economies: New Genetic Technologies for Global Food Production, eds. David J. Bennett and Richard C. Jennings. Published by Cambridge University Press. © Cambridge University Press 2013.

The availability of genetically improved seed is an important resource for farmers. However, seed is just one component of a sustainable farming system. Other approaches that will enhance the sustainability of our farms are also needed. These include more effective land and water use policies, integrated pest management approaches, as well as reductions in harmful inputs (Somerville and Briscoe, 2001).

Genetically improved seed must be integrated into ecologically based farming systems and evaluated in light of their environmental, economic and social impacts – the three pillars of sustainable agriculture. This chapter describes some lessons learned, over the last decade, of how genetically improved crops (with a particular focus on genetically engineered crops) have been integrated into agricultural practices around the world and discusses their current and future contribution to sustainable agricultural systems.

WHAT ARE GENETICALLY ENGINEERED CROPS?

Genetic engineering differs from conventional methods of genetic modification in two major ways: (1) genetic engineering introduces one or a few well-characterised genes into a plant species; (2) it can introduce genes from any species into a plant. In contrast, most conventional methods of genetic modification used to create new varieties (e.g. artificial selection, forced interspecific transfer, random mutagenesis, marker-assisted selection and grafting of two species, etc.) introduce many uncharacterised genes into the same species. Conventional modification can sometimes move genes between different genera, like wheat and rye or barley and rye.

SAFETY ASSESSMENT OF GENETICALLY ENGINEERED CROPS

There is broad scientific consensus that genetically engineered crops currently on the market are safe to eat. After 14 years of cultivation and a cumulative total of 2 billion acres planted, no adverse health or environmental effects have resulted from commercialisation of genetically engineered crops (Board on Agriculture and Natural Resources et al., 2002). The US National Research Council and the Joint Research Centre (the European Union's scientific and technical research laboratory and an integral part of the European Commission) have both concluded that there is a comprehensive body of knowledge that adequately addresses the food safety issue of genetically engineered crops (Committee on Identifying and Assessing Unintended Effects of Genetically Engineered

Foods on Human Health and National Research Council, 2004; European Commission Joint Research Centre, 2008). These and other recent reports conclude that the processes of genetic engineering and conventional breeding are no different in terms of unintended consequences to human health and the environment (European Commission Directorate-General for Research and Innovation, 2010).

This is not to say that every new variety will be as benign as the crops currently on the market. This is because each new plant variety (whether it is developed through genetic engineering or conventional approaches of genetic modification) carries a risk of unintended consequences. Whereas each new genetically engineered crop variety is assessed on a case-by-case basis by three governmental agencies, conventional crops are not regulated. To date, compounds with harmful effects on humans or animals have only been documented in foods developed through conventional breeding approaches. For example, conventional breeders selected a celery variety with relatively high amounts of psoralens to deter insect predators that damage the plant. (The psoralens are a family of naturally occurring toxic and mutagenic compounds, for example, used in Indonesia for catching fish.) Some farmworkers who harvested such celery developed a severe skin rash – an unintended consequence of this breeding strategy (Committee on Identifying and Assessing Unintended Effects of Genetically Engineered Foods on Human Health and National Research Council, 2004).

INSECT-RESISTANT CROPS

In the 1960s Rachel Carson brought to the attention of the wider public the detrimental environmental and human health impacts resulting from overuse or misuse of some insecticides. Even today, thousands of pesticide poisonings are reported each year (an estimated 300 000 deaths globally, around 1200 each year in California alone). This is one reason some of the first genetically engineered crops were designed to reduce reliance on sprays of broad-spectrum insecticides for pest control.

Maize and cotton have been genetically engineered to produce proteins from the soil bacterium *Bacillus thuringiensis* (Bt) that kill some key caterpillar and beetle pests of these crops. Bt toxins cause little or no harm to most non-target organisms including beneficial insects, wildlife and people (Mendelsohn *et al.*, 2003). Bt crops produce Bt toxins in most of their tissues. These Bt toxins kill susceptible insects when they eat Bt crops. This means that Bt crops are especially useful for controlling pests that feed inside plants and cannot be killed readily by sprays, such as the European corn-borer (*Ostrinia nubilalis*), which bores into stems, and the pink bollworm (*Pectinophora gossypiella*), which bores into bolls of cotton.

Bt toxins in sprayable formulations were used for insect control long before *Bt* crops were developed and are still used extensively by organic growers and others. The long-term history of use of *Bt* sprays allowed the US Environmental Protection Agency (EPA) and Food and Drug Administration (FDA) to consider decades of human exposure in assessing human safety before approving *Bt* crops for commercial use. In addition, numerous toxicity and allergenicity tests were conducted on many different kinds of naturally occurring *Bt* toxins. These tests and the history of spraying *Bt* toxins on food crops led to the conclusion that *Bt* maize is as safe as its conventional counterpart, and therefore would not adversely affect human and animal health or the environment (European Food Safety Authority, 2004).

Planting of *Bt* crops has resulted in the application of fewer pounds of chemical insecticides, and thereby has provided environmental and economic benefits that are key to sustainable agricultural production. Although the benefits vary depending on the crop and pest pressure, overall, the US Department of Agriculture (USDA) Economic Research Service found that insecticide use in the USA was 8% lower per planted acre for adopters of *Bt* corn than for non-adopters. Fewer insecticide treatments, lower costs and less insect damage led to significant profit increases when pest pressures are high (Fernandez-Cornejo and Caswell, 2006). When pest pressures are low, farmers may not be able to make up for the increased cost of the genetically engineered seed by increased yields. In Arizona, where an integrated pest management programme for *Bt* cotton continues to be effective, growers reduced insecticide use by 70% and saved more than $200 million from 1996 to 2008 (Naranjo and Ellsworth, 2009).

Planting of *Bt* crops has also supported another important goal of sustainable agriculture: increased biological diversity. An analysis of 42 field experiments indicates that non-target invertebrates (i.e. insects, spiders, mites and related species that are not pests targeted by *Bt* crops) were more abundant in *Bt* cotton and *Bt* maize fields than in conventional fields managed with insecticides. The conclusion that growing *Bt* crops promotes biodiversity assumes a baseline condition of insecticide treatments, which applied to 23% of maize acreage and 71% of cotton acreage in the USA in 2005 (Marvier *et al.*, 2007).

Benefits of *Bt* crops have also been well documented in less developed countries. For example, Chinese and Indian farmers growing genetically engineered cotton or rice were able to dramatically reduce their use of insecticides (Huang et al., 2002, 2005; Qaim and Zilberman, 2003; Bennett et al., 2006). In a study of pre-commercialisation use of genetically engineered rice in China, these reductions were accompanied by a decrease in insecticide-related injuries (Huang et al., 2005).

Despite initial declines in insecticide use associated with Bt cotton in China, a survey of 481 Chinese households in five major cotton-producing provinces indicates that insecticide use on Bt cotton increased from 1999 in 2004, resulting in only 17% fewer sprays on Bt cotton compared with non-Bt cotton in 2004 (Wang et al., 2008). A separate survey of 38 locations in six cotton-producing provinces in China showed that the number of sprays on all cotton fields dropped by about 20% from 1996 (before widespread cultivation of Bt cotton) to 1999 (2 years after widespread cultivation of Bt cotton) (Lu et al., 2010). This study also indicated a slight increase in insecticide use on all cotton fields from 1999 to 2008.

One limitation of using any insecticide, whether it is organic, synthetic or genetically engineered, is that insects can evolve resistance to it. For example, one crop pest, the diamondback moth (Plutella xylostella), has evolved resistance to Bt toxins under open field conditions. This resistance occurred in response to repeated sprays of Bt toxins to control this pest on conventional (non-genetically engineered) vegetable crops (Tabashnik, 1994).

Based partly on the experience with diamondback moth and because Bt crops cause season-long exposure of target insects to Bt toxins, some scientists predicted that pest resistance to Bt crops would occur in a few years. However, global pest monitoring data suggest that after more than a dozen years of widespread Bt crop use, Bt crops have remained effective against most pests (Tabashnik et al., 2008, 2009). Nonetheless, resistance to Bt crops has been reported in some field populations of at least four major species of target pests (Tabashnik et al., 2009; Bagla, 2010; Carriere et al., 2010; Storer et al., 2010).

Retrospective analyses suggest that the 'refuge strategy' – growing refuges of crop plants that do not make Bt toxins to promote survival of susceptible insects – has helped to delay evolution of pest resistance to Bt crops (Tabashnik et al., 2008, 2009). The theory underlying the refuge strategy is that most of the rare resistant pests surviving on Bt crops will mate with abundant susceptible pests from refuges of host plants without Bt toxins. If inheritance of resistance is recessive, the hybrid offspring produced by such matings will be killed by Bt crops, markedly slowing the evolution of resistance.

Despite the success of the refuge strategy in delaying insect resistance to Bt crops, this approach has limitations, including variable compliance by farmers with the requirement to plant refuges of non-Bt host plants. An alternative strategy, where refuges are scarce or absent, entails release of sterile insects to mate with resistant insects (Tabashnik et al., 2010). Incorporation of this strategy in a multi-tactic eradication programme in Arizona from 2006 to 2009 reduced pink bollworm abundance by more than 99%, while eliminating insecticide sprays against this pest. The success of such creative multidisciplinary integrated approaches, involving entomologists, geneticists, physiologists, biochemists and ecologists, provides a roadmap for the future of agricultural production and attests to the foresight of Rachel Carson.

HERBICIDE-TOLERANT CROPS

Weeds are a major limitation of crop production globally, as they compete for nutrients and sunlight. One method to control weeds is to spray herbicides that kill them. Many of the herbicides used over the last 50 years are classified as toxic or slightly toxic to animals and humans (classes I, II and III). Some newer herbicides, such as glyphosate (trade name 'Roundup'), are considered non-toxic (class IV). Roundup is essentially a modified amino acid that blocks a chloroplast enzyme (called 5-enolpyruvoyl-shikimate-3-phosphate synthetase, or EPSPS) required for plant, but not animal, production of tryptophan. Glyphosate has a very low acute toxicity, is not carcinogenic, breaks down quickly in the environment and thus does not persist in groundwater.

Some crop plants have been genetically engineered for tolerance to glyphosate. In these herbicide-tolerant crops, a gene, isolated from the bacterium Agrobacterium encoding an EPSPS protein resistant to glyphosate, is engineered into the plant. Growers of herbicide-tolerant crops can spray glyphosate to control weeds without harming their crop.

Although herbicide-tolerant crops do not directly benefit organic farmers, who are prohibited from using herbicides, or poor farmers in developing countries, who often cannot afford to buy the herbicides, there are clear advantages to conventional growers and to the environment in developed countries.

One important environmental benefit is that the use of glyphosate has displaced the use of more toxic (classes I, II and III) herbicides (Fernandez-Cornejo and Caswell, 2006). For example, in Argentina, soya bean farmers using herbicide-tolerant crops were able to reduce their use of toxicity class II and III herbicides by 83–100%. In North Carolina, the pesticide leaching was 25% lower in herbicide-tolerant cotton fields compared with those having conventional cotton (Carpenter, 2010).

Before the advent of genetically engineered soya bean, conventional soya bean growers in the USA applied the more toxic herbicide metolachlor (Class III) to control weeds. Metolachlor, known to contaminate groundwater, is included in a class of herbicides with suspected toxicological problems. Switching from metolachlor to glyphosate in soya bean production has had large environmental benefits as measured in environmental impact and likely health benefits for farmworkers (Fernandez-Cornejo and McBride, 2002).

In the Central Valley of California, most conventional alfalfa farmers use diuron (class III) to control weeds. Diuron, which also persists in groundwater, is toxic to aquatic invertebrates (EXTOXNET – Extension Toxicology Network, 1996). Planting of herbicide-tolerant alfalfa varieties is therefore expected to improve water quality in the valley and enhance biodiversity (Strandberg and Pederson, 2002). Further planting of herbicide-tolerant alfalfa varieties is pending a decision by USDA's Animal and Plant Health Inspection Service (APHIS), which has recently prepared a final environmental impact statement (EIS) evaluating the potential environmental effects of planting this crop (USDA APHIS, 2010).

Another benefit in terms of sustainable agriculture is that herbicide-tolerant maize and soya bean have helped foster use of low-till and no-till agriculture, which leaves the fertile topsoil intact and protects it from being removed by wind or rain. Thus, no-till methods can improve water quality and reduce soil erosion. Also, because tractor-tilling is minimised, less fuel is consumed and greenhouse gas emissions are reduced (Farrell et al., 2006; Committee on the Impact of Biotechnology on Farm-Level Economics and Sustainability and National Research Council, 2010). In Argentina and the USA, the use of herbicide-tolerant soya beans was associated with a 25–58% decrease in the number of tillage operations (Carpenter, 2010). Such reduced tillage practices correlate with a significant reduction in greenhouse gas emissions which, in 2008, was equivalent to removing nearly 6.9 million cars, equal to about 26% of all registered private cars in the United Kingdom (Brookes and Barfoot, 2010).

One drawback to the application of herbicides is that overuse of a single herbicide can lead to the evolution of weeds that are resistant to that herbicide. The evolution of resistant weeds has been documented for herbicide-tolerant traits developed through selective breeding, mutagenesis and genetic engineering. To mitigate the evolution of weed resistance and prolong the usefulness of herbicide-tolerant crops, a sustainable management system is needed. Such approaches require switching to another herbicide or mixtures of herbicides or employing alternative weed control methods (Committee on the Impact of Biotechnology on Farm-Level Economics and Sustainability and National Research Council, 2010). Implementation of a mandatory crop diversity strategy would also greatly reduce weed resistance. Newer herbicide-tolerant varieties will have tolerance to more than one herbicide, which will allow easier herbicide rotation or mixing, and, in theory, help to improve the durability of effectiveness of particular herbicides.

In addition to environmental issues, economic issues related to pollen flow between genetically engineered, non-genetically engineered and organic crops and to compatible wild relatives are also important to discussions of herbicide tolerance due to possible gene flow. These issues are addressed in the USDA report on genetically engineered alfalfa and are also discussed in other reviews (Ronald and Adamchak, 2008; McHughen and Wager, 2010; USDA APHIS, 2010).

VIRAL-RESISTANT CROPS

Although *Bt* and herbicide-tolerant crops are by far the largest acreage of genetically engineered crops on the market, other genetically engineered crops have also been commercialised and proven to be effective tools for sustainable agriculture. For example, in the 1950s, the entire papaya production on the island of Oahu in the Hawaiian chain was decimated by papaya ringspot virus (PRSV), a potyvirus with single-stranded RNA. Because there was no way to control PRSV, farmers moved their papaya production to the island of Hawaii where the virus was not yet present. By the 1970s, however, PRSV was discovered in the town of Hilo, just 20 miles away from the papaya-growing area, where 95% of the state's papaya was grown. In 1992, PRSV had invaded the papaya orchards and by 1995 the disease was widespread, creating a crisis for Hawaiian papaya farmers.

In anticipation of disease spread, Dennis Gonsalves, a local Hawaiian, and co-workers initiated a genetic strategy to control the disease (Tripathi *et al.*, 2006). This research was spurred by an earlier observation that transgenic tobacco expressing the coat protein gene from tobacco mosaic virus showed significant delay in disease symptoms caused by tobacco mosaic virus (Powell-Abel *et al.*, 1986). Gonsalves's group engineered papaya to carry a transgene from a mild strain of PRSV.

Conceptually similar (although mechanistically different) to human vaccinations against polio or smallpox, this treatment 'immunised' the papaya plant against further infection. The genetically engineered papaya yielded 20 times more papaya than the non-genetically engineered variety after PRSV infection. By September 1999, 90% of the Hawaiian farmers had obtained genetically engineered seeds, and 76% of them had planted the seeds. After release of genetically engineered papaya to farmers, production rapidly increased from 26 million pounds in 1998 to a peak of 40 million pounds in 2001. Today, 80–90% of Hawaiian papaya is genetically engineered. There is still no conventional or organic method to control PRSV. Funded mostly by a grant from the USDA, the project cost about US$60000, a small sum compared to the amount the papaya industry lost between 1997 and 1998, prior to introduction of the genetically engineered papaya.

GENETICALLY ENGINEERED CROPS ON THE HORIZON

Peer-reviewed studies of the genetically engineered crops currently on the market indicate that such crops have contributed to enhancing global agricultural sustainability. As reviewed here, benefits include massive reductions in insecticides in the environment (Qaim and Zilberman, 2003; Huang et al., 2005), improved soil quality and reduced erosion (Committee on the Impact of Biotechnology on Farm-Level Economics and Sustainability and National Research Council, 2010), prevention of destruction of the Hawaiian papaya industry (Tripathi et al., 2006), enhanced health benefits to farmers and families as a result of reduced exposure to harsh chemicals (Huang et al., 2002, 2005), economic benefits to local communities (Qaim et al., 2010), enhanced biodiversity of beneficial insects (Cattaneo et al., 2006), reduction in the number of pest outbreaks on neighbouring farms growing non-genetically engineered crops (Hutchison et al., 2010) and increased profits to farmers (Tabashnik, 2010). Genetically engineered crops have also dramatically increased crop yields – more than 30% in some farming communities (Qaim et al., 2010). As has been well documented for Bt cotton in Arizona, the ability to combine innovations in farming practice with the planting of genetically engineered seed has had a huge positive benefit/cost ratio, far beyond what could be achieved by innovating farming practices or planting genetically engineered crops alone. The benefit/cost ratio of Bt crops is the highest for any agricultural innovation in the last 100 years.

There are dozens of useful genetically engineered traits in the pipeline. These include nitrogen use efficiency (Arcadia Biosciences, 2010). Success of crops enhanced in these efficiencies would reduce water eutrophication caused by nitrogenous compounds in fertilisers and greenhouse gas emission resulting from the energy required to chemically synthesise fertilisers.

The USDA APHIS has developed a transgenic plum variety, the 'HoneySweet', which is resistant to plum pox, a plant disease that infects plum and other stone fruit trees, including peach, nectarine, plum, apricot and cherries. Although plum pox is very rare in the USA, and the few outbreaks are immediately eradicated, the 'HoneySweet' is a precautionary measure to avoid a major disruption in the availability of plums, prunes and other stone fruits should it become widespread as is already the case in Europe (USDA APHIS, 2009).

Other promising applications of genetic engineering are those that affect staple food crops. For example, rice is grown in more than 114 countries on six of the seven continents. In countries where rice is the staple food, it is frequently the basic ingredient of every meal. Thus, even modest changes in tolerance to environmental stress or enhanced nutrition can have a large impact in the lives of the poor. The development of food crops with enhanced nutrition is discussed by Adrian Dubock in Chapter 12 of this book, 'Nutritional enhancement by bio-fortification of staple crops'.

The development of submergence tolerant rice (Sub1-rice), through a non-genetically engineered process that involved gene cloning and precision breeding, demonstrates the power of genetics to improve tolerance to environmental stresses such as flooding, which is a major constraint to rice production in South and Southeast Asia (Xu et al., 2006). In Bangladesh and India, 4 million tons of rice, enough to feed 30 million people, are lost each year to flooding. Planting of Sub1-rice has resulted in three- to fourfold yield increases in farmers' fields during floods compared to conventional varieties. In 2011, 1 million farmers planted Sub1-rice. Although the Sub1 varieties provided an excellent immediate solution for most of the submergence-prone areas, a higher and wider range of tolerance is required for severe conditions and longer periods of flooding. With increasing global warming, unusually heavy rainfall patterns are predicted for rain-fed as well as irrigated agricultural systems. For these reasons, we and others have identified additional genes that improve tolerance (Seo et al., 2011). Such genes may be useful for the development of 'Sub1^{plus}' varieties.

In Africa, three-quarters of the world's severe droughts have occurred over the past 10 years. The introduction of genetically engineered drought-tolerant maize, the most important African staple food crop, is predicted to dramatically increase yields for poor farmers (African Agricultural Technology Foundation, 2010). Drought-tolerant maize will be broadly beneficial across almost any non-irrigated agricultural situation and in any management system. Drought-tolerance technologies are likely to benefit other agricultural crops for both developed and developing countries.

CONCLUSION

For hundreds of years, farmers have relied on genetically improved seed to enhance agricultural production. Without the development of high-yielding crop varieties over recent decades, two to four times more land would have been needed in the USA, China and India to produce the same amount of food. Looking ahead, without additional yield increases, maintaining current per capita food consumption will necessitate a near doubling of the world's cropland area by 2050. By comparison, raising global average yields to those currently achieved in North America could result in very considerable *sparing* of land (Waggoner, 1995; Green *et al.*, 2005). Because substantial greenhouse gases are emitted from agricultural systems, and because the net effect of higher yields is a dramatic reduction in carbon emissions (Burney *et al.*, 2010), development and deployment of high-yielding varieties will be a critical component of a future sustainable agriculture.

Thus, a key challenge is to raise global yields without further eroding the environment. Recent reports on food security emphasise the gains that can be made by bringing existing agronomic and food science technology and know-how to people who do not yet have it. These reports also highlight the need to explore the genetic variability in our existing food crops and to develop new genetic approaches that can be used to enhance more ecologically sound farming practices (Naylor *et al.*, 2007; World Bank, 2007; Royal Society, 2009).

Despite the demonstrated importance of genetically improved seed, there are still agricultural problems that cannot be solved by improved seed alone, even in combination with innovative farming practices. A premise basic to almost every agricultural system (conventional, organic and everything in between) is that seed can only take us so far. Ecologically based farming practices used to cultivate the seed, as well as other technological changes and modified government policies, clearly are also required.

In many parts of the world, such policies involves building local educational, technical and research capacity, food processing capability, storage capacity and other aspects of agribusiness, as well as rural transportation and water and communications infrastructure. The many trade, subsidy, intellectual property and regulatory issues that interfere with trade and inhibit the use of technology must also be addressed to assure adequate food availability to all. Despite the complexity of many of these interrelated issues, it is hard to avoid the conclusion that ecological farming practices using genetically engineered seed will play an increasingly important role in a future sustainable agriculture.

Fourteen years of extensive field studies (Carpenter, 2010) have demonstrated that genetically engineered crops are tools that, when integrated with optimal management practices, help make food production more sustainable. The vast benefits accrued to farmers, the environment and consumers explain the widespread popularity of the technology in many regions of the world.

The path towards a future sustainable agriculture lies in harnessing the best of all agricultural technologies, including the use of genetically engineered seed, within the framework of ecological farming.

ACKNOWLEDGEMENTS

I am grateful to Peggy Lemaux, Kent Bradford and Bruce Tabashnik for helpful discussions and critical review of the manuscript. This work was supported by NIH grant GM055962 and the US Department of Energy, Office of Science, Office of Biological and Environmental Research, through contract DE-AC02–05CH11231 between Lawrence Berkeley National Laboratory and the US Department of Energy. An earlier version of this manuscript was published in *Genetics*.

REFERENCES

African Agricultural Technology Foundation (2010). *Scientists Prepare for Confined Field Trials of Life-Saving Drought-Tolerant Transgenic Maize.* African Agricultural Technology Foundation, Nairobi, Kenya.

Arcadia Biosciences (2010). *Nitrogen Use Efficient Crops.* www.arcadiabio.com/nitrogen.php

Bagla, P. (2010). Hardy cotton-munching pests are latest blow to GM crops. *Science* **327**, 1439.

Bennett, R. M., Kambhampati, U. S., Morse, S., and Ismael, Y. (2006). Farm-level economic performance of genetically modified cotton in Maharashtra, India. *Review of Agricultural Economics* **28**, 59–71.

Board on Agriculture and Natural Resources, Committee on Environmental Impacts Associated with Commercialization of Transgenic Plants, National Research Council and Division on Earth and Life Studies (eds.) (2002). *Environmental Effects of Transgenic Plants: The Scope and Adequacy of Regulation.* National Academies Press, Washington, DC.

Brookes, G., and Barfoot, P. (2010). Global impact of biotech crops: environmental effects, 1996–2008. *AgBioForum* 13, 76–94.

Burney, J. A., Davis, S. J., and Lobell, D. B. (2010). Greenhouse gas mitigation by agricultural intensification. *Proceedings of the National Academy of Science USA* 107, 12052–12057.

Carpenter, J. E. (2010). Peer-reviewed surveys indicate positive impact of commercialized GM crops. *Nature Biotechnology* 28, 319–321.

Carriere, Y., Crowder, D. W., and Tabashnik, B. E. (2010). Evolutionary ecology of insect adaptation to Bt crops. *Evolutionary Applications* 3, 561–573.

Cattaneo, M. G., Yafuso, C., Schmidt, C., *et al.* (2006). Farm-scale evaluation of the impacts of transgenic cotton on biodiversity, pesticide use and yield. *Proceedings of the National Academy of Sciences USA* 103, 7571–7576.

Coakley, S. M., Scherm, H., and Chakraborty, S. (1999). Climate change and plant disease management. *Annual Review of Phytopathology* 37, 399–426.

Committee on Identifying and Assessing Unintended Effects of Genetically Engineered Foods on Human Health, and National Research Council (eds.) (2004). *Safety of Genetically Engineered Foods: Approaches to Assessing Unintended Health Effects.* National Academies Press, Washington, DC.

Committee on the Impact of Biotechnology on Farm-Level Economics and Sustainability, and National Research Council (eds.) (2010). *Impact of Genetically Engineered Crops on Farm Sustainability in the United States.* National Academies Press, Washington, DC.

European Commission Directorate-General for Research and Innovation (2010). *A Decade of EU-Funded GMO Research 2001–2010.* European Commission, Brussels, Belgium.

European Commission Joint Research Centre (2008). *Scientific and Technical Contribution to the Development of an Overall Health Strategy in the Area of GMOs.* European Commission, Brussels, Belgium.

European Food Safety Authority (2004). *Opinion of the Scientific Panel on Genetically Modified Organisms EFSA-Q-2003–121.* European Commission, Brussels, Belgium.

EXTOXNET – Extension Toxicology Network (1996). Pesticide information profile: diuron. In USDA Extension Service (ed.) *Pesticide Information Notebook.* Cornell University, Oregon State University, University of Idaho, University of California at Davis and Institute for Environmental Toxicology, Michigan State University.

Farrell, A. E., Plevin, R. J., Turner, B. T., *et al.* (2006). Ethanol can contribute to energy and environmental goals. *Science* 311, 506–508.

Fernandez-Cornejo, J., and Caswell, M. (2006). *The First Decade of Genetically Engineered Crops in the United States,* Economic Information Bulletin No. EIB-11. US Department of Agriculture, Washington, DC.

Fernandez-Cornejo, J., and McBride, W. D. (2002). *Adoption of Bioengineered Crops,* Agricultural Economic Report No. AER810. US Department of Agriculture, Washington, DC.

Green, R. E., Cornell, S. J., Scharlemann, J. P. W., and Balmford, A. (2005). Farming and the fate of wild nature. *Science* 307, 550–555.

Huang, J., Rozelle, S., Pray, C., and Wang, Q. (2002). Plant biotechnology in China. *Science* 295, 674–676.

Huang, J., Hu, R. Rozelle, S., and Pray, C. (2005). Insect-resistant GM rice in farmers' fields: assessing productivity and health effects in China. *Science* 308, 688–690.

Hutchison, W. D., Burkness, E. C., Mitchell, P. D., *et al.* (2010). Areawide suppression of European corn borer with *Bt* maize reaps savings to non-*Bt* maize growers. *Science* 330, 222–225.

Lobell, D.B., Schlenker, W., and Costa-Roberts, J. (2011). Climate trends and global crop production since 1980. *Science* 333, 616–620.

Lu, Y., Wu, K., Jiang, Y., *et al.* (2010). Mirid bug outbreaks in multiple crops correlated with wide-scale adoption of *Bt* cotton in China. *Science* 328, 1151.

Marvier, M., McCreedy, C., Regetz, J., and Kareiva, P. (2007). A meta-analysis of effects of *Bt* cotton and maize on nontarget invertebrates. *Science* 316, 1475–1477.

McHughen, A., and Wager, R. (2010). Popular misconceptions: agricultural biotechnology. *New Biotechnology* 27, 724–728.

Mendelsohn, M., Kough, J., Vaituzis, Z., and Matthews, K. (2003). Are *Bt* crops safe? *Nature Biotechnology* 21, 1003–1009.

Naranjo, S. E., and Ellsworth, P. C. (2009). Fifty years of the integrated control concept: moving the model and implementation forward in Arizona. *Pest Management Science* 65, 1267–1286.

Naylor, R., Falcon, W., and Fowler, C. (2007). *The Conservation of Global Crop Genetic Resources in the Face of Climate Change*. Rockefeller Foundation, Bellagio, Italy.

Powell-Abel, P., Nelson, R. S., De, B., *et al.* (1986). Delay of disease development in transgenic plants that express the tobacco mosaic virus coat protein gene. *Science* 232, 738–743.

Qaim, M., and Zilberman, D. (2003). Yield effects of genetically modified crops in developing countries. *Science* 299, 900–902.

Qaim, M., Subramanian, A., and Sadashivappa, P. (2010). Socioeconomic impacts of *Bt* (*Bacillus thuringiensis*) cotton. *Biotechnology in Agriculture and Forestry* 65, 221–224.

Ronald, P., and Adamchak, R. (2008). *Tomorrow's Table: Organic Farming, Genetics and the Future of Food*. Oxford University Press, New York.

Royal Society (2009). *Reaping the Benefits: Science and the Sustainable Intensification of Global Agriculture*. The Royal Society, London.

Schlenker, W., and Roberts, M. J. (2009). Nonlinear temperature effects indicate severe damages to US crop yields under climate change. *Proceedings of the National Academy of Sciences USA* 106, 15594–15598.

Seo, Y. S., Chern, M., Bartley, L. E., *et al.* (2011). Towards a rice stress response interactome. *PloS Genetics* 7, e1002020.

Somerville, C., and Briscoe, J. (2001). Genetic engineering and water. *Science* 292, 2217.

Storer, N. P., Babcock, J. M., Schlenz, M., *et al.* (2010). Discovery and characterization of field resistance to *Bt* maize: *Spodoptera frugiperda* (Lepidoptera: Noctuidae) in Puerto Rico. *Journal of Economic Entomology* 103, 1031–1038.

Strandberg, B., and Pederson, M. B. (2002). *Biodiversity of Glyphosate Tolerant Fodder Beet Fields. Timing of Herbicide Application.* National Environmental Research Institute, Silkeborg, Sweden.

Tabashnik, B. E. (1994). Evolution of resistance to *Bacillus thuringiensis*. *Annual Review of Entomology* 39, 47–79.

Tabashnik, B. E. (2010). Communal benefits of transgenic corn. *Science* 330, 189–190.

Tabashnik, B., Gassman, A. J., Crowder, D. W., and Carriere, Y. (2008). Insect resistance to *Bt* crops: evidence versus theory. *Nature Biotechnology* 26, 199–202.

Tabashnik, B. E., Van Rensburg, J. B. J., and Carrière, Y. (2009). Field-evolved insect resistance to Bt crops: definition, theory, and data. *Journal of Economic Entomology* **102**, 2011–2025.

Tabashnik, B. E., Sisterson, M. S., Ellsworth, P. C., *et al.* (2010). Suppressing resistance to Bt cotton with sterile insect releases. *Nature Biotechnology* **28**, 1304–1307.

Tripathi, S., Suzuki, J., and Gonsalves, D. (2006). Development of genetically engineered resistant papaya for papaya ringspot virus in a timely manner: a comprehensive and successful approach. *Methods in Molecular Biology* **354**, 197–240.

USDA APHIS (2009). *HoneySweet Plum Trees: A Transgenic Answer to the Plum Pox Problem*. USDA, Washington, DC. www.ars.usda.gov/is/br/plumpox/

USDA APHIS (2010). *USDA Announces Final Environmental Impact Statement for Genetically Engineered Alfalfa*. USDA, Washington, DC. www.aphis.usda.gov/biotechnology/alfalfa_documents.shtml

Waggoner, P. E. (1995). How much land can ten billion people spare for nature? Does technology make a difference? *Technology in Society* **17**, 17–34.

Wang, S., Just, D. R., and Pinstrup-Anderson, P. (2008). Bt-cotton and secondary pests. *International Journal of Biotechnology* **10**, 113–120.

World Bank (2007). Agriculture for development. In *World Development Report*, The World Bank, Washington, DC.

Wu, K.-M., Lu, Y.-H., Feng, H.-Q., *et al.* (2008). Suppression of cotton bollworm in multiple crops in China in areas with Bt toxin-containing cotton. *Science* **321**, 1676–1678.

Xu, K., Xu, X., Fukao, T., *et al.* (2006). Sub1A encodes an ethylene responsive-like factor that confers submergence tolerance to rice. *Nature* **442**, 705–708.

KEY FURTHER RESOURCES

Ronald, P., and Adamchak, R. (2008). *Tomorrow's Table: Organic Farming, Genetics and the Future of Food*. Oxford University Press, New York, NY.

Brand, S. (2010). *Whole Earth Discipline: Why Dense Cities, Nuclear Power, Transgenic Crops, Restored Wildlands, and Geoengineering Are Necessary*. Penguin, London.

Specter, M. (2009). *Denialism: How Irrational Thinking Hinders Scientific Progress, Harms the Planet, and Threatens Our Lives*. Penguin, London.

McWilliams, J. (2009). *Just Food: Where Locavores Get It Wrong and How We Can Truly Eat Responsibly*. Little, Brown, Boston, MA.

Ronald, P. (2008). The new organic, *The Boston Globe*, 16 March. http://www.boston.com/bostonglobe/ideas/articles/2008/03/16/the_new_organic/?page=full

12

Nutritional enhancement by biofortification of staple crops

If undernutrition were a disease, such as H1N1, and unprocessed food were a drug or a vaccine, both would have the full attention of the entire international community.

Lancet **374**, 1473 (31 October 2009)

HISTORICAL CONTEXT

It is widely accepted that the human population's transition from hunter–gatherers to cultivators of plants started to occur in Mesopotamia (now Iraq) about 10000 years ago. From then until now, farmers have been selecting the best crop plants from which to breed and produce better plants. More recently, until about 1940, maize (corn) yields increased only slowly in the USA, and area cultivated increased, raising overall production (Figure 12.1). Since about 1940, increasingly more sophisticated technology has been applied to crop production in most regions of the world.

In the mid-1960s Norman Borlaug in the Green Revolution helped many countries through plant breeding to achieve higher staple crop yields and thereby offset the effects of fast population growth. The very easy international transfer of seed between countries facilitated the rapid progress of breeding programmes and benefit-sharing between the collaborating countries (Hesser, 2006). This is as it should be for a 'public good'. In the mid-1960s the public good was a dwarf wheat variety from Mexico resistant to fungal disease. We need to be able to

Successful Agricultural Innovation in Emerging Economies: New Genetic Technologies for Global Food Production, eds. David J. Bennett and Richard C. Jennings. Published by Cambridge University Press. © Cambridge University Press 2013.

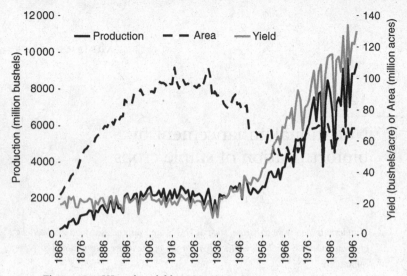

Figure 12.1 US maize yield, 1866–1996.

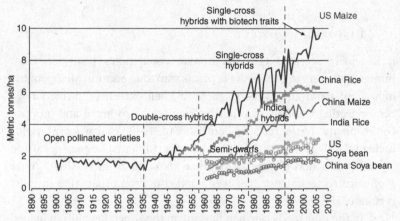

Figure 12.2 Science and innovation as drivers of yield improvement.
(Source Data from USDA, IRRI & FAO © PHII)

internationally cooperate as easily now as in the 1960s to biofortify staple foods, whatever the technology used to create them.

The new cultural practices introduced since the 1940s include the use of herbicides, insecticides, fungicides and fertiliser as well as selective breeding and, in the last 15 years, genetic tools and manipulation, including, but not limited to, genetic engineering. Genetic improvement is considered responsible for about 50% of US maize yield increase per hectare. In the UK, the breeder's contribution to wheat yield may be as much as 90% (Pers.commun. Stephen Smith, Research Fellow at Pioneer).

Table 12.1 *Per capita food production as a share of per capita production in 1961–5, by world region*

Continent	1961–5	1971	1981	1991	2001
Africa	100	103	94	90	90
Asia	100	104	114	134	173
South America	100	100	115	118	144
World	100	107	112	115	126

These technologies have been applied in a number of countries and crops (Figure 12.2).

However, not all regions have benefited from these technologies. In contrast to other regions, in Africa with the lowest use of improved seed and fertiliser (and presumably other agricultural inputs), per capita food production declined in the 40 years to 2001 (Table 12.1, Southgate and Graham, 2006).

Until recently the preferred crop improvements have focused on harvestable yield. In grain crops, such as wheat, maize and rice, and many other staple crops such as cassava and plantain (cooking banana), yield has been mainly characterised by the carbohydrate content of the seed or plant part consumed. Yield is important for macronutrient nutrition and is easily measured by plant breeders: yield is the usefully harvested sum effect of all the variables, biotic and abiotic (that is, caused by living organisms or not) affecting a plant during its growth and maturation before harvest. (Post-harvest losses of up to 30%, and in some societies food wastage after preparation, are also important, but another story.)

Carbohydrates in the diet are an important source of the energy needed for life. However, animals, including humans, need more than carbohydrates alone. Proteins, fats and micronutrients are also important. And often staple crops contain little or no essential micronutrients in the storage tissues which are used as human food.

MICRONUTRIENTS

As early as 1881 in Switzerland the existence of what are now known as micronutrients in food were postulated following animal experiments with milk constituents. In the early part of the twentieth century science started to investigate and understand the role of vitamins in human health resulting in the Nobel Prize for Medicine to two scientists, Eijkman and Hopkins, in 1929 in connection with vitamins (Hopkins, 1929).

Figure 12.3 Goitre on a gargoyle (Freiburg, Germany).

Humans obtain their nutrients from the food they consume. Micro-nutrients – vitamins (if they are of organic origin) and minerals (when inorganic) – are essential for optimal human development and health, although the levels at which they are needed are so low as to contribute essentially nothing as an energy source. Plants synthesise vitamins, or their precursors, necessary for human nutrition. Minerals such as iron and zinc are absorbed from the environment in which the plant grows. Both vitamins and minerals may be accumulated and stored in specific plant tissues, and their concentration may be related to the growth stage of the plant or its seeds. Of course the plants produce or absorb these chemicals to suit the plants' life needs, but humans (and other animals) have evolved to take advantage of this chemical diversity for their separate needs.

Iodine insufficiency adversely impacts physical and mental development, those affected being known as cretins. In the Middle Ages goitre, involving enlargement of the thyroid gland due to iodine insufficiency (as under the neck of the female in Figure 12.3) was common in middle Europe until sea salt with traces of iodine became common.

The early micronutrient deficiency diseases that were recognised and addressed included beriberi (deficiency in thiamine, vitamin B_1), scurvy (vitamin C deficiency), rickets (vitamin D deficiency) and pellagra (deficiency in vitamin B_3). Connections with dietary deficiencies were noted, and these diseases make few headlines today. However, there is evidence that rickets is becoming of concern again in some societies as a result of too little sunlight, due to cultural influences especially (Robinson et al., 2006).

Sufficient folic acid (vitamin B_9) in the diet in very early pregnancy, even before the mother is aware she is pregnant, is very important to prevent neural tube defects in the developing fetus which may result in the many deformities in children. The best source of folate is green vegetables, often inaccessible to impoverished and marginalised populations. Many countries have problems with this deficiency.

Iron deficiency is the most prevalent and serious global deficiency, affecting about 50% of the global population and a greater percentage in the developing countries. Its effects on physical and mental development may be debilitating, and usually irreversible. Amelioration remains challenging despite the promise of some fortification and conventional breeding and genetic engineering approaches for biofortified staple crops.

Vitamin A deficiency is also extremely widespread in developing countries, and much data has been gathered about its impact as well as the reduction in its impact by current interventions (WHO, 2005; West et al., 2010). The most recent World Health Organization estimates (2009 and 2012) suggest 190 to 250 million people, including young mothers but predominantly children less than 5 years old, are vitamin A deficient. The deficiency, caused by insufficient coloured fruit, and/or vegetables and/or animal products in the diet, is the largest cause of childhood blindness (Whitcher, Srinivasan and Upadhyay, 2001), and separately increases mortality due to common diseases as a result of immune function suppression. Careful meta-analysis of the impact of vitamin A supplementation has proven how useful it is, reducing all-cause child mortality under 5 years by 24% (Mayo-Wilson et al., 2011) to 34% (Fawzi et al., 1993). With the latest (2010) estimate of global child deaths at nearly 8.1 million in 2009 (You et al., 2010), vitamin A availability for all children in undernourished settings could potentially prevent around 1.9–2.7 million child deaths annually (Tang et al., 2012). By any measure vitamin A deficiency is a serious public health problem (Table 12.2).

And the direct mortality due to vitamin A deficiency, not to mention its impact as the largest cause of childhood blindness, is of

Table 12.2 *Vitamin A deficiency disorders (numbers affected annually)*

Children ~170–230 million
Pregnant women ~20 million
Xerophthalmia
children ~3 million
pregnant women ~3 million
Blindness ~1/2 million (of which ~2/3 die within a few months)
Preventable deaths ~1.9–2.7 million

Source: Adapted from A. Sommer (pers. comm.) following Tang *et al.* (2012).

Table 12.3 *Global population mortality*

Cause	Annual mortality (millions)
Vitamin A deficiency	1.9–2.7[a]
HIV/AIDS[b]	1.7
Tuberculosis[b]	1.4
Malaria[b]	0.75

[a] From Tang *et al.* (2012).
[b] WHO 2010 data from website.

the same order of magnitude as other important public-health problems of the developing world (Table 12.3).

Dietary diversity including foods from both plant and animal sources is the ideal for all populations to avoid nutritional deficiencies. Most people in wealthy and educated populations enjoy ample dietary diversity. Additionally, many members of these populations choose to take dietary supplements of vitamins or minerals, perhaps influenced by marketing as much as by nutritional science. Conversely, many people – around 2 billion of the world's 7 billion population – are deprived of sufficient variety in their diets. Such deprivation can arise for a number of reasons, but is often associated with educational, and especially financial, impoverishment. Poor people are less likely in general to have a well-balanced diet, and developing countries have a higher percentage of poor in the population. Accordingly, malnutrition – both macronutrient and micronutrient – tends to be associated today with developing countries.

A regular meeting of Nobel Laureate economists – the Copenhagen Consensus – has consistently ranked in each of their three meetings over the past eight years micronutrient provision as the highest or second highest priority for cost-effective solutions to the 30 most pressing problems of mankind. Indeed addressing micronutrient malnutrition has usually been listed as two or three of the top five or six priorities on each occasion (Copenhagen Consensus, 2004, 2008, 2012). The micro-nutrients are essential for human development and health in different ways, and there is some data for synergistically beneficial interrelationships between them.

ALLEVIATION OF MICRONUTRIENT DEFICIENCY IN POPULATIONS

Several micronutrient deficiencies cause serious population health problems and related individual and family misery, and significantly reduce productivity. Anticipated high child mortality caused in part by micronutrient deficiencies may influence parents to have more children, increasing strain on already stretched resources. Several approaches, additional to encouraging dietary diversity, have been adopted or have potential for addressing micronutrient deficiencies of populations.

Micronutrient fortification

Industrial fortification (for example iodine in salt, vitamin A and D in margarine, fluoride in toothpaste, folic acid in flour) has been used successfully in both industrialised and developing countries to ensure sufficiency of micronutrients to populations. This type of fortification depends on the industrial processing of food and the distribution infrastructure for the processed food and has to be paid for by someone. However, it can be very effective: in Lao People's Democratic Republic palpable goitre reduced from 40% to 9% of children aged 6 to 12 years in about 10 years after salt was routinely iodised. This result was achieved despite many villages in Lao not being connected by roads. In the USA folate-deficiency-induced neural tube deformities in children were dramatically reduced following folic acid fortification of wheat flour starting in 1998 (Honein, 2001).

Micronutrient supplementation

Supplementation involves provision and consumption of tablets or capsules containing micronutrients.

In Lao PDR, iron tablets as supplements are provided to women during antenatal care visits. In follow-up visits within a year of giving birth less than 14% of mothers had taken the iron tablets, although higher success (~35%) was achieved in urban areas.

To prevent clinical vitamin A deficiency, chemically synthesised vitamin A supplement (globally around 500 million vitamin A in gelatin capsules per year) has been administered periodically since the early 1990s to 'at risk' populations to reduce child and maternal mortality and reduce vision problems. The capsules themselves cost approximately US$0.02 each, but the infrastructure to deliver them costs US$1–2 each. So the whole programme costs $500 million to $1 billion annually (Tang et al., 2012). After almost 20 years of such distribution UNICEF found that vitamin A supplementation programme coverage had stagnated at 58% in 103 countries but coverage was still improving, to 77% by 2009.

Currently significant proportions of the costs of vitamin A supplementation are paid for by US and Canadian government international aid programmes. Another problem for vitamin A supplementation is donor fatigue: after more than 20 years these aid programmes are suggesting to developing countries that they pay for their own vitamin A supplementation programmes. There are calls to abandon vitamin A supplementation programmes and their protection. Undoubtedly improvements in sanitation, water supply, vitamin A supplementation coverage and measles immunisation have reduced the mortality due to vitamin A deficiency. But with more than 5000 preventable deaths every day, it still remains startlingly high.

Vitamin A supplementation does not markedly change the determinants that keep populations deficient. Based on sporadically done serum retinol surveys, there remains no clear indication that vitamin A status has improved through supplementation. If this supplementation were withdrawn tomorrow mortality would almost certainly rise. Current global vitamin A deficiency burden estimates do not consider the tenuous nature of the protection provided by supplementation, and therefore the risks of ceasing funding, in the absence of other deficiency mitigations (K. P. West, pers. comm.).

Apart from difficulties in co operation of nutritionally disadvantaged people with supplementation programmes such programmes depend on an infrastructure for delivery, which has a cost. Often the costs of this infrastructure are reduced by sharing distribution costs with childhood immunisation programmes such as for measles and polio. But some of these diseases are, thankfully, being eliminated by the success of immunisation and eventually can cease, which will increase the cost of micronutrient supplementation.

A concern of conventional supplementation programmes, or even fortification programmes, is that they may miss the most needy, the most inaccessible and marginalised members of society. Even

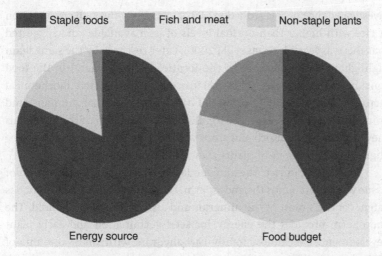

Figure 12.4 Share of energy source and food budget in rural Bangladesh. (www.harvestplus.org/content/food-crisis)

individuals living in the countryside often do not have the resources to access a diverse variety of foods, and are overdependent on the staple food in the area, which offer minimal or none of the essential micronutrients. Non-staple foods, often the most expensive, are the source of most micronutrients in the diet (Figure 12.4 http://www.harvestplus.org/content/food-crisis accessed November 8, 2012).

Biofortification

From the 1940s to the 1970s the international nutritional focus was on macronutrient deficiency, particularly carbohydrates. Since the 1970s there has been an increasing global awareness of the preventable impact of dietary micronutrient deficiency on human health and mortality, and related reduction in the economic productivity of populations. Many meetings of the United Nations and other supranational organisations have addressed the problem of nutrition. Most notably the Johannesburg Millennium summit brought together 129 heads of state to ratify the Millennium Development Goals, many of which relate significantly to better nutrition for the world's poor (United Nations, 2010).

Almost in parallel over the same time period there has been increased attention to plant breeding for increased plant and human nutrition (in the case of iron and zinc) and human nutrition (vitamin A). An important early pilot study with nuns in the Philippines (which has had government policies targeting malnutrition since the mid-1940s) investigated the possible reduction of iron deficiency through consumption of rice with higher than normal levels of iron available, and suggested that this was feasible (Haas et al., 2000). Later research by the same team highlighted the importance of the location of the mineral in the food consumed (Haas et al., 2005). Plants are sophisticated biochemical factories producing or storing all the chemicals that humans and . animals need for life (Beyer, 2010). But the distribution of the chemicals they synthesise or absorb and store within the plant tissues may not be aligned with the parts of plants that humans usually eat.

In the case of rice, the alurone layer of the grain, a thin sheath of tissue which surrounds the endosperm, the part of polished rice which is eaten, contains most of the minerals and also a trace of carotenoid. The endosperm provides the energy for seed germination and early plant development. It is almost totally carbohydrate with small amounts of fat and protein. The germinating rice seed and rice seedling can absorb from the alurone layer the minimal amounts of minerals it needs for its development before its root and leaf systems are developed. Rice has to be polished for storage as a human food or fats in the alurone layer surrounding the endosperm become rancid, tasting bad. Polishing removes the alurone layer. In the study of nuns (Haas et al., 2005), differential polishing was used so that the high-iron rice actually retained more of the relatively iron-rich alurone layer. With the control group of nuns (on a low-iron rice diet) the rice was polished to remove the entire alurone layer.

The nuns study was important research to demonstrate the potential for biofortification. Even modest increases (1.42 mg/day) in consumption of iron with the high-iron rice improved the vitamin A status of individuals in the controlled and randomised trial. But at the same time a challenge was created for the rice breeders (Brooks, 2010). The challenge was that the difference in iron content of the high-iron and normal rice had to be achieved by differential removal of the alurone layer by differential polishing. More alurone layer meant higher iron, but less storage capability for the rice. So for polished rice with normal storage characteristics, the breeders need to target higher iron in the endosperm, the part of rice which is eaten.

The result of these trials and related developments subsequently led to the establishment of the Harvest Plus programme hosted by the International Food Policy Research Institute (IFPRI) in Washington, DC, and the International Centre for Tropical Agriculture (CIAT) in Cali, Columbia. These two are part of the 16-centre research grouping which concentrates on improvements in agriculture for the poor, the Consultative Group for International Agricultural Research (CGIAR).

NUTRITION AND AGRONOMY

Harvest Plus is attempting through breeding to increase levels of nutrients already present in low amounts in the edible parts of staple crops whilst maintaining all the beneficial crop production and management characteristics. The intention is to *'move millions over the threshold from malnourishment to micronutrient sufficiency'* (Bouis et al., 2011). Currently Harvest Plus is breeding iron, zinc and provitamin-A nutrient enriched – biofortified – sweet potato, bean, pearl millet, cassava, maize, rice and wheat, and variously targeting (with the different crops) Uganda, Mozambique, Rwanda, Democratic Republic of Congo, Nigeria, Zambia, Bangladesh, India and Pakistan. Biofortified versions of these Harvest Plus target crops have been or are scheduled for release to growers in the lead countries between 2007 and 2013.

Normally, growers adopt a crop variety for cultivation based on its potential for their economic gain, usually measured by the value of yield to the processor or the wholesaler, retailer or consumer. But a biofortified crop is designed to be consumed by micronutrient-deficient populations. Adoption requires adoption by both growers and consumers, and more challengingly still, for obvious reasons almost simultaneously by both sectors.

What is clear, and was also noted in the nun trials, is that growers do not want better nutrition at the expense of yield. All biofortification projects are well aware that the nutritional traits need to be carried by high-yielding, agronomically normal if not excellent crop varieties. The characteristics of the crop plant cannot suffer without potentially damaging the adoption of the biofortified variety by the consumers it is designed to assist. Fortunately in many developing countries most growers are also consumers, or at least live with them. Urban resource-poor people will need government programmes to get the biofortified crops to them.

For nutritional traits which are invisible, for example iron and zinc, so long as the increased nutrient concentrations in biofortified varieties do not alter the processing, storage or eating qualities of the crop at the appropriate stage of its use, distribution to the population should be straightforward, depending on high-yielding varieties being used to carry the nutritional trait, varieties conforming to local practices, and government policy endorsing the use for population benefit.

Crops that can provide a source of vitamin A are usually coloured and as a result the tissue consumed is often yellow or orange because it contains the provitamin A. Harvest Plus has experience with the orange-fleshed sweet potato in Mozambique (Low *et al.*, 2007) and Uganda (Hotz *et al.*, 2012). Consumers may need a little education which is a challenge to achieve at the very low per capita cost required at a population level, especially with multi-ethnic, possibly multi-dialect, largely illiterate and dispersed populations. Experience with this crop, and studies with other coloured biofortified crops (for example Golden Rice – of which more later) both suggest that neither the colour of the crop nor its method of production are of overriding importance to the consuming public who are much more interested in the nutritional benefit and lack of incremental cost.

WHAT IF CONVENTIONAL BREEDING IS NOT POSSIBLE?

Now, what if conventional breeding is not possible, or there is no source of variation of the targeted nutritional trait from which to breed a biofortified improved variety?

Some crops do not depend on seed for reproduction, but reproduce vegetatively, plantain and cassava being two well-known examples. Their way of reproduction has also inhibited breeding even for agronomically important traits. Nevertheless, these crops are important crops for food security in the developing world.

Also, in some seed-based crops there is none of the desired nutrient at any level in the part of the plant which is eaten by humans. Rice is a good and important example, rice being the most important source of calories for the world's population. Excluding the second most important, wheat, the calories provided by rice to the world's human population are greater than from all the other crops together. Rice provides more than 80% of the calories for half the world's population every day.

Rice has been genetically engineered to produce folate with the potential to address folic acid deficiency in early pregnancy and reduce related neural tube deformities in children (Storozhenko *et al.*, 2007). Regrettably, so far there appears to be no interest in advancing the proof of concept research to practical application in developing countries, which should be straightforward.

In some crops, including rice, there may be some iron, for example, but it may naturally occur in the wrong place for optimal human nutrition without changing processing or consumption behaviour. There is some progress and expectation that sufficiently elevated levels of zinc to match nutritional targets may be achieved through conventional plant breeding of rice.

For reasons of biology, achieving this aim for iron is not so certain. Until recently it has also been considered that the genetic control of iron uptake and storage within the plant is very complex, involving perhaps 12 different genes, and beyond the capabilities of conventional plant breeding and therefore requiring genetic engineering. However, in recent years at least two different groups in Switzerland and Australia have made some progress in raising iron levels of rice endosperm through genetic engineering approaches (Wirth *et al.*, 2009; Johnson *et al.*, 2011). Similarly, high-iron bananas have been created through the adoption of genetic engineering approaches, and the technology is being transferred from Australia to India.

The Australian group has also demonstrated the potential for increasing levels of both iron and zinc in rice by continuously transcribing a gene at a high level rather than it being transcribed at the normal level as needed (Johnson *et al.*, 2011).

GOLDEN RICE

In 1999, just prior to the pilot study for the high-iron study of nuns in the Philippines, two German scientists (Ingo Potrykus and Peter Beyer) made the breakthrough they had been working on for a decade. They and their teams had brought together two, at the time extremely novel, pieces of science (Ye *et al.*, 2000). Beyer's team had figured out what genes were needed to switch on the pre-existing synthetic pathway in the rice endosperm and moreover had targeted the expression to be tissue-specific to endosperm. Potrykus' team had previously managed for the first time to achieve genetic transformation by introducing new genes to *Oryza indica* rice, the species most commonly grown and cultivated in Asia (Datta *et al.*, 1990). Working together and using initially three new genes (now two) the two teams added them to the approximately 30000 genes in the rice genome. The rice endosperm was thus caused to express carotenoids, the precursors which animals, including humans, use to synthesise vitamin A. (At the time all the early commercialised genetically engineered crops from the private sector contained only one introduced gene.) The research team realised their success early in 1999, when yellow-coloured polished grains, indicating the presence of carotenoids, were observed for the first time.

The original idea for a yellow rice to combat vitamin A deficiency came from a rice breeder, Peter Jennings, in a 1984 meeting. Subsequently Potrykus and Beyer started research to try to create such a rice. Other tissues of rice contained carotenoids, but not the grain. As there was no rice variety which contained carotenoids in its endosperm, there was no variation to improve on using conventional breeding technologies. At the time, most people considered the quest would be unsuccessful and so funding for the research was difficult to find, eventually coming from Potrykus' own institute, the Eidgenössische Technische Hochschule Zürich (ETH Zurich), Swiss Federal Funds, the European Commission and the Rockefeller Foundation who had hosted the 1984 meeting referred to above.

The inventors of Golden Rice were always clear from the outset that they wanted to donate their technology free of charge to benefit resource-poor people in countries where vitamin A deficiency was a problem. That was what had motivated their decade of research since the early 1990s. Nevertheless they recognised early on that their scientific skills, great as they were, were insufficient alone to achieve their vision, and that they needed a partner experienced in commercial international agricultural product development.

At the time I was employed by Zeneca and experienced in negotiation and suggested to the inventors that it was possible to integrate the inventors' altruistic and humanitarian purpose with the commercial interests of Zeneca (which became Syngenta). This could be done by assigning the 'market' for humanitarian applications involving resource-poor farmers in developing countries to the inventors, with commercial rights being retained by the company for all markets. In return for formal undertakings to the inventors and their subsequent licensees, being included in licences with Syngenta, the inventors licensed the company with their rights in the Golden Rice technology in a costless, royalty-free transaction. Syngenta obligations entered into included supporting regulatory submissions and donating technology improvements to the inventors for the carefully but generously defined humanitarian uses (Paine et al., 2005; Krattiger and Potrykus, 2007; Potrykus, 2010). Resource-poor farmers do not export rice, and under the terms of their agreement with Syngenta the inventors are not able to grant licences for large-scale commercial exploitation of Golden Rice. The technology improvements obligation is actually a two-way obligation, with the expectation at signature of the agreement that improvements would flow from the inventors and their teams to Syngenta. But later, in 2004 and 2005, Syngenta scientists in pursuit of commercial goals modified the genetic engineering approach leading to higher levels of expressed carotenoids, and a higher proportion of the important carotenoid beta-carotene. Pursuant to the earlier agreement this was supplied for humanitarian purposes (Paine et al., 2005).

Subsequent to agreeing the licence terms the three of us then established a network of public-sector Golden Rice licencees in developing countries wishing to collaborate in the humanitarian aims of the project. We also formed a Humanitarian Board of volunteers from different fields of expertise to assist us in guiding the strategy for Golden Rice and to make the Golden Rice trait available free of charge within national regulations (Potrykus, 2010). A website (www. goldenrice.org) was established to facilitate communication about the project to a very interested global audience and to reduce the onerous task for the inventors of repeatedly answering similar questions from different interested individuals and the press.

Careful agronomic work followed with the public-sector institution licencees of Professor Potrykus (Professor Beyer can also grant licences) at the Indian Agricultural Research Institute, Delhi, and the International Rice Research Institute in the Philippines. The Golden Rice trait was bred into the *Oryza indica* varieties of rice preferred in Asia where most vitamin A deficiency occurs, and the production and management characteristics of the transformed rice were carefully checked. The expression levels of beta-carotene, the most important source of vitamin A (Grune *et al.*, 2010) in the different rice varieties used and from the different genetic transformation events, were also measured.

Much other detailed research was conducted, including, most importantly, research to understand the efficiency of the bioconversion of beta-carotene to vitamin A in the human body (Tang *et al.*, 2009, 2012). This research was led by collaborators from the US Department of Agriculture based in medical schools with an interest in nutrition and carotenoids at Tufts University, Boston and Baylor College of Medicine. The research was mostly funded by the US National Institutes of Health, and also involved, for important aspects involving adults and children, investigators and institutions in China. This research proved that the bioconversion of Golden Rice's beta-carotene to vitamin A is significantly better than any other plant source of beta-carotene so far investigated, and not significantly different from the bioconversion of pure beta-carotene dissolved in oil, so is unlikely to be improved upon. This result is important in reducing the amount of Golden Rice which needs to be consumed to improve the vitamin A status of an individual.

Based on all this research the Golden Rice Humanitarian Board were able in 2009 to choose one lead genetic transformation event to be provided as a public good to all public-sector licencee institutions in countries that wanted access to assist the nutrition of their people – free of any cost, and subject only to applicable regulations and commitment

Figure 12.5 Petri dish of Golden Rice in front of Golden Rice plants.

to the terms of the Humanitarian Licence. One transformation event, and the regulatory package of data accompanying it, being provided free of charge to the countries has long been a Humanitarian Board strategy to facilitate adoption at lowest cost. Most rice is consumed close to where it is grown and Golden Rice is targeted at the resource-poor. Of course countries will wish to introduce the trait into appropriate rice varieties agronomically adapted to the locations of and preferred by the target consumers in their countries.

We are excited by the demonstrated potential for Golden Rice in its current form to significantly reduce vitamin A deficiency in rice consuming societies. Data have been generated proving that Golden Rice can be as effective as preformed vitamin A in capsules or milk or eggs as a source of vitamin A, and suggesting that only 40 g/day of dry Golden Rice – about the amount in the Petri dish in Figure 12.5 – when cooked and consumed will save lives and sight.

More research is planned to demonstrate that daily consumption of Golden Rice will improve the vitamin A status of individuals.

ACCEPTANCE OF GOLDEN RICE

Despite the demonstrated, *ex-ante*, potential of Golden Rice, and the transparency of its not-for-profit humanitarian vision, significant delays to the project are, and have been, caused by a number of factors. These mostly arise because of human suspicion of genetically engineered crops, anti-globalisation sentiments, suspicion of private-sector motives for support, and in some cases fear of activist attack. Much of these sentiments arise in 'rich Europe' with little appreciation of, and less empathy for, continuing agricultural, food and nutrition issues in developing countries.

Vitriolic verbal attacks have occurred, initially from Greenpeace and Friends of the Earth and related groups. Campaigners for the 'Keep Wales GM Free' campaign accused the scientists involved and their institutions as well as the whole Golden Rice Humanitarian Board of 'crimes against humanity' and transgressing the Nuremburg Code because of the research involving Chinese children which was done to establish the bioconversion efficiency of the beta-carotene in Golden Rice to vitamin A. This was in spite of the facts that animal models were not suitable, that beta-carotene at the levels found in food matrices is safe and that the management of such clinical trials follows accepted international regulation and practice by the clinicians and medical institutions involved! We were confident of our actions, but took comfort that two independent ethical reviews agreed with us and disagreed with the anti-GM activists. Even Patrick Moore, the founder of Greenpeace, has accused his former colleagues of crimes against humanity for their opposition to Golden Rice (Moore, 2010).

The suspicion of genetically engineered crops arises from the idea that genetic engineering is a greater danger to environmental and human health than other forms of plant breeding. As the first genetically engineered (as well as the first purposefully created) biofortified crop, Golden Rice is likely to bear more of the burden of opposition than subsequent biofortified crops created with the same technology. There is, of course, zero evidence that crops created using genetic engineering are, as a result of the technology used, any worse or better for human or environmental health than crops bred using other technologies. But this topic is dealt with elsewhere in this book.

We have conducted social marketing research with resource-poor growers and consumers. When the question is: 'Look, you can have this rice, it's yellow, it grows the same as the other rice, it processes, stores, cooks and tastes the same. It doesn't cost more than white rice. And when you or your children eat this much every day, you won't die from lack of vitamin A or go blind. Would you like it?' They say, 'Yes please, we'd like to try it.' They are very interested in nutrition and their family's welfare. They are pleased the trait will not cause the rice to cost more, and not concerned how it is created. Even the colour is unlikely to be a major obstacle to adoption as long as nutritionists and health and education professionals join their agricultural colleagues in designing locally appropriate introduction programmes to inform growers and consumers.

Of course, initially seed has to be delivered to the farmers by the public-sector-managed seed system as for new rice varieties. After this, communities – especially sharing of seed by farmers – can be relied on to disseminate the nutritional benefit. As the technology is in the seed, no factory is needed. No road infrastructure. No 'cold chain' of refrigeration needed, at the time of writing, with all vaccines for distribution from point of manufacture to point of use. No more money for cultivation or purchase than for other rices. No special processing. No special packaging. No aid agencies, philanthropic or government funding. Farmers earning less than about US$10 000 per annum from agriculture – which includes all resource-poor farmers in developing countries – will be free to sow, harvest, save seed, replant, and sell locally without any farmers or government or anyone else needing to pay any royalty or technology fee of any kind. Golden Rice will be available to the people who can benefit from it and be completely sustainable.

Our vision for Golden Rice is that eventually people will regard white rice as something old-fashioned and unusual, much as the original white or purple carrot is regarded today in comparison with its variant, the orange one, which is so familiar but was introduced by Dutch carrot breeders to honour their royal house of Orange hundreds of years ago.

CONCLUSION

The *ex-ante* economic impact analyses of Golden Rice by Anderson *et al.* (2004; Anderson and Pohl Nielsen, 2004) forecast a significant impact on the wealth of Asian countries: conservative Golden Rice consumption could add from US$4 billion to US$18 billion annually to Asian GDP. By enhancing the effectiveness of their immune systems, the well-being of the poorest individuals in society would be significantly improved, and agricultural productivity increased very markedly.

A World Bank study (Galloway and McGuire, 1994) assigned a present value of $US45 to each annual case of iron deficiency averted and US$96 for each annual case of vitamin A deficiency averted for preschool children (Bouis *et al.*, 2011).

Ex-ante studies (Stein *et al.*, 2006), which assumed a 12:1 bioconversion of beta-carotene to vitamin A, combined with the later demonstrated, much more highly efficient human bioconversion research results 2.3:1 (Tang *et al.*, 2012) suggest that it will cost much less than $US3.0 per person, adult or child, to prevent mortality and morbidity

Table 12.4 *Global expenditure on research in public health problems*

	US$ 2008	Disease burden in low- and middle-income countries
HIV/AIDS/malaria/tuberculosis	8.18 bn	23% (combined)
Maternal and child health (excluding biofortified staple crops)	3.17 bn	
Non-communicable diseases (surgery, diabetes, circulatory disease)	0.1215 bn (122 m)	45%
Biofortified staple crops (not from referenced source, approximation)	0.015 bn (15 m)	Very large

Source: Gates (2012).

caused by vitamin A deficiency with Golden Rice. Similar *ex-ante* cost benefits have been calculated for putative multi-biofortified rice in China (De Steur *et al.*, 2012). These estimates derive from the knowledge that once a biofortified crop is created and is stable, there is no additional cost associated with its cultivation and use: all the costs of a biofortified crop are in the initial development, making the initial seed available and related education and social marketing programmes.

By the time it is in use a biofortified crop should cost no more than the variety from which it is derived. Thus in an effort to provide long-term and sustainable prevention of micronutrient deficiencies, biofortified staple foods have the potential to provide a safe and effective additional approach to current interventions.

Biofortified staple crops are also an excellent pragmatic fit with the 1992 United Nations nutritional recommendation for locally available food-based strategies to alleviate micronutrient deficiencies (http://www.fao.org/docrep/V7700T/v7700t02.htm accessed 9 November 2012). Therefore biofortified foods also have the potential to sustainably complement other nutritional deficiency interventions with less cost and logistical problems and so better population coverage.

Biofortification work in rice has, like other advances described in this book, been facilitated by rapid progress over the past few decades in the mapping of plant genomes and the development of sophisticated molecular techniques which has led to extraordinary progress in gene discovery and the elucidation of biosynthetic pathways. As all organisms have many genes and associated metabolic pathways in common, conclusions with relevance to crop species can be drawn from research with organisms with shorter life cycles.

All the biofortification programmes mentioned in these pages, whether through genetic engineering or conventional plant breeding approaches, are however dramatically underfunded (Table 12.4).

More than twice as much money is spent globally on research to combat HIV/AIDS, tuberculosis and malaria annually than is spent on the whole of agricultural research for the seven most important crops by the private, public and philanthropic sectors, estimated at a little more than US$3.0 billion annually (Gates, 2012). It is time to start funding biofortified crops from the health budgets which should be promoting them, and from the very significant funds – US$40 billion – available in support of the Millennium Development Goals for 5 years from 2010 (United Nations, 2010).

The first medicine is food!

REFERENCES

Anderson, K., and Pohl Nielsen, C. (2004b). *Golden Rice and the Looming GMO Trade Debate: Implications for the Poor*, CEPR Discussion Paper No. 4195. Centre for Economic Policy Research, London.

Anderson, K., Jackson, L. A., Pohl Nielsen, C. (2004a). *Genetically Modified Rice Adoption: Implications for Welfare and Poverty Alleviation*, Discussion Paper No. 0413. Centre for International Economic Studies, Adelaide, Australia.

Beyer, P. (2010). Golden Rice and 'golden crops' for human nutrition. *New Biotechnology* 27, 478–481.

Bouis, H. E., Hotz, C., McClafferty, B., Meenakshi, J. V., and Pfeiffer, W. (2011). Biofortification: a new tool to reduce micronutrient malnutrition. *Food and Nutrition Bulletin* 32, S31–S40.

Brooks, S. (2010). *Rice Biofortification: Lessons for Global Science and Development*. Earthscan, London.

Copenhagen Consensus (2004). www.copenhagenconsensus.com/Home-1.aspx

Copenhagen Consensus (2008). www.copenhagenconsensus.com/Default.aspx?ID=953

Copenhagen Consensus (2012). www.copenhagenconsensus.com/Projects/CC12/Outcome.aspx

Datta, S. K., Peterhans, A., Datta, K., and Potrykus, I. (1990). Genetically engineered fertile *indica*-rice plants recovered from protoplasts. *BioTechnology* 8, 736–740.

De Steur, H., Gellynck, X., Blancquaert, D., *et al.* (2012). Potential impact and cost-effectiveness of multi-biofortified rice in China. *New Biotechnology* 29, 432–442.

Fawzi, W. W., Chalmers, T. C., Herrera, M. G., and Mosteller, F. (1993). Vitamin A supplementation and child mortality: a meta-analysis. *Journal of the American Medical Association* 269, 898–903.

Gates, B. (2012). *Annual Letter*. www.gatesfoundation.org/annual-letter/2012/

Galloway, R., and McGuire, J. (1994). Determinants of compliance with iron supplementation: supplies, side effects, or psychology? *Social Science and Medicine* 39, 381–390.

Grune, T., Lietz, G., Palou, A., *et al.* (2010). β-Carotene is an important vitamin A source for humans. *Journal of Nutrition* 140, 2268S–2285S.

Haas, J. D., del Mundo, A. M., and Beard, J. L. (2000). A human feeding trial of iron-enhanced rice. *Food and Nutrition Bulletin* 21, 440–444.

Haas, J. D., Beard, J. L., Murray-Kolb, L. E., *et al.* (2005). Iron-biofortified rice improves the iron stores of non-anemic Filipino women. *Journal of Nutrition* **135**, 2823–2830.

Hesser, L. (2006). *The Man Who Fed the World*. Durban House, Dallas, TX.

Honein, M. A., (2001). Impact of folic acid fortification of the US food supply on the occurrence of neural tube defects. *Journal of the American Medical Association* **285**, 2981–2986.

Hopkins, F. (1929). *The Early History of Vitamin Research, Nobel Lecture 1929*. www. nobelprize.org/nobel_prizes/medicine/laureates/1929/hopkins-lecture.html

Hotz, C., Loechl, C., Abdelrahman, L., *et al.* (2012). Introduction of β-carotene-rich orange sweet potato in rural Uganda results in increased vitamin A intakes among children and women and improved vitamin A status among children. *Journal of Nutrition*, published online 8 Aug 2012.

Johnson, A. A. T., Kyriacou, B., Callahan, D. L., *et al.* (2011). Constitutive over-expression of the *OsNAS* gene family reveals single-gene strategies for effective iron and zinc biofortification of rice endosperm. *PLoS ONE* **6**, e24476.

Krattiger, A., and Potrykus, I. (2007). Golden Rice: a product-development partnership in agricultural biotechnology and humanitarian licensing. In A. Krattiger, R. T. Mahony, L. Nelsen, *et al.* (eds.) *Executive Guide to Intellectual Property Management in Health and Agricultural Innovation: A Handbook of Best Practices*. www.ipHandbook.org

Low, J. W., Arimond, M., Osman, N., *et al.* (2007). A food-based approach introducing orange-fleshed sweet potatoes increased vitamin A intake and serum retinol concentrations in your children in rural Mozambique. *Journal of Nutrition* **137**, 1320–1327.

Lynch, S. R. (2011). Why nutritional iron deficiency persists as a worldwide problem. *Journal of Nutrition* **141**, 763S–768S.

Mayo-Wilson, E., Imdad, A., Herzer, K., Yakoob, M., and Bhutta, Z. (2011). Vitamin A supplements for preventing mortality, illness and blindness in children aged under 5: systematic review and meta-analysis. *British Medical Journal* **343**, d5094.

Moore, P. (2010). *Confessions of a Greenpeace Dropout: The Making of a Sensible Environmentalist*. Beatty, San Francisco, CA.

Paine, J. A., Shipton, C. A., Chaggar, S., *et al.* (2005). Improving the nutritional value of Golden Rice through increased pro-vitamin A content. *Nature Biotechnology* **23**, 482–487.

Potrykus, I. (2010). The private sector's role for public sector genetically engineered crop projects. *New Biotechnology* **27**, 578–581.

Robinson, P. D., Hogler, W., Craig, M. E., *et al.* (2006). The re-emerging burden of rickets: a decade of experience from Sydney. *Archives of Disease in Childhood* **91**, 564–568.

Save the Children (2012). *A Life Free From Hunger: Tackling Child Malnutrition*. www.savethechildren.org/site/c.8rKLIXMGIpI4E/b.7980641/k.C98/Nutrition_Report_2012.htm

Southgate, D., and Graham, D. (2006). *Growing Green: The Challenge of Sustainable Agricultural Development in Sub-Saharan Africa*. International Policy Press, London.

Stein, A. J., Sachdev, H., and Qaim, M. (2006). *Can Genetic Engineering for the Poor Pay Off? An Ex-ante Evaluation of Golden Rice in India*, Discussion Paper No. 05/2006. Centre for Agriculture in the Tropics and Subtropics, University of Hohenheim, Germany.

Storozhenko, S., De Brouwer, V., Volckaert, M., *et al.* (2007). Folate fortification of rice by metabolic engineering. *Nature Biotechnology* **25**, 1277–1279.

Tang, G., Qin, J., Dolnikowski, G. G., Russell, R. M., and Grusak, M. A. (2009). Golden Rice is an effective source of vitamin A. *American Journal of Clinical Nutrition* 89, 1776–1783.

Tang, G., Hu, Y., Yin, S.-A., *et al.* (2012). Golden Rice β-carotene is as good as β-carotene in oil in providing vitamin A to children. *American Journal of Clinical Nutrition* 96, 658–664.

United Nations (2010) Summit. www.un.org/en/mdg/summit2010/

West, K. P. Jr, Klemm, R. D. W., and Sommer, A. (2010). Vitamin A saves lives: sound science, sound policy. *World Nutrition* 1, 211–229.

Whitcher, J. P., Srinivasan, M., and Upadhyay, M. P. (2001). Corneal blindness: a global perspective. *Bulletin of the World Health Organization* 79, 214–221.

WHO (2009). *Global Prevalence of Vitamin A Deficiency in Populations at Risk 1995–2005*. World Health Organization, New York.

WHO (2012). *Nutrition: Micronutrient Deficiencies*. World Health Organization, New York. www.who.int/nutrition/topics/vad/en/

Wirth, J., Poletti, S., Aesclimann, B., *et al.* (2009). Rice endosperm iron biofortification by targeted and synergistic action of nicotianamine synthase and ferritin. *Plant Biotechnology Journal* 7, 631–644.

Ye, X.-D., Al-Babili, S., Kloti, A., *et al.* (2000). Engineering the provitamin A β-carotene biosynthetic pathway into (carotenoid-free) rice endosperm. *Science* 287, 303–305.

You, D., Jones, G., and Wardlaw, T. (2010). *Levels and Trends in Child Mortality: Estimates Developed by the United Nations Inter-Agency Group for Child Mortality Estimation*. United Nations Children's Fund (UNICEF), New York.

KEY FURTHER RESOURCES

Mayer, J. E., Pfeiffer, W. H., and Beyer, P. (2008). Biofortified crops to alleviate micronutrient malnutrition. *Current Opinion in Plant Biology* 11, 166–170.

www.goldenrice.org.

www.harvestplus.org.

LARRY MURDOCK, IDAH SITHOLE-NIANG AND
T. J. V. HIGGINS

13

Transforming the cowpea, an African orphan staple crop grown predominantly by women

AN IMPORTANT FOOD LEGUME

Cowpea – also known as black-eyed pea, niébé, southern pea and Augen-bohne among other names – is the archetypal legume crop of Africa (Figure 13.1). It thrives in the Sahel and the dry and moist Guinea savannahs in a great continental arc stretching from Senegal in the west to Sudan and Somalia in the east and thence south through western Kenya and on into Tanzania, Zimbabwe and northern South Africa. Across this vast reach of land, where the typical soils are poor in organic matter and nutrients and rainfall is erratic, cowpea thrives. It is the crop of women above all, and of the impoverished and the resource-deprived. It provides the crutch of subsistence for tens of millions of rural poor in Africa. Although millions of hectares of it are grown across the continent, it has been sufficiently neglected by donors and national research programmes to have been included in the book *Lost Crops of Africa* (Anon, 2006).

Cowpea, *Vigna unguiculata* Walp., is a drought-tolerant, heat-adapted annual herbaceous legume. It is predominantly self-pollinating, although a slight amount of outcrossing occurs – on the order of 1%, depending on cultivar and location (Fatokun and Ng, 2007). Root-associated rhizobia fix nitrogen at rates ranging from 25–179 pounds of nitrogen per hectare (Sprent *et al.*, 2010). It can be cultivated where annual rainfall is as low as 300 mm (Silva *et al.*, 2010). Experts disagree about exactly where cowpea was first domesticated but there is no question about its African origin (Vaillancourt and Weeden, 1992).

Successful Agricultural Innovation in Emerging Economies: New Genetic Technologies for Global Food Production, eds. David J. Bennett and Richard C. Jennings. Published by Cambridge University Press. © Cambridge University Press 2013.

Figure 13.1 Cowpea grain, rich in protein and carbohydrate, is an important food for tens of millions of rural and urban Africans, and an important source of cash income for millions of resource-poor farmer families. (Photo: Dan Pikler).

African cowpea can have a predominantly bushy upright habit or take the form of a prostrate, crawling plant. It can serve as excellent ground cover, preserving moisture while adding nitrogen. Cowpea hay is valuable as fodder, typically gathered after the mature pods are harvested.

Women are the primary cultivators of cowpea in many parts of Africa, both for the excellent nutrition it offers their families as well as for the income it generates when they sell it in the local markets or to travelling traders. Commonly grown for its grain, its foliage is rich in protein, dried leaves (25–45% protein) being comparable in protein content to the grain (20–30% protein) (Ohler *et al.*, 1996). Fresh foliage is consumed as a salad, or cooked, and dried foliage is sold in markets, especially in East Africa. During the hungry time – when last year's grain has been exhausted and next year's crop is not yet ready to harvest – farm families gather fresh tender leaves of the growing cowpea crop. In Senegal, green cowpea pods are boiled and shelled, and the soft beans eaten during this period. The roadsides are dotted with women offering fresh pods for sale.

From Africa (annual production in 2010 was 5.2 million tonnes, 94% of the annual world production) (http://faostat.fao.org/) cowpea spread around the world, reaching southern Europe two thousand years ago and South Asia a thousand years later. In recent centuries, north-western Brazil and the southern USA became cowpea producers. Cowpea is the most economically important legume crop of Africa (Langyintuo et al., 2003). In West and Central Africa much of the cowpea grain produced in the northern savannahs and Sahel moves south via centrally located aggregation centres such as Kano, Nigeria. The grain – once carried by camels and horses along ancient trade routes – now reaches the populous coastal urban centres like Lagos and Accra by truck.

Not only is most of the world's cowpea produced in Africa, most of it is consumed there (Figure 13.2). There are potential markets in Europe as well as in Brazil and India, but inadequate production, quality issues and the lack of standards, plus high internal demand has thus far kept the crop on the continent where it is produced. It is noteworthy that Nigeria, the world's largest producer of cowpea, is also the world's largest importer, gathering much of the crop of its neighbour to the north, Niger, as well as Cameroon and other contiguous cowpea-producing nations like Burkina Faso and Benin. Economic studies show that African markets can benefit from increased production – as might, for example, result from the introduction of insect-resistant Bt cowpea – with increases in trade and only modest effects on market price of the grain (Langyintuo et al., 2003). Greater production could easily be absorbed by the existing African markets. This could happen without major price disruptions or changes in the pattern of trade if increased production is spread over the region and not limited to one or two countries.

INSECTS CONSUME TOO MUCH

Over the last 30 years cowpea breeders in Africa have worked to develop high-yielding cowpea producing large, white seeds with rough seed coats – the preferred type for regional trade. In many ways they have succeeded, producing cultivars with yield potentials as high as 1500–2000 kg/ha. But such yields are rarely achieved on farmers' fields, where, depending on the location and year, yields typically are more like 150–400 kg/ha. Why the huge yield gap?

The principal reasons are the biological constraints that plague this otherwise hardy crop. The main biological constraints are insects. The crushing pressure that insects represent is proven by comparisons of yields with and without insecticide treatments. Yields can increase 20-fold when effective insecticides are applied. Farmers know that in some years there will be no grain yield at all unless they use insecticides – in those years all they harvest without insecticides is hay. Insecticides

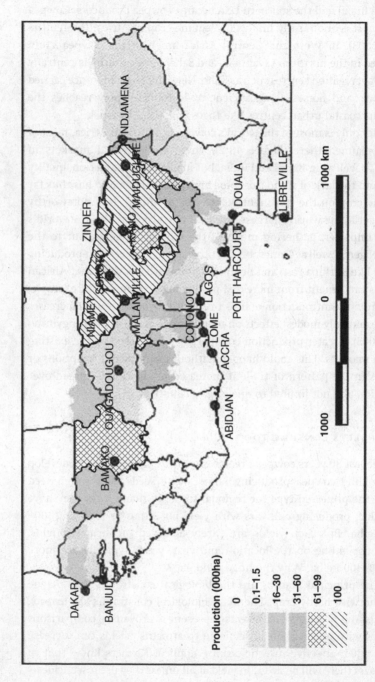

Figure 13.2 Distribution of cowpea-growing areas in West and Central Africa. (Reproduced with permission from Langyintuo et al., 2003).

are in any case not a good solution to the problems of field-grown cowpeas. Insecticides are often not available, expensive when they are, of low quality, not approved for use on cowpeas, and farmers may not have spray equipment or know-how to use them safely.

Among the principal insect pests in the field are the *Maruca* pod-borer (MPB), thrips, several species of pod-sucking bugs and aphids. After the crop is harvested, cowpea is plagued by seed-feeding beetles of the subfamily Bruchinae. The cowpea bruchid (*Callosobruchus maculatus* Fabricius), which has a high reproductive potential, is capable of destroying the entire hard-won crop after only a few months of storage. As a result farmers commonly sell their crop at harvest when the price is at its low point of the year because they can't store it. This situation is happily changing, with the advent of the Purdue Improved Cowpea Storage (PICS) technology and its wide dissemination in Africa's main cowpea-growing regions (www.ag.purdue.edu/ipia/pics/Pages/Home.aspx; see also Baributsa *et al.*, 2010).

In addition to insect pests, cowpea in the field suffers from other biological constraints that can be locally serious in some years, such as bacterial and viral diseases. Parasitic plants, especially *Striga gesneroides*, appear to be increasingly troublesome in some areas, and *Alectra vogelli* reduces yields as well.

There are numerous social, economic and policy constraints hindering the advance of cowpea as a crop. One is the underinvestment in agricultural research that has plagued African nations in their eagerness to develop their cities and manufacturing infrastructure. Agriculture is simply not sexy and exciting, and cowpea – being a minor crop and a crop of the poor and of women – has a particularly bad association. A second reason is that there is little international trade in cowpea outside of Africa, so investments are not directed to crops that merely feed people and do not promise to bring in serious amounts of foreign exchange. Third, while the tools of biotechnology have begun to be brought to bear on cowpea, the slowness with which African governments have passed biosafety laws and created regulatory processes and bureaucracies has held back the testing of biotech-based crops. The former colonial masters in Europe, being outspoken opponents of biotechnology, have had a particularly regressive effect on their former colonies – where they retain a certain grudging influence. Fourth, international institutions have given all too little of their attention to this minor crop. The International Institute of Tropical Agriculture (IITA), although it was enthusiastically involved in the early efforts to develop insect-resistant cowpea through genetic engineering, has largely, but, we hope, temporarily, disappeared from the scene. It makes sense for the IITA to be actively involved in promoting GM cowpea because it has the world-mandate for cowpea with this improvement. Fortunately, another African institution, the African Agriculture Technology Foundation (AATF), shaped under the pioneering leadership of Dr Eugene Terry, has had the foresight to take Bt cowpea into its portfolio of projects.

GENE TECHNOLOGY FOR COWPEA

More than 20 years ago, Murdock, Hardin, Bressan and colleagues argued that there was a need for a genetic transformation system for cowpea. With encouragement and input from IITA's B. B. Singh and S. R. Singh and African cowpea breeders, they recognised that existing germplasm does not contain sufficient levels of resistance to some of the major biotic and abiotic stresses that constrain productivity of this important crop. Purdue brought several teams together to tackle the problem and with very limited resources made excellent progress which Kononowicz has summarised (Kononowicz *et al.*, 1997). Following a meeting organised by Purdue in Dakar in 2001 and the establishment of the Network for the Genetic Improvement of Cowpea for Africa (NGICA), the Rockefeller Foundation, USAID, the Bean/Cowpea Collaborative Research Support Program (CRSP) and AATF gave new impetus to the search for a cowpea transformation system. Transformation has now been achieved and is working in several laboratories (Popelka *et al.*, 2006; Adesoye *et al.*, 2008; Solleti *et al.*, 2008).

This new technology is being used mainly to introduce resistance to *Maruca* pod-borer, a devastating pest that frequently reduces cowpea yield by more than 50% and to the bruchid *Callosobruchus maculatus*, which, as already mentioned, can decimate the grain during post-harvest storage. Sprays of *Bacillus thuringiensis* (Bt) are known to control MPB (Taylor, 1968). We will illustrate the approaches that are being taken with our current work on using a gene that encodes a crystal protein from Bt which has been synthesised and reconstructed for expression in cowpea. This Bt protein is Cry1Ab and is known to control MPB, whose caterpillars mostly attack floral organs and the developing pod (Taylor, 1967). The later-stage larvae also attack the leaves and petioles. Thus, the gene was designed so that it is active in vegetative and reproductive parts of the plant. The reconstructed gene was linked to a second gene which encodes a protein that allows selection of plant cells that survive in the presence of the antibiotic kanamycin. The two genes were incorporated into a circular DNA vector, derived from *Agrobacterium tumefaciens* and

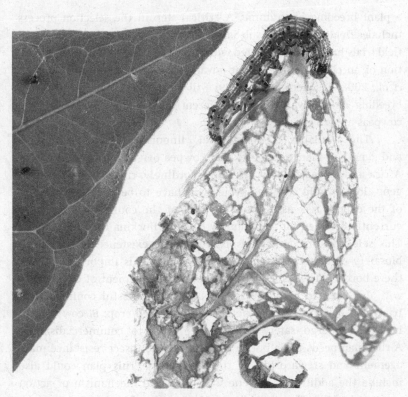

Figure 13.3 In vitro bioassay of transgenic cowpea using *Helicoverpa armigera* larvae. On the right is shown part of a non-transgenic leaf, while on the left is part of a leaf from a transgenic cowpea expressing the *cry1Ab* gene, after 10 days' exposure to the larvae. (Photo: Carl Davies).

capable of transferring both genes into plants cells. Cells that survived in the presence of kanamycin were propagated by the appropriate combination of sugars, amino acids and plant hormones until whole plants were regenerated. They contained the genes integrated in the plant chromosome and were stably transmitted from generation to generation. This process of gene transfer and subsequent plant regeneration is called plant transformation. Cowpea transformation can be divided into several steps and has been described in detail on the Network for the Genetic Improvement of Cowpea for Africa (NGICA) website (www.entm.purdue.edu/NGICA/cowpea.html). Hundreds of independent transgenic events containing the *Bt* gene have been generated and have been tested for effectiveness in caterpillar bioassays using *Helicoverpa armigera* (Figure 13.3). They are now going through a stringent selection procedure to find lines with potential to act as parents in a plant breeding programme. A critical step in the selection process includes field trials in the sub-Saharan African environment. Four such field trials have been conducted with promising results. Following selection of an elite line, the recent advances in cowpea genomics (Timko et al., 2008; Muchero et al., 2009) will be invaluable in speeding the breeding of new varieties for different agro-ecological zones where cowpeas are grown.

The available ethnobotanical, linguistic, phytogeographical and genetic evidence suggests that cowpea originated in the Horn of Africa and spread to West Africa. Accordingly, the issue of potential gene flow to wild cowpea species will have to be addressed as part of the overall risk assessment in each of the countries where it is currently destined for use, namely Nigeria, Burkina Faso and Ghana. This selection of countries is based on the existence of functional biosafety regulatory systems, although cowpea is important beyond these borders. The issue of trans-boundary movement of GM cowpea will also need to be addressed. Following successful confined field trials and like any other genetically engineered crop, Bt cowpea will have to undergo safety assessments prior to commercialisation. A roadmap needs to be put in place to ensure insect resistance management and stewardship of the technology. This plan could also include the addition of a gene with a second mechanism of action against the MPB. This would have the benefit of both enhancing the resistance of the Bt plants but also increasing the durability of the resistance – resistant MPB populations would be much less likely to develop against two genes versus one. In sum, there needs to be a sound regulatory system under which the Bt cowpea will be deployed. If this were regionally harmonised it would certainly go a long way towards addressing the trans-boundary issues that will emerge. Strides have been made towards dealing with stewardship issues associated with deployment of Bt cowpea in West Africa thanks to an expert panel convened at the Danforth Foundation to address issues arising including (1) gene flow; (2) non-target arthropods; and (3) insect resistance management (Huesing et al., 2011).

PUBLIC AWARENESS

Ambivalence towards GM crops in Africa has not gone away. This is despite the fact that African heads of states through the esteemed High Level Panel on Biotechnology has in fact embraced the technology (www. nepadst.org/doclibrary/pdfs/biotech_africarep_2007.pdf). The European legacy on the African continent still leaves an indelible mark on GM crops and their adoption. As mentioned, African countries are still particularly sensitive to the leanings of their former colonial masters. This is despite the fact that a highly publicised Regional Approach to Biotechnology and Biosafety Policy in Eastern and Southern Africa (RABESA) study showed that there is more significant trade amongst African countries than with Europe (www.asareca.org/paap/uploads/publications/RABESA). Some of that trade is in fact in benign commodities such as teas and coffee that do not have GM counterparts at present (Paarlberg, 2008). Needless to say, there has been much dialogue and public awareness around specific GM crops undergoing confined field trials on the continent. The Open Forum for Agricultural Biotechnology, through its various country chapters in Nigeria, Kenya, Uganda, Egypt and to a lesser extent Tanzania, has been particularly effective in addressing some of these public concerns. According to Clive James (James, 2012) the adoption of genetically engineered crops in the next 5 years will largely depend on three things: *'the timely implementation of appropriate, responsible, and cost/time-effective regulatory systems; strong political will and support and a continuing wave of improved crops that will meet the priorities of industrial and developing countries in Asia, Latin America and Africa'*. For Africa, Bt cowpea is high on that list.

SEED SYSTEMS

While effective seed systems will play a critical role is the wide diffusion of *Bt* cowpea in West Africa, to date those systems remain largely weak and undeveloped. For the deployment of *Bt* cowpea to succeed in West Africa it will need a concerted effort by all concerned to establish a workable system that will ensure seed access, quality control and the necessary stewardship for the technology to perform. This therefore dictates that the improved cowpea be profitable, reduce poverty, and increase food and income security amongst smallholder farmers in sub-Saharan Africa. While donors and governments may be found who will help a *Bt* cowpea seed system develop, the best system will ultimately be one that is profit-driven and self-sustaining. The PICS project mentioned earlier, which has had success in developing a value chain for cowpea storage technology, has learned many lessons about how to work effectively with a variety of stakeholders in Africa to create a system that reaches and benefits farmers in a way that may be sustainable. Some of these lessons will be useful in fostering the development of a cowpea seed system.

THE ROAD AHEAD

Numerous challenges to implementing the cowpea value chain remain. These include the participatory development and diffusion of improved cowpea varieties adapted to climate change, environmentally sound pest control, value addition and market access (Murdock *et al.*, 2008).

While significant achievements have been made in bringing Bt cowpea to where it is now, the goal of delivering improved seeds to African farmers still needs to be achieved. There are intermediate steps along the path that need to be taken, namely continued product development, commencement of product deployment activities, implementation of sound regulatory systems as well as establishment of a workable seed delivery system in those countries. For maximum benefit, it is also vital that deployment takes into account the regional impact of adoption. If some countries adopt and others do not, the relative shifts in trade will benefit the adopters and work to the disadvantage of non-adopters (Langyintuo and Lowenberg-DeBoer, 2006).

FINAL THOUGHTS

It has been said somewhere that every technology has a downside, and there are no exceptions. This is as true for genetically modified cowpea as it is for a shiny new Mercedes automobile. But the risk of the automobile is real, ever-present and substantial. We accept it every day of our lives. Every one of us knows people – friends, relatives, acquaintances – who have died in automobile crashes. By comparison, the risks of genetically modified crops, while real, are microscopic, if not submicroscopic. Clever people, of course, can think of imagined dangers all day long, dangers that are irrefutable because they are vague and untestable. We hear, for example, people declaring that they would accept GM foods when they are proven safe. They forget that science can never prove such a thing – it can only approach the truth, never absolutely get there. For that matter, you can't *prove* that any food is safe – the best you can do is say that, so far, there is no reason to believe there is harm associated with it. When benefits of Bt cowpea technology are put on the balance pan, the weight of those benefits depends on the perspective of the person loading the balance. We believe that the right perspective, the humanitarian perspective, puts the children and the good people of Africa in front of the balance. They deserve and need better food, more food, and cleaner food. More abundant cowpea, at lower cost – the result of greater production throughout the main cowpea-growing regions of West and Central Africa thanks to deployment of *Bt* cowpea – would go a long way to making for a better life for millions of poor Africans. The risk of *Bt* cowpea, theoretically real but vanishingly small, is far outweighed by the benefits it offers to people in need.

REFERENCES

Adesoye, A., Machuka, J., and Togun, A. (2008). *CRY 1AB* trangenic cowpea obtained by nodal electroporation. *African Journal of Biotechnology* 7, 3200–3210.
Anon (2006). *Lost Crops of Africa*, vol. 2, *Vegetables*. National Academies Press, Washington, DC.
Baributsa, D., Lowenberg-DeBoer, J., Murdock, L. L., and Moussa, B. (2010). Profitable chemical-free cowpea storage technology for smallholder farmers in Africa: opportunities and challenges. *10th International Working Conference on Stored Product Protection*, Estoril, Portugal.
Fatokun, C., and Ng, Q. (2007). Outcrossing in cowpea. *International Journal of Food, Agriculture and Environment* 5, 334–338.
Huesing, J., Romeis, J., Ellstrand, N., *et al.* (2011). Regulatory considerations surrounding the deployment of *Bt*-expressing cowpea in Africa: report of the deliberations of an expert panel. *GM Crops* 2, 211–224.
James, C. (2012). *Global Status of Commercialized Biotech/GM Crops: 2011*, ISAAA Brief No. 43. International Service for the Acquisition of Agri-biotech Applications, Ithaca, NY.
Kononowicz, A. K., Cheah, K. T., Narasimhan, M. L., *et al.* (eds.) (1997). *Developing a Transformation Systems for Cowpea (*Vigna unguiculata *[L.] Walp.)*. International Institute of Tropical Agriculture, Ibadan, Nigeria.
Langyintuo, A. S., and Lowenberg-DeBoer, J. (2006). Potential regional trade implications of adopting *Bt* cowpea in west and central Africa. *AgBioForum* 9, 111–120.
Langyintuo, A. S., Lowenberg-Deboer, J., Faye, M., *et al.* (2003). Cowpea supply and demand in West and Central Africa. *Field Crops Research* 82, 215–231.
Muchero, W., Diop, N. N., Bhat, P. R., *et al.* (2009). A consensus genetic map of cowpea [*Vigna unguiculata* (L) Walp.] and synteny based on EST-derived SNPs. *Proceedings of the National Academy of Sciences USA* 106, 18159–18164.
Murdock, L., Coulibaly, O., Higgins, T., *et al.* (2008). Cowpea. In C. Kole and T. C. Hall (eds.) *Compendium of Transgenic Crop Plants: Transgenic Legume Grains and Forages*, pp. 23–36. Blackwell Publishing, Oxford.
Ohler, T., Nielsen, S., and Mitchell, C. (1996). Varying plant density and harvest time to optimize cowpea leaf yield and nutrient content. *HortScience* 31, 193–197.
Paarlberg, R. (2008). *Starved for Science: How Biotechnology Is Being Kept Out of Africa*. Harvard University Press, Cambridge, MA.
Popelka, J. C., Gollasch, S., Moore, A., Molvig, L., and Higgins, T. J. V. (2006). Genetic transformation of cowpea (*Vigna unguiculata* L.) and stable transmission of the transgenes to progeny. *Plant Cell Reports* 25, 304–312.
Silva, V. D. P. R., Campos, J. H. B. C., Silva, M. T., and Azevedo, P. V. (2010). Impact of global warming on cowpea bean cultivation in northeastern Brazil, *Agricultural Water Management* 97, 1760–1768.

Solleti, S., Bakshi, S., Purkayastha, J., Panda, S., and Sahoo, L. (2008). Transgenic cowpea (*Vigna unguiculata*) seeds expressing a bean α-amylase inhibitor 1 confer resistance to storage pests, bruchid beetles. *Plant Cell Reports* **27**, 1841–1850.

Sprent, J. I., Odee, D. W., and Dakora, F. D. (2010). African legumes: a vital but under-utilized resource. *Journal of Experimental Botany* **61**, 1257–1265.

Taylor, T. A. (1967). The bionomics of *Maruca testulalis* Gey. (Lepidoptera: Pyralidae), a major pest of cowpeas in Nigeria. *Journal of the West African Science Association* **12**, 111–129.

Taylor, T. A. (1968). The pathogenicity of *Bacillus thuringiensis* var. *thuringiensis* Berliner for larvae of *Maruca testulalis* Geyer. *Journal of Invertebrate Pathology* **11**, 386–389.

Timko, M., Rushton, P., Laudeman, T., *et al.* (2008). Sequencing and analysis of the gene-rich space of cowpea. *BMC Genomics* **9**, 103.

Vaillancourt, R. E., and Weeden, N. F. (1992). Chloroplast DNA polymorphism suggests Nigerian center of domestication for the cowpea, *Vigna unguiculata* (Leguminosae). *American Journal of Botany* **79**, 1194–1199.

14

Transgenic marine algae for aquaculture: a coupled solution for protein sufficiency

INTRODUCTION

Many developing economies are in or near the tropics and have desert areas bordering the sea. These areas are ripe for development with a dual interrelated industry that can solve many problems for the local populace as well as contribute to local and world food security: this is mass land-based culture of microalgae, coupled in part with aquaculture, using the algae as a major feedstock for aquaculture of fin-fish, shellfish, and crustaceans. This requires that the algae be domesticated to be a robust crop having the highest value-added possible, which can only be performed using genetic engineering. Additional systems' engineering is needed to further lower production costs. The populace in many of these areas is both underemployed and undernourished, and aquaculture is labour-intensive. Fish and other seafoods are the most efficient converters of plant protein into animal protein, far superior to poultry and livestock. Only 1.5 kg (dry weight) of feed is required to produce 1 kg (fresh weight) of salmon fillets, whereas 5 and 8 kg of feed are required to produce 1 kg of chicken and pork fillets, respectively. The biggest impediment to aquaculture is the availability of appropriate sources of protein and of oil for feeding. Aquaculture has depended on fishmeal and fish oil to supply these needed components, but their supply is diminishing due to depletion of ocean fisheries coupled with an increase in both demand and price (Naylor et al., 2009) (Figure 14.1) due to growing markets for animal protein in China and India. Huge efforts are being made to replace fishmeal with other protein sources, to some success with livestock and poultry, but less so

Successful Agricultural Innovation in Emerging Economies: New Genetic Technologies for Global Food Production, eds. David J. Bennett and Richard C. Jennings. Published by Cambridge University Press. © Cambridge University Press 2013.

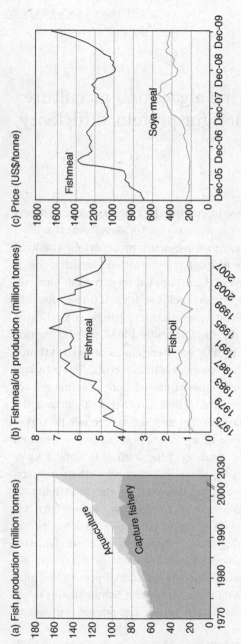

Figure 14.1 Depleting supplies and increasing need for fishmeal and fish-oil. (a) Aquaculture, requiring large amounts of fishmeal and fish-oil, is continually increasing. (Redrawn and condensed from www.aquaculture.ca/gifs/world-fish-production.jpg) (b) Fishmeal production is decreasing due to less fish catch at sea. (Redrawn and condensed from www.blueharvest.com.au/Aquafeed/tabid/152/Default.aspx) (c) The price of fishmeal is continually increasing, especially relative to soya bean meal. (Redrawn and condensed from www.hammersmithltd.blogspot.com)

with aquaculture. Soya, the major plant protein source in most animal feeds, acts as an antifeedant reducing all-around digestibility of feed pellets with many species of fish, especially the more carnivorous ones (Gatlin *et al.*, 2007). In addition, soya is missing key amino acids such as methionine, which must be added as a specially formulated synthetic product.

Microalgae have long served as an ideal food for fish larvae in hatcheries and have been shown to be an adequate at least partial replacement for fishmeal for later stages of growth (Reitan, 2011), but the cost of production of microalgae has been much too high, and the availability too low for large-scale feeding trials. Cultured microalgae have been looked at mainly as a replacement of high omega-3 fish oils (Patnaik *et al.*, 2006). Algal meal can at least partially replace fishmeal on a 1:1 basis (Walker and Berlinsky, 2011), yet the algal meal is only about 70% digested whereas fishmeal is almost completely digested. There is some literature about the use of cultivated and wild-harvested macroalgae (seaweed) as aquaculture feed but the yields are more variable and less intensive than obtainable with microalgae. Some such systems have environmental justification, such as macroalgae cultivated on ropes surrounding penned aquaculture in the sea, which purifies waste from the fish, but the amounts yielded per fish cannot sustain an aquaculture industry. Single-celled cyanobacteria (blue–green algae) in many ways are superior to microalgae (growth rate, and ability to be cultivated at higher temperatures), but their lipids are in the form of glyco- or phospholipids and cannot be easily extracted like the triglycerides stored in microalgae. The multicellular cyanobacterium *Spirulina* has been more widely tested as a source of feed for aquaculture, but it grows more slowly than microalgae, and the present production costs only allow it to be marketed as a 'health food' supplement. Freshwater algae have also been tested as a source of feed for aquaculture, but the limited availability of fresh water in the desert areas under consideration restricts its use for producing feed. Thus, this chapter focuses on domesticating marine microalgae as a feed source, domesticated to replace fishmeal and fish-oil for intensive aquaculture penned at sea or in ponds on land.

The potential for algal meal to fill the feeding gap is immense. Marine microalgae grow on seawater, so they would not compete with conventional crops for fresh water. Algal yields per unit area can easily be more than 20 times those of soya bean, and with adequate management they can utilise nearly 100% of the sunlight all year round, being continuously kept in a constantly increasing exponential rate of growth, unlike crops. With soya beans only the seeds are used, with algae the 'whole plant'. Crops are typically carbon-dioxide deficient, as despite the world's problems, the carbon dioxide concentration in air is still low; industrial waste carbon dioxide maximises algal growth such that sunlight is limiting; thus the preference of ±30° from the equator to obtain maximum light all year round for algal culture. Unlike conventional agriculture where a large proportion of applied fertiliser is washed away in run-off, is leached down to aquifers or remains in parts of plants that are not used, nearly 100% of applied fertiliser is inside cultured algae, due to rapid uptake as part of evolving a competitive advantage over other algae. Unlike cereal crops, algae do not sequester phosphate, iron, copper and zinc in phytic acid, limiting their availability. The 25% oil naturally found in algae is high in omega-3 fatty acids and can replace fish-oil as an energy source for fish, and the resulting fish will be high in omega-3 fatty acids. Fish fed other vegetable oils lack this healthy nutrient. Thus microalgae could be a major feed source for an aquaculture industry, providing employment, inexpensive protein and oil, replacing at least in part meal and oil derived from fish, as well as supporting an export industry, both of algae but especially aquaculture products.

LIMITATIONS TO ACHIEVING THIS UTOPIA

There have never been large enough quantities of inexpensive algae to perform full-scale feedings trials to assess the complete value of the algae as fishmeal/fish-oil replacements, and to ascertain what could be done to achieve optimal feeding efficiency from the algae. This is due to inadequacies in present production systems as well as inherent problems in the algae.

High production and harvest costs

At present, the cost of producing algae is exorbitant for use as a bulk feed source due to high production costs of the algae themselves and the high cost of harvesting the tiny algae. Until recently two production choices were presented to potential growers – open circulating 'raceway' ponds, with continuous pumping and mixing of large volumes of dilute cultures. These are cheap to build but expensive to run, and can easily become contaminated with other algal species (of which some are poisonous), microorganisms, protozoans, various algal-eating microfauna, etc. Much of the carbon dioxide pumped in (at high energy costs) is lost. Despite being dilute, most of the algae are statistically in the dark due to light penetration problems. Evaporative cooling is cheap, but the lost water must be replaced by fresh water to prevent its becoming over-saline.

Closed systems, usually clear pipes or vertical or angled panels, are capitally expensive, but use far less water because algae can be more concentrated. These are used for high-value products but cannot be designed for cheap production. Closed systems are coming on line made of clear plastic film, floating on cooling water in fields similar to rice paddies, or at sea and mixed by wave motion (Meiser and Verhein, 2011). These could be a starting point for developing systems that can produce algae cost-effectively for fishmeal replacement, and maybe eventually for biofuels. Algae for biofuels are presently worth about one-third of algae for fishmeal and fish-oil (unless the biofuels are subsidised).

Harvesting microalgae with high-speed centrifuges, the current technology, is far too expensive both in capital and running costs. Inexpensive, non-toxic flocculation systems have recently been developed that de-water at a fraction of the cost of centrifugation (Schlesinger et al., 2012). These are predicated on the algae being more concentrated than the typical algal culture, but fit the densities achieved in some of the closed systems. The flocculated algae can be further de-watered by filtration; the flocs, unlike individual algae, do not block the filters. Drying (where algae will not be used fresh) can be performed utilising cylindrically concave mirrors focusing sunlight on algae moving on a belt. Algae would only be harvested when growing rapidly, so solar drying energy is not needed at night, and less on cloudy days.

Algae, like all other crops, need domestication

Some naively think that one can take algae from the sea and cultivate them, unless one wishes to perpetuate the mistakes admitted by our predecessors (Sheehan et al., 1998). Algae require a suite of 'platform' traits to be appropriate for wide-scale culture: the ability to grow densely, without cultures 'crashing' (suddenly dying at a critical high density due to the phenomenon known as quorum sensing); resistance to contamination by other algae, microorganisms and herbivores; and an engineered inability to succeed in natural ecosystems as a biosafety measure (Table 14.1). In addition, they must be domesticated to have value-added traits to become a more highly digestible, optimal feed source. They should possess the most cost-effective balance of essential amino acids, fatty acids and carbohydrate for mixing with other, far less

Table 14.1 *Platform domestication traits for algae, and examples of transgenes for domestication*

Trait	Gene or trait	Comments
Herbicide resistance for resistance to algal contamination		
Glyphosate	Modified EPSP synthase	1000 µM herbicide required
Glufosinate	*bar*	30 µM herbicide required
Fluorochloridone	Mutant phytoene desaturase	<1 µM herbicide required
Butafenacil	Mutant protoporphyrinogen oxidase	<1 µM herbicide required
Resistance to microorganisms		
Bacteria/fungi		
Antimicrobial proteins	E.g. lactoferricin	
Viruses		
RNAi or overexpression	Specific pieces of viral DNA or cDNA	Untested
Resistance to zooplankton		
Protozoans	Antimicrobial peptides	Untested
Sea lice	Avermectins	Untested
No quorum sensing	Anti-apoptosis genes	Achieve dense cultures
Maximum growth		
Smaller PSII antennae	*tla*1 gene	
Systems/synthetic biology	New light reactions	Untested
	New dark reactions	Untested
Heat tolerance	*psbA* double mutant and/or polygenes	Works in cyanobacteria
Inability to grow in nature		
	Δ carbonic anhydrase	Carbon starved in nature
	Δ nitrate/nitrate reductase	Nitrogen starved in nature
	Partially suppressed RuBisCO	Carbon starved in nature

Δ, deleted section of gene resulting in inactivity.
Further information and references can be found in Gressel (2013).

expensive feed material. In addition, more value can be obtained from algae if they can replace the many expensive microadditives put in standard fish diets (enzymes that assist obtaining maximum feed efficiency from cheaper grains, minerals, oral vaccines, replacements for antibiotics and hormones, etc.).

Domestication took thousands of years to get to our present crops from puny, low-yielding wild species, which were full of anti-nutrients, attacked by insects and diseases, that dropped most of their seeds before harvest, were not responsive to fertiliser, and could not be cultured in dense monocultures. Crop domestication was initially a function of inadvertent selection of biodiversity within the species assisted by sexual reproduction that allowed novel trait combinations. Breeding later accelerated domestication using an understanding of genetics. Most algal species lack a sexual cycle, so selection would be wholly dependent on inherent mutations, limited by the genes present in each species's genome. Whereas various laboratories and companies have selected for some adaptive traits (e.g. heat tolerance) in mass culture, it is not known even whether the adaptations are via mutation or are epigenetic (i.e. not directly affecting its DNA). Crops need novel traits that are not inherent within the species, and these are being introduced by genetic engineering (Gressel, 2008). If we wish to domesticate algae rapidly by suppressing unwanted genes, increasing the expression of other inherent genes and adding traits not found in the algae, genetic engineering is required for obtaining the ideal algae for feeding in aquaculture. One strain will probably not fit all uses, and possibly different traits will have to be introduced into different microalgae for the different species of fin-fish, shellfish, and crustaceans being fed.

DOMESTICATING ALGAE AS AN OPTIMAL FEED SOURCE FOR AQUACULTURE

The use of genetic engineering for domesticating algae is discussed at length in a recent review (Gressel, 2013). The genes that are needed as platform traits are summarised in Table 14.1, and those traits that can add value in Table 14.2.

It is exceedingly important to keep cultures from becoming contaminated as many potentially contaminating microalgae and cyanobacteria are poisonous to cultivated aquaculture species and, if not, to the humans who later eat them (e.g. Ianora and Miralto, 2010). As can be seen in Table 14.1, four different genes conferring herbicide resistance have been transformed into different algal species but the large

Table 14.2 *Some possible value-added domestication traits for algae used as an enhanced feed in aquaculture*

Trait	Gene or trait	Comments
Enhancing digestibility	Antisense or RNAi of cell wall glycosyl transferases	Algal genes unknown
Enhancing digestibility	Introduce vacuolar sequestered carbohydrases	
Increasing methionine content	Modified cystathionine synthase + zein peptide	Requires high level of promotion
Increasing lysine content	Feedback insensitive dihydrodipicolinate synthase	Requires high level of promotion
Enriching/modifying omega-3 fatty acids	ALA, EPA and elongases	Suppression or overexpression, depending on need
Release bound phosphate, Fe, Zn	Phytase	
Increase iron content	Inactive ferritin	
Increase Cu and Zn	Inactive CuZn superoxide dismutase	
Remove fishy odour	Express trimethylamine oxidase	
Replace feed efficiency-enhancing antibiotics	Antimicrobial peptides	
Controlling sea lice	Avermectins	
Vaccines	Various genes	
Increase growth rate of fish	Fish growth hormone	Hard to pass into bloodstream

Further information and references can be found in Gressel (2013).

amounts required of two of them render their use impractical. There is a vast array of known antimicrobial peptides that can suppress bacterial, fungal, and some protozoan infections. These are not antibiotics used in medicine and indeed kill antibiotic-resistant bacteria (Hancock and Sahl, 2006). It is improbable that resistance will rapidly evolve to an antimicrobial peptide, as this would require mutations having a negative effect on several different characters in cell membrane structure (Alberola *et al.*, 2004). The mechanisms that provide pathogens with multi-drug antibiotic resistance do not confer resistance to these antimicrobial peptides; thus, there can be little fear that they will increase the presence of pathogens resistant to conventional antibiotics (Park *et al.*, 2011).

The genes encoding the natural avermectin insecticides from broths fermented by the soil bacterium *Streptomyces avermitilis* are known and could be useful in controlling herbivorous zooplankton (Qui *et al.*, 2011). Both the antimicrobial peptides and avermectin-synthesising traits could also add value in animal feed, increasing feed efficiency in the same manner as banned antibiotics, and might control fish parasites such as sea lice (Table 14.2).

Too often wild-type cultures of algae grew well until stressed, and then the populations crashed. In nature, the evolutionarily desirable reaction to stress is for most of a population to die, leaving resources for the chosen few. This is governed by 'quorum sensing', which sets off a controlled cell death response. Such culture crashes are the dread of commercial production and some of the genes that prevent such chain reactions are becoming known (Rajamani *et al.*, 2011).

Various strategies are being used to increase the photosynthetic efficiency of algae being cultivated in situations so different from nature, where carbon dioxide and minerals are not limiting and light is the limiting factor. The strategies include reducing the chlorophyll content in the antennae that collects light so that light reaches more cells, to a complete revamping of the light and the dark reactions of photosynthesis.

It is imperative that the inevitable spill from large-scale algal systems does not unbalance natural ecosystems, whether the algae are genetically engineered or not. Strategies have been elucidated to mitigate such effects by suppressing genes unneeded in culture but necessary for reproduction in nature. Because of the high carbon dioxide environment of cultivation, the enzyme carbonic anhydrase that traps and conveys carbon dioxide to the chloroplast is unnecessary and can be deleted, and part of the RuBisCO, the inefficient chloroplast enzyme that fixes carbon dioxide into its first organic product, is redundant and can be suppressed. Likewise algae cultivated on urea or ammonia do not need nitrate or nitrate reductases. Suppressing these genes relegates escaped algae to oblivion in natural ecosystems (Chen *et al.*, 2011).

The production of various vaccines in algae, typically in *Chlamydomonas* chloroplasts, has been widely discussed (Siripornadulsil *et al.*, 2007). While expression can be high, there is little evidence of actual biological effect. Many peptides that are part of immunisation regimes must be coated with sugar molecules (glycosylated) in the cytoplasm after they are synthesised, or they are not active. There is no glycosylation of chloroplast-expressed proteins. Marine algae that have adequate nuclear gene expression and proper subcellular targeting allowing glycosylation may prove superior. Gut-active vaccines will be easier to develop than those such as fish growth hormone, which must penetrate the intestinal mucosa to get into the bloodstream.

Other possible value-added traits are outlined in Table 14.2 and in Gressel (2013).

MUST ALL THESE TRAITS BE ENGINEERED INTO ALGAE?

The price of fishmeal and fish-oil have both recently stabilised at about $US1400/ton, but is bound to go up when the present economic slump is over due to the inelastic supply of fishmeal and the increasing demand for the products of aquaculture (Figure 14.1). If the algae truly replace fishmeal 1:1, they are worth $US1400/ton; if only 70% of the feed value of fishmeal, then $US980/ton. Thus, if one can currently produce wild-type algae profitably at less than $US1.00/kg there is an immediate market for algae to replace at least part of the fishmeal. Some of those developing new cultivation systems claim production costs to be much lower (on the internet). Thus, one could immediately start production with large-scale feeding trials with whatever genes are presently available and obtain results that will provide the feedback on what genes, beyond those needed for a robust platform, will be needed, and will be most valuable.

IS THE SUPPLY OF CARBON DIOXIDE LIMITING FOR CULTURING ALGAE IN DEVELOPING COUNTRIES?

Many experts have excluded the tropical desert areas as being inappropriate for algal culture. Their logic is clear: the advantage of micro-algae is that very high growth rates can only be achieved by supplying adequate amounts of carbon dioxide. The pumping costs would be too high if air was the source of carbon dioxide. Despite the doubled levels of carbon dioxide since the Industrial Revolution, the concentration is still very low. The obvious solution is to use purified flue gas from coal-fired power plants as a source of carbon dioxide. There are very few such power plants in tropical desert areas, so these areas are written off.

There is one forgotten source of carbon dioxide in some tropical desert areas, and carbon dioxide could surprisingly be cheaply shipped to others. Much of the supply of natural gas (methane) now comes from the desert or near-desert areas of the Persian Gulf, West and North Africa, the South Pacific and Northern Australia. This natural gas comes out of the ground mixed with carbon dioxide, which is separated from the methane before compressing to liquid natural gas (LNG). This pure carbon dioxide could then be used for algal culture.

What about carbon dioxide for the poorer areas where there is no natural gas? First, even in some desert areas where there is natural gas there is poverty and high unemployment, e.g. North and West Africa. Second, if the developed countries are serious about carbon sequestration, there is another source of cheap carbon dioxide for poor countries. Presently the talk is to liquefy carbon dioxide from flue gases and store it underground or in the deep sea. The ships that bring LNG to the developed world return with seawater as ballast. Instead they could return with liquid carbon dioxide at near the same shipping cost and drop it off at a new aquaculture alga-culture area. This seems a far safer and cost-effective use of carbon dioxide than storing it at the bottom of the sea or in abandoned mines. Some of the energy released in vaporising the LNG could be used to pre-cool the carbon dioxide, lowering the cost of liquefaction. On arrival in the tropics, the energy released in vaporising the liquid to gaseous carbon dioxide for alga-culture could be used to flash freeze aquaculture products for export, as well as for air conditioning to make life more bearable in these hot areas. The saving in fossil fuel should at least partially offset the carbon credits that would be received from underground/underwater sequestration.

CONCLUDING REMARKS

It is clear that marine algae cultured on seawater with waste fossil carbon dioxide can soon serve as a major feed source of protein and lipid for aquaculture. Culturing algae is probably already near being cost-effective, and will remain more cost-effective than cultivating algae as an unsubsidised biofuel feedstock. When the algae are optimally genetically engineered with many value-added traits and the production systems honed to maximal efficiency, there is a good chance that algae will be competitive with feed grains, allowing marginal agricultural areas to revert to nature, and to compensate for urban encroachment on farmlands. Such algae might represent the next quantum leap in productivity that keeps Malthus' predictions from coming true, despite population growth.

Such algal culture coupled with aquaculture could change the economies of tropical developing countries with two new exportable cash crops: algal meal and aquacultural products. Unlike cash crops such as cotton, which often leave the locals starving, algae would be a cash crop that could solve some food sufficiency problems in the countries where they are grown.

ACKNOWLEDGEMENTS

The author gratefully acknowledges the scientific team at TransAlgae Ltd, with whom he interacted as chief scientist and learnt together about algae for 3 years. He especially appreciates the co-learning experience with Drs Shai Ufaz, Ofra Chen, Daniella Schatz and Doron Eisenstadt. Neither they nor TransAlgae are responsible for the views stated herein.

REFERENCES

Alberola, J., Rodriguez, A., Francino, O., et al. (2004). Safety and efficacy of antimicrobial peptides against naturally acquired leishmaniasis. *Antimicrobial Agents and Chemotherapy* **48**, 641–643.

Chen, O., Ufaz, S., Eisenstadt, D., et al. (2011). Transgenically mitigating the establishment and spread of transgenic algae in natural ecosystems by suppressing the activity of carbonic anhydrase. *US Patent Application* 2011/0045593.

Gatlin, D. M., Barrows, F. T., and Brown, P. (2007). Expanding the utilization of sustainable plant products in aquafeeds: a review. *Aquaculture Research* **38**, 551–579.

Gressel, J. (2008). *Genetic Glass Ceilings: Transgenics for Crop Biodiversity.* Johns Hopkins University Press, Baltimore, MD.

Gressel, J. (2013). Transgenic marine microalgae as an enriched fishmeal replacement. In A. Richmond and Q. Hu (eds.) *Handbook of Microalgal Culture*, 2nd edn. Blackwell, Chichester, UK.

Hancock, R. E., and Sahl, H. G. (2006). Antimicrobial and host defense peptides as new anti-infective therapeutic strategies. *Nature Biotechnology* **24**, 1551–1557.

Ianora, A., and Miralto, A. (2010). Toxigenic effects of diatoms on grazers, phytoplankton and other microbes: a review. *Ecotoxicology* **19**, 493–511.

Li, M. H., Robinson, E. H., Tucker, C. S., Manning, B. B., and Khoo, L. (2009). Effects of dried algae *Schizochytrium* sp., a rich source of docosahexaenoic acid, on growth, fatty acid composition, and sensory quality of channel catfish *Ictalurus punctatus. Aquaculture* **292**, 232–236.

Meiser, A., and Verhein, M. (2011). Photobioreactor. *US Patent Application* 2011/0124087.

Naylor, R. L., Hardy, R. W., Bureau, D. P., et al. (2009). Feeding aquaculture in an era of finite resources. *Proceedings of the National Academy of Sciences USA* **106**, 15103–15110.

Park, S.-C., Park, Y., and Hahm, K.-S. (2011). The role of antimicrobial peptides in preventing multidrug resistant bacterial infections and biofilm formation. *International Journal of Molecular Sciences* **12**, 5971–5992.

Patnaik, S., Samocha, T. M., Davis, D. A., Bullis, R. A., and Browdy, C. L. (2006). The use of HUFA-rich algal meals in diets for Litopenaeus vannamei. Aquaculture Nutrition **12**, 395–401.

Qui, J., Zhuo, Y., Zhu, D., et al. (2011). Overexpression of the ABC transporter AvtAB increases avermectin production in Streptomyces avermitilis. Applied Microbiology and Biotechnology **92**, 337–345.

Rajamani, S., Teplitski, M., Kumar, A., et al. (2011). N-Acyl homoserine lactone lactonase, aiia, inactivation of quorum-sensing agonists produced by Chlamydomonas reinhardtii (Chlorophyta) and characterization of aiia transgenic algae. Journal of Phycology **47**, 1219–1227.

Reitan, K. I. (2011). Digestion of lipids and carbohydrates from microalgae (Chaetoceros muelleri Lemmermann and Isochrysis aff. galbana clone T-ISO) in juvenile scallops (Pecten maximus L.). Aquaculture Research **42**, 1530–1538.

Schlesinger, A., Eisenstadt, D., Bar-Gil, A., et al. (2012). Inexpensive non-toxic flocculation of microalgae contradicts theories; overcoming a major hurdle to bulk algal production. Biotechnology Advances **30**, 1023–1030.

Sheehan, J., Dunahay, T., Benemann, J., and Roessler, P. (1998). A Look Back at the U.S. Department of Energy's Aquatic Species Program: Biodiesel from Algae. National Renewable Energy Laboratory, Golden, CO.

Siripornadulsil, S., Dabrowski, K., and Sayre, R. (2007). Microalgal vaccines. In R. León, A. Gaván, and E. Fernández (eds.) Transgenic Microalgae as Green Cell Factories. Landes Bioscience, Austin, TX.

Walker, A. B., and Berlinsky, D. L. (2011). Effects partial replacement of fish meal protein by microalgae on growth, feed intake, and body composition of Atlantic cod. North American Journal of Aquaculture **73**, 76–83.

Part 3   Lessons learned about implementing new genetics crops in policy

Introduction

BRIAN HEAP

It is said that A. E. Housman once began a book review, *'There is much in this book that is true and much that is new. But that which is true is not new and that which is new is not true'* (Naiditch, 1988). Are there any lessons to be learned about implementing the genetics of crop production into policy that are new or true?

Picked up in a recent report titled *The Future of Food and Farming: Challenges and Choices for Global Sustainability* (Foresight, 2011) is the growing international unease that global action is required to avert future global food security crises, and that we will need the best science, technology and innovation to address the challenge. The remarkable successes of the modern era have come from scientific research and the talents and dedication of some of the very best minds and research teams in the world. The lesson for today's citizens is to grasp a communal vision for an adventure in the pursuit and wise application of knowledge. It is essential to create a climate of well-justified confidence in scientific and technological progress as described by Jack Bobo and Roger Beachy in Chapter 19, restore self-assurance in the processes of accountability and equity as outlined by Mark Cantley and Drew Kershen in their Chapter 16, and enhance the opportunities that exist for rewarding collaboration that can promote synergy and economies of scale if managed positively.

Successful Agricultural Innovation in Emerging Economies: New Genetic Technologies for Global Food Production, eds. David J. Bennett and Richard C. Jennings. Published by Cambridge University Press. © Cambridge University Press 2013.

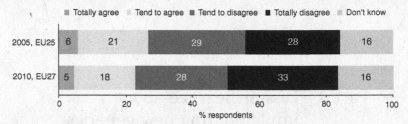

Figure 1 Support for GM food EU25 2005 vs. EU27 2010. (Gaskell, *et al.* 2010)

The principal worry is that, with the global population expected to rise from about 6 billion to about 9 billion over the next 40 years, the growth in demand for food will outstrip the growth in nutritional supply, leading to increased hunger and deteriorating international security. In particular, there is concern for the sustainability of current food systems in Africa – a region anticipated to double in size from about 1 billion inhabitants to close to about 2 billion inhabitants by 2050 (see Burckhardt *et al.*, Chapter 15). Unlike other developing regions, such as Asia and Latin America which have both experienced significant uplifts in productivity and food security over the past 50 years, sub-Saharan Africa has largely stood still. To make matters worse, opportunities to increase production through land expansion are proving increasingly challenging and costly, and consequently the most likely scenario is that in the future Africans will need to produce more food from the same amount of (or even less) land.

In Europe, the idea of turning good scientific evidence about GM crops into policy has been dictated by the interests of politicians responding to a sceptical public. Here there are on average about three times more opponents to GM than supporters. Roughly a third of all people totally disagree with using GM food and it is only about 5% that strongly support the use of GM food. In spite of being used for decades in the USA, in Europe GM food is seen as being potentially unsafe, unnatural and not particularly beneficial (see Figure 1). It is notable that the negative perception of GM use in Europe is mainly linked to food, while regenerative medicine methods, including embryonic stem cell research, have average approval levels of more than 60%. One chink of light is with cisgenic GM crops (with genes added only from the same species or from plants crossable under conventional breeding methods) which are deemed more acceptable in Europe with an average support rate of 55% (Eurobarometer, 2006).

Sir David King, until 2007 the UK's Chief Scientific Advisor, once said that as many as 700 000 lives were lost every year because of malnutrition and unhygienic food and water which could be saved by implementing advanced farming methods, such as GM (Alleyne, 2008). The immediate reaction from a blog entry on GMWatch was a discussion of 'the many lies of David King' (GM Watch, 2010). Likewise, the Norwegian Council for Africa refers to agro-multinationals such as Monsanto, Dupont or Syngenta as forms of 'genetic colonialism' that cause *desertification, deforestation, water depletion, loss of income and hunger'* (Sharife, 2008).

Science needs openness and the lesson of the genetics of Lysenko-ism in the former Soviet Union shows what can happen to science in a society where political orthodoxy determines what is acceptable and what is not, or where political/industrial imperatives are more highly valued than the best international scientific evidence which points to a different solution. The pressures now are greater than ever: bigger population, tougher international competition, warmer climate, and rapid consequences of taking the wrong decisions. Decision-makers (not only in government) can now make gross mistakes that are bigger than ever, and the technology of instant communication means that they are immediately exposed to universal critique.

Finally, three lessons for policy-makers emerge in the following chapters which are true, if not new. First, you need good entrepreneurs to translate new knowledge into benefit (see Jack Bobo and Roger Beachy, Chapter 19). Second, decision-makers must, and increasingly do, recognise when they are dealing with science-related issues rather than policy-related priorities. And third, we all need to distinguish good practices and regulations from bad ones, and say so. The decision-maker has to understand the options available and the foreseeable consequences of each. Lord May of Oxford, when Chief Scientific Advisor at the Cabinet Office, stressed the need for a transparent and objective system that ensured scientific advice is thoroughly comprehensible and rigorously tested. Governments need to be forewarned of potentially controversial issues at an early stage to ensure that the best scientific advice can be marshalled, that consultations are broadly based involving all interested parties and that new knowledge is translated into publicly acceptable policies (Government Guidelines, 2010).The belief expressed in 1997 was that the UK had relied too much in the past for analysis of complex issues on character and experience, thereby consolidating power in relatively few individuals, rather than on the rationality of their views. *'In the USA, on the other hand, trust rests in formal processes and style of reasoning that ensure transparency and objectivity of governmental decisions'* (Jasanoff, 1997). As Thomas Babington Macaulay (1800–59) wrote, *'We know no spectacle so ridiculous as the British public in one of its periodical fits of morality'* (Macaulay, 1831). However, in both the UK and the USA the dangerous drift away from the world that people experience and inhabit now demands an even greater dialogue between decision-makers and critics so that uncertainties in science and policy are debated openly and honestly.

REFERENCES

Alleyne, R. (2008). *Europe's GM food fear 'exacerbates famine'*, Daily Telegraph, 7 September. www.telegraph.co.uk/news/worldnews/2700176/Europes-GM-food-fear-exacerbates-famine.html

Eurobarometer (2006). *GM Food: Europeans Still See More Risks than Benefits*, GMO Compass. www.gmo-compass.org/eng/news/stories/227.eurobarometer_europeans_biotechnology.html, page 37

Foresight (2011). *The Future of Food and Farming: Challenges and Choices for Global Sustainability*, Final Project Report. Government Office for Science, London.

Gaskell, G., Stares, S., Alansdothir, A., et al. (2010). *Europeans and Biotechnology in 2010: Winds of Change?* European Commission, Brussels, Belgium. http://ec.europa.eu/public_opinion/archives/ebs/ebs_341_winds_en.pdf, page 37

GMWatch (2010). Lack of GMOs costs lives, claims leading scientist. GM Watch. www.gmwatch.org/latest-listing/1-news-items/11857-lack-of-gmos-costs-lives-claims-leading-scientist

Government Guidelines (2010). *The Government Chief Scientific Adviser's Guidelines on the Use of Scientific and Engineering Advice in Policy Making*. Government Office of Science and Technology, Department for Business, Innovation and Skills, London.

Jasanoff, S. (1997). Civilisation and madness: the great BSE scare of 1996. *Public Understanding of Science* 6, 221–232.

Macaulay, T. B. (1831). In *Museum of Foreign Literature, Science and Art, No. 19*, New Series vol. 12, p. 431. Littell, Philadelphia, PA.

Naiditch, P. G. (1988). *A. E. Housman at University College, London: The Election of 1892*. E. J. Brill, Leiden, The Netherlands.

Sharife, K. (2008). *Genetic Colonialism: From Mendel to Machiavelli*. The Norwegian Council for Africa, Oslo. www.afrika.no/Detailed/17800.html

SAMUEL BURCKHARDT, CLAUDIA CANALES HOLZEIS,
JULIAN GRAY AND BRIAN HEAP

15

Enabling factors for an innovation-ready agricultural landscape in African countries

INTRODUCTION

Food security is a pressing issue on the African continent and, in view of projected population growth, the problem is likely to worsen unless addressed immediately (Godfray *et al.*, 2010). Although no single strategy will provide the full solution, genetically modified (GM) crops are widely considered an important tool to increase agricultural productivity. Despite an 80-fold increase globally in planted hectares between 1996 and 2009, only three African nations have fully operational procedures for the adoption of GM crops to early 2012. They are South Africa (with 2.1 million hectares of GM crops), Burkina Faso (0.2 million ha) and Egypt (0.1 million ha). In February 2009 a Biosafety Bill was signed by Kenya opening the way towards the commercialisation and release of GM crops. Ghana took a similar step in December 2011 signalling the interest that exists among several African nations in the legislative and regulatory requirements for the introduction of GM crops.

Criticism and controversy still surround the use of GM crops. Concerns recur about human safety, environmental risks and fears of smallholder exploitation by multinational GM seed manufacturers. In addition, severe underfunding of agricultural research in sub-Saharan African (SSA) nations, slow and problematic privatisation of seed distribution sectors, weak agricultural extension services and a lack of infrastructure and credit facilities can hinder the adoption of the technology.

Successful Agricultural Innovation in Emerging Economies: New Genetic Technologies for Global Food Production, eds. David J. Bennett and Richard C. Jennings. Published by Cambridge University Press. © Cambridge University Press 2013.

Table 15.1 *Potential applications of GM crop technology*

Current
Tolerance to broad-spectrum herbicide
Maize, soya bean, brassica
Resistance to chewing insects
Maize, cotton, oilseed, brassica
Nutritional biofortification
Staple cereal crops, sweet potato
5–10 years
Resistance to fungus and virus pathogens
Potato, wheat, rice, banana, fruits, vegetables
Resistance to sucking insects
Rice, fruits, vegetables
Improved storage and processing
Wheat, potato, fruits, vegetables
10–20 years
Drought and salinity tolerance, improved nitrogen use, high
temperature tolerance
Staple cereal and tuber crops
Over 20 years
Apomixis, nitrogen fixation, conversion to perennial habit
Staple cereal, tuber crops

Source: After Godfray *et al.* (2010).

A number of European Union countries have shown strong opposition to GM crops and this has sent a potent signal of consumer resistance to developing nations (Paarlberg, 2008). Many international observers have been heavily critical of the EU for its role, as a major trading partner, in influencing anti-GM policy in many African countries. In particular, reference is made to instances in the past when Zambia (2002, 2010) and Angola (2004) rejected food aid targeted at starving citizens because of its GM origin.

A further criticism is that, to date, commercialised GM crops are mostly suited to temperate climates, although cotton and maize constitute important exceptions. Crops targeted to Africa are currently being developed largely through private/public partnerships (cowpea, cassava, banana and maize), and several initiatives are at the field trial stage. And very promising applications of GM technology are afoot (Table 15.1). However, lessons learned from other continents tell of certain requirements for a dynamic and sustainable GM crop adoption regime. It is not just a question of access to the technology or to improved seeds, but of the enabling factors that facilitate an agricultural landscape capable of adopting innovation (Juma, 2011).

ENABLING FACTORS

Ghana and Tanzania have been selected with four considerations in mind – geography, governance, willingness to adopt new technologies, and corruption level. Located in West and East Africa respectively they are close to nations that have either commercialised the technology (Burkina Faso) or are very near to doing so (Kenya and Uganda). Ghana has made more rapid progress than Tanzania, and some of the important enabling factors will now be examined under seven headings.

Current production of modifiable crops

According to data for 2009, from the Food and Agriculture Organization of the United Nations (see FAO, 2012 for latest data) for the two nations, in Ghana (Figure 15.1) yams represent the main agricultural product (about $US1.5 billion) followed by cassava (about $US1.3 billion) and plantains (about $US0.7 billion). Of the crops currently grown elsewhere in the globe as GM varieties, only maize and tomatoes feature in Ghana's top 20 commodities by value. If the scope is broadened to GM crops currently in development, then yams, cassava, rice, sorghum and coconuts could be

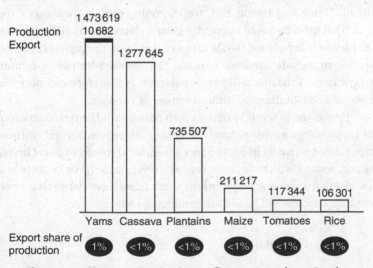

Figure 15.1 Ghana: production (upper figure, open columns) and export (lower figure black column) share of crops with GM potential (2009, in US$ × 1000).

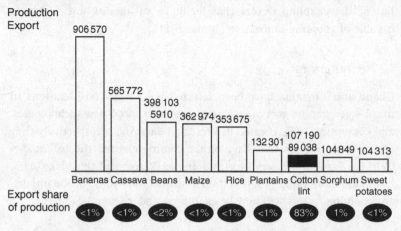

Figure 15.2 Tanzania: production (upper figure, open columns) and export (lower figure black column) share of crops with GM potential (2009, in US$ × 1000).

added to the list. A crude measure of Ghana's GM potential can be calculated by considering the cumulative value/quantity of modifiable produce – up to $US3 billion/18 million tonnes of produce.

In Tanzania (Figure 15.2), bananas are the main agricultural product (about $US0.9 billion in 2009), followed by cassava (about $US0.6 billion), beans (about $US0.4 billion) and maize (about $US0.4 billion). Maize and cotton lint are the only crops in Tanzania's top 20 commodities by value currently grown elsewhere as GM varieties. If the scope is broadened to GM crops currently in the pipeline, the list expands to include bananas, cassava, rice, sweet potatoes, sorghum, and potatoes. Tanzania's GM crop potential is therefore estimated to be about $US2.2 billion/20 million tonnes of produce.

Hence, the potential exists in both nations for the introduction of GM crops across a wide range of products. Moreover, several of these crops cannot be easily improved by conventional means of plant breeding (e.g. banana which can only reproduce vegetatively) or because of a lack of suitable genetic variation for some traits (cowpea) so that alternative approaches will have to be countenanced.

Laws and regulations

What is the state of the requisite legal and regulatory framework which must be in place for the commercial release of such crops? The global legal and regulatory framework surrounding GM crops is built upon two international United Nations commitments – the Convention of Biological Diversity (CBD) of 1993 and the Cartagena Protocol on Biosafety (CPB) of 2000. The former is an international commitment to the conservation of biological diversity, sustainable use of biological resources, and sharing equitably the benefits arising from the use of genetic resources. The latter is an international regulatory framework that reconciles the respective needs of trade and environmental protection with respect to trans-boundary movement of genetically modified organisms (GMOs). Countries that sign and ratify the Cartagena Protocol accept being bound by the provisions of the Protocol. All 51 sub-Saharan African countries are parties to the CBD; however, a detailed study by the African Agricultural Technology Foundation in 2006 found that only 36 African countries had signed and ratified the Cartagena Protocol.

Ghana can be considered to have a positive stance towards GM crops and progress has been made in respect of an enforceable regulatory framework for GMOs. Ghana's parliament approved a Legislative Instrument on Biosafety in 2008, and in December 2011 the Biosafety Bill was signed. The legislation paves the way for confined field trials of GM crops such Bt cowpea and GM sweet potato, and will establish a framework for the commercial cultivation of GM crops.

The government of Tanzania has been less transparent about the development of policy for GM crops than Ghana, notwithstanding several public announcements of support for the technology to increase food production. While the planting of GM crops is currently prohibited, the situation surrounding field trials is uncertain. For example, despite several reports in the press and quotes from government ministers stating that GM crop field trials have taken place in Tanzania for cotton, tobacco and maize, the current legislation contains a strict liability and redress clause that is likely to hamper not only adoption but also the confined field-testing of GM crops. These confusing positions seem to mirror the country's stance towards GM crops, namely, an ambiguity about political commitment to the adoption of the technology.

Both countries are aware of the needs for regulation and legal enforcements and at the present time more needs to be done in Tanzania compared with Ghana. This is not to overlook the place of national regulations that relate to the environment, food, drugs and plant quarantine; in some instances they may appear to be disabling rather than enabling factors in the adoption of GM crops.

Trade flows

An important factor in the adoption of GM crops is whether doing so would affect a SSA nation's international trade. According to WTO (2012a) Africa's exported agricultural products in 2010 were worth approximately $US55 billion. Europe was the largest importer with $US20 billion of product imported, followed by Asia ($US9 billion), the Middle East ($US6 billion) and North America ($US3 billion). Trade within Africa accounted for $US11 billion; SSA nations have a share of 80% of trade with partners outside the region, a share that could be even higher given the economic potential of the continent. Nevertheless, Africa still suffers from a negative trade balance in food trade.

In 2010, Ghana exported $US5.8 billion of agricultural goods, which was about 73.4% of its total exports (WTO, 2012b). The main target destinations of total exports were South Africa (53.5%), the European Union (17.9%) and United Arab Emirates (6.8%). According to FAO and WTO data, exports represented a very small share of the crops with GM potential. Tanzania exported $US1.2 billion of agricultural goods to other countries which represented about 32% of its total exports. In Tanzania, exports of crops with GM potential was slightly higher than in Ghana, but the share of total production remained fairly small. Cotton-related products with 74% had the highest share of exports in relation to total production (GM cotton is not segregated from conventional cotton in international markets).

The data indicate, therefore, that there is no clear trade-related reason against the adoption of GM crops in either Ghana or Tanzania because the crops under development are, as yet, not important food export commodities for Europe.

Agricultural research capacity

An important pillar for the adoption of agricultural innovations is an adequately funded and skilled system for agricultural research in both the developed and the developing world (FAO, 2012). Local agricultural research capacity plays a fundamental role in strengthening extension services and these are key for the effective dissemination of technological advances to farmers, the ultimate end users.

Agricultural research and development (R&D) in SSA was deeply neglected during the last two decades of the last century, but this trend has since been reversed. The increased recognition by African policymakers that agricultural development is an engine of economic growth has been reflected in an increase of about 20% in investments in agricultural R&D in SSA between 2001 and 2008, after more than a decade of stagnation (Beintema and Stads, 2011). However, over one-third of this growth is attributable to a $US110 million increase in spending in Nigeria. Ghana, Sudan, Tanzania and Uganda also experienced relatively high increases in total spending of between $US25 million and $US56 million each. The 'Big Eight' – Nigeria, South Africa, Kenya, Ghana, Uganda, Tanzania, Ethiopia and Sudan – accounted for 70% of regional public agricultural R&D spending and 64% of all researchers in 2008 in SSA. Regrettably, during the same 2001–8 period, 13 out of 29 SSA countries studied experienced negative yearly growth in public agricultural R&D spending, and five of these countries also experienced negative economic growth. Therefore, because of the relative size and economic importance of the 'Big Eight', developments in capacity or spending have a significant impact on regional trends.

Accompanying increases in public spending on agricultural R&D in SSA, contributions by international donors and funding agencies have rebounded in more recent years with the establishment of regional projects funded through World Bank loans as part of the West Africa Agricultural Productivity Program (WAAPP) and the East Africa Agricultural Productivity Program (EAAPP). Activities focus on generating and disseminating improved agricultural technologies that address both national and regional priorities. In relative terms, Ghana ranks highly for agricultural research capacity in SSA, but in absolute terms it lags far behind South Africa due to a combination of lower investment, poorer regulatory frameworks and a historically more sceptical stance towards the perceived benefits of investing in research. This is changing. Agricultural R&D spending in Ghana more than doubled during 2000–8 with staff numbers showing steady but less dramatic growth. However, most of this investment went to increased salary expenditure at the Council for Scientific and Industrial Research (CSIR), which after years of underfunding was incommensurably low and uncompetitive; rather than to expansion of research activities or to improve equipment or infrastructure. Nonetheless, Ghana has already committed an investment of 1% of the national GDP to support science and research in 2012, an indication of the renewed political support to this area. Ghana has at least six institutes with agricultural biotechnology research capacity and an estimated 28 agricultural biotechnology projects – only one of which involved GM technology. Non-profit and for-profit companies do not play a major role in agricultural R&D in Ghana and government research organisations account for about

85% of all agricultural R&D spending, the remainder being accounted for by higher education organisations.

A very positive development for the whole region and continent is the renewed interest in studying the genomes of 'orphan' food crops which are very important nationally, but have traditionally been neglected because they are not main commodities in international markets. An international consortium has recently announced a plan to raise $US40 million to sequence the genomes of 24 orphan crop species. Part of the initiative is to set up a Plant Breeding Academy in Accra to train African scientists on how to use this information effectively in breeding programmes.

Nonetheless, significant challenges remain that potentially threaten Ghana's ability to maintain its current research capacity. The average qualifications of agricultural research staff have changed very little since the 1990s and many research agencies are faced with an ageing pool of scientists while a concurrent ban on recruitment limits new hires.

The overall level of agricultural research capacity in Tanzania is not too dissimilar from that of Ghana. There are at least four institutes that possess agricultural biotechnology research capacity and an estimated 22 agricultural biotechnology projects were undertaken in the same year – with one involving GM technology. Tanzania has five government agencies involved in agricultural research that collectively account for about 70% of total R&D spending. Public agricultural R&D has traditionally been highly dependent on donor funding – a characteristic not limited to Tanzania and one that has resulted in a high level of funding volatility. The Tanzanian government has therefore now prioritised research and the aim is to increase public research investment to 1% of GDP across all research sectors – equating to about $US45 million for agriculture and livestock research – and to increase researchers' salaries by more than 80%.

Clearly, both Ghana and Tanzania recognise the significance of agricultural research as an enabling factor in the adoption of new technologies, not least in the development of indigenous research and in training for expert extension services.

Intellectual property protection

It will be some time before most African nations are in a strong position to file their own GM crop patents. Only South Africa is mentioned in the top international countries. The African Regional Intellectual Property Organization (ARIPO) is an inter-governmental organisation that has the capacity to hear applications and counts 17 members, mainly in SSA

(Ghana and Tanzania are members). In 2008, ARIPO recorded 435 applications; the European Patent Office, as a comparison, filed 146150 applications in the same year (WIPO, 2010). Ghana and Tanzania were granted two patents between 1995 and 2008, both filed in the US Patent Office (World Patent Report, 2010). The numbers of patent applications are likely to increase with the improved funding of research initiatives in the national agricultural R&D centres of both nations. A key, and presently unknowable, factor of course, is the proportion of patents which result in beneficial application as against those which do not and remain as 'sleeping' patents.

Transport, power infrastructure and communications

An efficient transport system is of vital importance especially for seed distribution and access to markets. Indicators include road density (km of roads per 100 km^2 of land area), share of paved roads (% of total roads), quality of port infrastructure (1 = extremely underdeveloped, to 7 = well developed) and number of motor vehicles per 1000 people for transport infrastructure. SSA has, compared with the rest of the world, a very high share of unpaved roads, which makes transport generally more demanding and less efficient.

In Ghana the government has been active with infrastructure improvement programmes in the last 30 years. With a score of 4.0, the perceived quality of port infrastructure in Ghana is above the SSA average. In terms of electric power consumption (kWh per capita), another indicator of the readiness to adapt to technological innovation, Ghana, at 259 kWh per capita, consumes around half the SSA average of 549 kWh per capita (data taken from African Development Bank Group, 2005 and World Bank, 2010).

Tanzania's roads are in a slightly worse condition than Ghana with 71% of the roads in a fair or good condition. The road density, at 9 km per 100 km^2 of land area, is lower than in comparable SSA nations, and only 8.6% of the roads are paved (SSA average of 12.1%). The ranking of port infrastructure scores Tanzania at 2.8 (SSA average, 3.7). With 82 kWh electric power consumed per capita, Tanzania sits far below the SSA average of 550 kWh per capita. This is an indication that large parts of the Tanzanian population, and in particular the rural population, have only limited access to electricity (World Bank, 2010).

The progress of technology-driven communication can be most simply measured by the mobile phone coverage or internet penetration. The number of mobile phone subscribers in Africa soared from

16 million in 2000 to 376 million in 2008, with 60% of the population using them in 2008 compared with 10% in 1999 (CIA World FactBook, 2010). Although the highest demand is coming from major cities, there is also an increasing coverage of rural areas pointing to the potential for ICT-driven agricultural extension systems. However, the full potential of mobile communication will not be realised if the development of other critical infrastructure such as roads, education and health facilities are neglected (Aker and Mbiti, 2010).

In terms of internet users, the penetration of 13.5% in the continent as a whole (139 million users on 1 December 2011; nearly 650 million in November 2012) is lower than the world average of 32.7%, but this has been characterised by a substantial growth over the past 10 years, suggesting that the tipping point in some African countries may have been reached (Internet World Stats, 2011). More recent figures, however, report a figure of nearly 650 million on November 2012. An analysis of the potential for ICT-driven agricultural extension systems within both nations indicates that the use of mobile phone technologies is currently more viable than the internet with Ghana ahead of Tanzania.

Transport, power infrastructure and communications are crucial elements in the development of sustainable agricultural systems that include the adoption of new technologies such as the use of modern genetic techniques for plant breeding. While Ghana may lead Tanzania in some respects, both nations are aware of the need for future investment in these areas.

Macroeconomic and political stability

The stability of the macroeconomic environment is important for the overall competitiveness of a country because the government cannot provide services efficiently if it has to make high interest payments on its past debts. Meissner (2010) has summarised the impact of the current international financial crisis on the many drivers of African growth that have been endangered including the prices of and demand for primary commodities, capital flows, foreign direct investment and regional integration. The global recession threatens the gains achieved in the last decade in the fight against poverty as the economic growth in Africa overall contracted from an average of 5% between 2003 and 2008 to about 2.8% in 2009. Ghana and Tanzania have seen a GDP growth of about 5% though inflation has been above or close to double-digit figures but is expected to slow. However, both countries have scored weakly in terms of macroeconomic stability with a relatively poor central government balance and high inflation rates (World Bank, 2010).

A relatively low macroeconomic stability is not necessarily an unsurpassable hurdle for the adoption of new technologies such as GM. Burkina Faso has already proven that success with GM crop adoption is possible despite a comparably low level of macroeconomic stability. Nonetheless, on the financial stability front, uncertainties persist about the willingness of banks to lend to private investors and in this respect the major financial stability risks will lie outside the continent (Meissner, 2010).

The corruption perception index (CPI) is another factor to consider since the continent as a whole suffers from corruption levels which have a negative effect on inward investment. The CPI score for SSA averages 2.84 (1 = very corrupt; 10 = absence of corruption) against a world average of 4.0 across 178 countries. In terms of political stability, while there is a trend towards a smaller number of political incidents in Africa, the State Fragility Index indicates that African nations still appear more fragile than many other countries (Transparency International, 2010).

Ghana scores relatively well on low corruption. Its regional rank in SSA is at 7 out of 47 countries, and it is seen as a comparatively very secure nation. Although it is organised as an institutionalised democracy, economic effectiveness is still relatively low, which is typical for many SSA countries. Tanzania is average in terms of perceived corruption, ranking 20th place in the SSA stakes. However, it is seen as a relatively stable country with a moderate state fragility score that ranks it among the top nations in Africa. The security level is comparable with Ghana and very high for regional standards.

The Global Competitiveness Report (Swab, 2009) summarised the two countries in the following way (selected). Ghana displays excellent public institutions and governance indicators, but there has been a deterioration of the country's macroeconomic stability. Overall there was reasonable public trust of politicians, relative judicial independence, and corruption levels lower than in most other countries of the region. Some aspects of the country's infrastructure are also good by regional standards, particularly roads and ports, financial markets are relatively sophisticated, but education levels continue to lag behind international standards at all levels, and the country is not harnessing new technologies for productivity enhancements (ICT adoption rates are very low). The country is characterised by high and increasing macroeconomic instability, the government is running high and increasing fiscal deficits, debt levels are high, inflation is well into the double digits, and high interest rate spreads point to inefficiencies in the financial system.

Tanzania benefits from public institutions that are characterised by reasonable public trust of politicians, relative government even-handedness in its dealings with the private sector, and a security situation that is good by regional standards. In addition, some aspects of the labour markets lend themselves to efficiency, such as the high female participation in the labour force. There has also been a measurable improvement in the sophistication of financial markets, but there are weaknesses in infrastructure with poor-quality roads, ports and electricity supply, and few telephone lines. And although primary education enrolment is commendably high, enrolment rates at the secondary and university levels are among the lowest in the world. Related to the education level of the workforce, the adoption of new technologies is low in Tanzania, with very low uptake of ICTs such as the internet and mobile telephony. In addition, the quality of the educational system receives a poor assessment. And the basic health of the workforce is also a serious concern with poor health indicators and high levels of diseases such as malaria, tuberculosis and HIV.

Therefore, both nations are classified qualitatively in the 'average readiness' category in terms of the adoption of GM technologies (Figure 15.3).

PAVING THE WAY

The implementation of agricultural innovation, which includes improved planting material and enhanced agricultural practices, is the most important constraint to date in developing countries for raising agricultural production. Agricultural performance is location-dependent, and therefore the combination of tools most suited to raise productivity and income for farmers will vary in different regions. For this reason, an agricultural landscape that is ready for innovation is not defined by the adoption (or not) of a specific technological advance. Rather, the defining characteristic is the ability to assess existing technologies and subsequently make the most suitable of these available to farmers. For instance, significant increases in agricultural productivity can be achieved by planting the best available seed adapted to local soil conditions, the use of appropriate nitrogen fertiliser applications and the adoption of management practices that combat poor soil quality. GM crops, although not a panacea, can in certain instances overcome some of the limitations faced by conventional plant breeding techniques, and therefore deserve careful consideration. Establishing a national debate that implicates all relevant actors and is free from the interference of non-African outsiders is essential to pave the way for an informed and nuanced deliberation of the relative merits and risks of a technology.

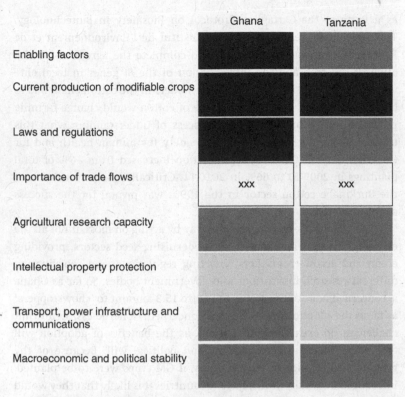

Figure 15.3 Enabling factors for the adoption of GM crops, and a qualitative assessment of innovation-readiness in Ghana and Tanzania. Dark grey, high readiness; light grey, average; stippled, low; xxx, not graded.

The enabling factors listed in the previous section and summarised in Figure 15.3 will impact how a country embraces, and is able to efficiently deploy, novel technologies. The fact that a nation does not embrace all the enabling factors listed in Figure 15.3 should not be interpreted as an impediment to the adoption of new technologies of crop production, but there is one essential requirement – the political commitment and support from national governments.

The experience of Burkina Faso is revealing: a small, landlocked country with a very poor economy managed to complete a national project that made it the second country in Africa to adopt a GM crop (the first being South Africa). What is more, it is very likely that Burkina Faso will have a decade advance on the next SSA countries to adopt the technology. The Burkinabe government had to create an enabling regulatory framework and prepare its institutions for the review and approval of the new technology in compliance with its commitments as a Party of the Cartagena Protocol on Biosafety in Biotechnology. The agricultural research institute, Institut de l'Environnement et de Recherches Agricoles (INERA), had to complete the agronomic evaluation of the crop and the introgression of the *Bt* genes in local elite varieties of cotton. The different stakeholders in the cotton chain had to be informed about how this new type of cotton would change farming practices and be involved in the process of understanding what this technology would mean in terms of safety for human health and for the environment. GM cotton production increased from 16% of total plantings in 2009/10 to 66% in 2010/11. Critically, the privatisation of the Burkinabe cotton sector in the 1990s was pivotal for the success of this initiative.

Political back-up can pave the way by acting on most, if not all the enabling factors whether it is by modernising seed sectors, providing credit and access to markets, enacting regulations, and coordinating different research institutions and government bodies. So far as Ghana is concerned none of the scores in Figure 15.3 amount to 'showstoppers' as far as the adoption of GM crops is concerned. Nonetheless, significant challenges do exist for both nations as the benefits of adoption will take time, financial commitment and political will. Because of the geographical location of both nations, if GM crops were to be planted/introduced internally or in adjacent countries it is likely that they would unofficially spread to neighbouring countries (as was the case in South America) mediated by farmer-to-farmer exchange due to the permeable nature of the borders. Appropriate biosafety regulatory systems need to be in place to regulate this exchange and prevent the surge of illegal seed markets.

A regional approach to modernising agriculture is essential, and drawing on our analysis of the enabling elements, it is unsurprising that Ghana is ahead of Tanzania in being the more ready and likely to move to its first full-scale GM implementation programme. In terms of timescales it is probable that Ghana will draft its first GM crop development plan within the next 5 years, but there is much less confidence that Tanzania will be ready so quickly. Considering a minimum development period of 6–8 years, the first commercialised GM crops are not likely to be planted in Ghana much before 2020 and in Tanzania much before 2023. Table 15.2 summarises our view as to the most likely crops to be planted as GM variants for each country.

A recurrent topic throughout this analysis has been the role of the agricultural extension services and how they can provide services to

Table 15.2 *GM crops most likely to be adopted*

	Probable	Possible
Ghana	Cotton	Cassava
	Maize	Rice
	Sweet potato	
	Cowpea	
Tanzania	Maize	Banana
	Cotton	Cassava
		Rice

subsistence smallholders. The focus has to be on crop type, partnerships involving private, public and non-governmental agencies, and on investments on incentives for farmers. It has to respect decisions taken at a local level and funding channelled through a variety of sources which are most suited to the rapid evaluation and adoption of agricultural innovations. External organisations have to recognise that in funding projects they will have to operate within the constraints of target countries, and be aligned to government priorities.

Communication about the modern advances in plant breeding will improve knowledge of risk assessment and management modalities, measures for mitigating smallholder risk as in Advanced Market Commitments (AMCs) used for vaccine development (Elliot, 2010; GAVI Alliance, 2010), micro-insurance schemes (Roth and McCord, 2008), and value-based input pricing schemes. It will also assist outcome assessments of new technologies on human welfare evaluated through indicators of food security, health and education participation levels; finances through indicators of net economic gains for smallholders; and safety through indicators such as biodiversity, trait resistance and crop contamination (Qaim, 2010). While this chapter has focused on Ghana and Tanzania, the enabling framework as well as many of the implementation actions can be used to evaluate and improve the innovation-readiness and GM preparedness of any SSA nation, including those nations that are considering GM technology such as Uganda and even Zimbabwe. For nations that follow this path they will need to be equipped to meet the obligatory legal and regulatory framework and to make independent evaluations on three critical dimensions. Does smallholder demand for an agricultural innovation such as GM crops exist? Are the nation and its people ready, informed and able to fulfil smallholder demand? If not, what improvements are required for readiness to be achieved?

ACKNOWLEDGEMENT

This publication was made possible through the support of a grant from the John Templeton Foundation. The opinions expressed are those of the authors and do not necessarily reflect the views of the John Templeton Foundation.

REFERENCES

African Agricultural Technology Foundation (2010). *Project 4: Water Efficient Maize for Africa (WEMA)*. AATF, Nairobi, Kenya. www.aatf-africa.org

African Development Bank Group (2005). *Ghana: Review of Bank Assistance to the Transport Sector*. Abidjan, Côte d'Ivoire.

Aker, J. C., and Mbiti, I. M. (2010). Mobile phones and economic development in Africa. *Journal of Economic Development* **24**, 207–232.

Beintema, N., and Stads, G. (2011). *African Agricultural R&D in the New Millennium: Progress for Some, Challenges for Many*, IFPRI Food Policy Report. http://bch.cbd.int/

CIA World FactBook (2010). *Mobile Phone User Statistics*. https://www.cia.gov/library/publications

Elliot, K. A. (2010). *Pulling Agricultural Innovation and the Market Together*, Working Paper No. 215. Center for Global Development, Washington, DC.

FAO (2012). www.fao.org/countries/55528/en/

GAVI Alliance (2010). Advance market commitments for vaccines: creating markets to save lives. *AMC Fact Sheet 2010*. www.vaccineamc.org

Godfray, H. C. J., Beddington, J. R., Crute, I. R., *et al.* (2010). Food security: the challenge of feeding 9 billion people. *Science* **327**, 812–818.

Internet World Stats (2011). www.internetworldstats.com/stats.htm

Juma, C. (2011). *The New Harvest: Agricultural Innovation in Africa*. Oxford University Press.

Meissner, R. (2010). Africa's macroeconomic and financial stability prospects: the cases of South Africa and Nigeria. *Consultancy Africa Intelligence*. www.consultancyafrica.com/

Paarlberg, R. (2008). *Starved for Science: How Biotechnology Is Being Kept out of Africa*. Harvard University Press, Cambridge, MA.

Qaim, M. (2010). Benefits of genetically modified crops for the poor: household income, nutrition, and health. *New Biotechnology* **27**, 552–557.

Roth, J., and McCord, M. J. (2008). Agricultural microinsurance: global practices and prospects. *Microinsurance Centre*. www.microinsurancenetwork.org/

Swab, K. (ed.) (2009). *The Global Competitiveness Report 2009–2010*. World Economic Forum, Geneva, Switzerland. www.weforum.org/

Transparency International (2010). *Corruption Perceptions Index*. www.transparency.org/

WIPO Patent Report (2010). *Patent Grants by Patent Office, Broken Down by Resident and Non-Resident (1883–2008)*. www.wipo.int/freepublications/

World Bank (2010). *Annual Report: World Development Indicators (WDI)*. World Bank, Washington, DC.

World Patent Report (2010). *Patent Grants by Country of Origin and Patent Office (1995–2008)*. www.wipo.int/ipstats/

WTO (2012a). www.wto.org/english/res_e/statis_e/its2011_e/its2011_e.pdf

WTO (2012b). http://stat.wto.org/CountryProfiles/GH_e.htm

MARK F. CANTLEY AND DREW L. KERSHEN

16

Regulatory systems and agricultural biotechnology

A SKELETAL DESCRIPTION OF THE REGULATION OF AGRICULTURAL BIOTECHNOLOGY

From the very beginnings of modern biotechnology in the early 1970s, scientists, governmental regulators and the general public have engaged in debate about the appropriate level of regulatory oversight. Scientists themselves early imposed a several-year moratorium on the use of certain recombinant DNA (rDNA) techniques until they could reach a consensus about the risks involved with modern biotechnology. With the Asilomar Conference (1974), they expressed confidence in modern biotechnology while recommending that the US National Institutes of Health (NIH) establish guidelines for research using rDNA techniques. Scientists then lifted their voluntary moratorium. The scientific consensus was that modern biotechnology could be used safely and effectively to create microorganisms, plants and animals with genes engineered into their cells to address practical problems in medicine, agriculture, food processing, environmental remediation and industrial production (e.g. biofuels and starches).

In light of this scientific consensus, commentators began to proclaim a biological revolution in the way that human beings would interact with their world, comparable to the revolutions of physics and chemistry in the nineteenth and twentieth centuries respectively. Scientific and technical optimism abounded among the scientific community about the coming era of the biological revolution. Alas, the optimism, well founded in science, ultimately collided with the reality of other societal forces – scientific ignorance and misunderstanding,

Successful Agricultural Innovation in Emerging Economies: New Genetic Technologies for Global Food Production, eds. David J. Bennett and Richard C. Jennings. Published by Cambridge University Press. © Cambridge University Press 2013.

technological phobias and antagonism, domestic and international politics, inherent expansionist tendencies of regulators, competing companies' self-interest in survival and market power, and a determined protest industry comprising groups with a variety of ideological agendas. Each of these oppositional forces worked vigorously to counter the scientific and technical optimism about modern biotechnology.

In the 1980s, the World Health Organization (WHO, 1982), the Organisation for Economic Cooperation and Development (OECD, 1986) and the US National Academy of Sciences (NAS, 1987) published careful studies of modern biotechnology and reached the following conclusions:

- Modern biotechnology is an extension and refinement of prior genetic engineering in plant breeding. Genetic engineering is not new. Indeed, all but a very small number of food crops are the product of genetic selection and breeding due to human intervention.
- Modern biotechnology allows for more precise, better understood, more predictable, and more limited genetic modifications than traditional breeding techniques that also introduce genes from distant (beyond species) relatives.
- The scientific basis for regulation focuses on risks, not speculative hazards, and balances the benefits of the new microorganism, crop or animal with identifiable risks known from comparable organisms. Science urges that regulators think carefully about benefits and risks. Science does not support a regulatory structure that insists upon no risks.
- The scientific basis for regulation focuses on the product (the crop or the food) and not the process by which the product is produced. A regulatory structure rooted in science would be a product-based, not process-based, regulatory structure.

Despite these WHO, OECD and NAS pronouncements, and similar pronouncements from other organisations and scientific societies, the opposing forces unceasingly promoted what Swiss botanist Klaus Ammann (2011) calls the 'genomic misconception', i.e. that modern biotechnology, especially rDNA technology, created new, unique and unprecedented risks to human and animal health, the environment and to social structures in farming and agricultural development. The 'genomic misconception' manifests itself in the regulatory structures that governments have adopted around the world.

In the United States, under a Coordinated Framework for Biotechnology, the government adopted a policy to use already existing legal authority and administrative agencies to regulate agricultural biotechnology. In addition, the Coordinated Framework adopted a product, not process, approach to regulation – that is to say, an approach which focuses on the product, not the means by which it was produced. Consequently, three US regulatory agencies have been assigned responsibility towards biotechnology: the US Department of Agriculture, Animal Plant Health Inspection Service (USDA-APHIS) for plant pests and noxious weeds (the agronomic impacts); the Environmental Protection Agency (EPA) for pesticides and chemicals (the environmental impacts); and the Food and Drug Administration (FDA) for food products (food safety issues). However, upon closer inspection as to how the US regulatory system actually operates, one realises that USDA-APHIS, EPA and FDA have invoked triggers for mandatory and voluntary regulatory review based upon the use of rDNA techniques and the transfer of genes from beyond the genus botanical classification. For example, USDA-APHIS created a new category of 'regulated article' for field trials and commercial release of transgenic crops; the EPA created a new category of 'plant-incorporated protectants' (PIPs) for transgenic insect-resistant crops; the FDA adopted a policy of voluntary consultation for foods containing ingredients from transgenic crops and subjected a transgenic salmon to the laws relating to the approval of veterinary drugs. Each regulatory agency thus subjects agricultural biotechnology to extra and intensive regulatory oversight.

In the European Union (EU) beginning in 1990 and continuing to the present (2012), the European Commission and the EU Member States created new laws and regulations and new regulatory pathways applicable only to modern biotechnology, predominantly as used in agriculture. (Other applications were addressed, but could escape new oversight if existing EU regulation already provided adequately severe oversight – as was and is the general case for pharmaceuticals.) The EU required prior approval for field trials and commercial release through this specialised regulatory structure. In addition, the EU created legal and regulatory requirements with regard to food, feed and seed labels, and imposed standards of traceability and liability upon biotech products. As a consequence, the EU regulatory structure focused precisely on the process by which an agricultural product was produced and focused upon both risks and speculative hazards, demanding close to 'no risk' before granting approval and downplaying, if not completely ignoring, the benefits of agricultural biotechnology. Finally, the EU imposed a 'zero tolerance' standard for any agricultural biotechnology crop or animal that had not gained EU regulatory approval. One gets an inkling of the complexities created by the EU regulatory structure when one realises that a significant distinction is made between products made 'from' agricultural biotechnology (the underlying commodity) and products made 'with' agricultural biotechnology (using enzymes and other processing aids).

As for international law, the genomic misconception appeared in the Convention on Biological Diversity (CBD) (1992) in a clause calling for the establishment of a binding agreement related to trade focused solely on biotechnological products. The negotiations for the binding agreement formally began in 1996 and agricultural biotechnology soon became the primary focus. When the Cartagena Protocol on Biosafety (CPB) became a binding international agreement in 2003, the CPB set forth an advance informed agreement procedure built upon extensive and mandatory risk assessment and risk management provisions for the commercial release and the commodity trade in agricultural biotechnology products. Moreover, CPB Article 27 called for the creation of an additional process for the negotiation of a further protocol related to liability and redress for agricultural biotechnology products. In 2010, negotiators presented to the world a Nagoya–Kuala Lumpur Supplemental Protocol on Liability and Redress (a document that, as of January 2012, has not attained sufficient country ratifications to enter into force). Perhaps the general attitude of these three international negotiations can best be presented by stating that the CPB uses the term 'benefit(s)' twice (both times in the CPB Introduction) and the term 'risk(s)' 67 times. From 2003 forward, the United Nations Environmental Program – Global Environment Facility (UNEP-GEF) has travelled the globe promoting the 'risk' attitude towards agricultural biotechnology among the developing nations of Africa, Asia and Latin America.

IMPACTS OF THE REGULATORY STRUCTURES UPON AGRICULTURAL BIOTECHNOLOGY

For agricultural biotechnology, the consequences of the regulatory structures described in the preceding history are predictable and unfavourable – though each structure has impacted agricultural biotechnology in unique ways.

In the United States, the three primary regulatory agencies have approved, and continue to approve, transgenic crops for agricultural production and the foods and feeds from these crops for human and animal consumption. As of 2012, the United States has approved 86 petitions for commercial release (deregulation) of transgenic traits in eight different major crops – maize, soya bean, cotton, canola, sugar beet, alfalfa, papaya and squash. In 2011, the US regulatory agencies approved a nutritionally approved (high oleic) soya bean and a drought-tolerant maize for commercial release that will probably be marketed in 2012 or 2013 – both first biotechnology products with nutritional benefits and abiotic stress traits. In light of these US approvals, it is clear that the US regulatory structure is not an impermeable barrier to agricultural biotechnology innovation. In fact in 2010, US farmers planted about 66.8 million hectares using transgenic seeds, with a very high percentage of production of the initial six listed crops being transgenic.

However, the approval process has become increasingly time-consuming, costly and effort-intensive. Transgenic products require ten times more money (in the multi-millions of dollars), many more years (5–7 being usual) and much greater documentation (reams and reams of data) than non-transgenic products. More discouraging, the US regulatory system has become desultory with some biotechnology products in regulatory limbo without any decision other than additional requests for information and time to study. Transgenic salmon has been in the US system for more than 10 years; a rice for producing a proven rehydrating medicine for children with diarrhoea truly became an orphan drug when its creator abandoned it because of the regulatory costs, delays and demands.

Moreover, the US regulatory structure creates a litigation risk that is relatively unique to the United States legal system. Groups opposed to regulatory actions have brought many lawsuits based on the National Environmental Policy Act (NEPA), the Plant Pest Act (PPA), the Endangered Species Act (ESA) and other statutes to contest the approval decisions of the agencies. Regardless of the outcome, these lawsuits impose delays, litigation expenses and great uncertainty upon transgenic products. In addition, due to the strict regulatory oversight for agricultural biotechnology, any agricultural biotechnology trait found in unapproved crops or uses immediately gives rise to US class action lawsuits for economic damages to domestic and international markets. Several class action lawsuits have settled for hundreds of millions of dollars in damages payable to the class action litigants and their attorneys. In other words, a transgenic product must not only successfully circumnavigate the US regulatory structure, the transgenic product must also successfully circumnavigate the US judicial system.

Unsurprisingly, the complex and intricate regulatory structure in Europe has primarily resulted in regulatory paralysis. In light of the decision-making structure of the system, particularly the risk management stage, the EU has been unable to reach affirmative decisions. Rather for each product, after years of risk analysis and lengthy debate among the EU members, the final result is stalemate. In certain instances, the European Commission then has the legal authority to approve the agricultural biotechnology product by default – but even then the European Commission often dithers rather than rendering a favourable decision rooted in the risk analysis evidence supporting regulatory approval. The European Commission has approved approximately 40 agricultural biotech products for importation for food and feed uses but has approved only two agricultural biotechnology crops (a potato and a maize) for commercial cultivation on European farmlands. Consequently, EU officials and others often describe the EU system as dysfunctional.

Without political support from the EU, ideological groups and member governments antagonistic to agricultural biotechnology have dominated the debate in the European press and the minds of Europeans. Retailers have purposefully avoided agricultural biotechnology products anytime the use of those products requires a mandatory label. Universities have shied away from undertaking research in agricultural biotechnology. Protestors routinely vandalise field trials and farmers' transgenic fields without fear of significant penal or civil consequences in the European courts. As a result, agricultural seed and chemical companies (e.g. BASF in early 2012) have basically abandoned Europe for the United States and Brazil – more favourable climes for agricultural biotechnology product approval and public acceptance. Europe is the only continent where agricultural biotechnology is declining as opposed to increasing in use.

Despite the risk aversion permeating the CPB and the UNEP-GEF programmes about biosafety, developing nations grew 71 million hectares of transgenic crops in 2010 (48% of the worldwide hectarage) with 14.4 million small resource-poor farmers (90% of all transgenic farmers) using agricultural biotechnology to improve their yields, the productivity of their labour, their safety, their income and their food security. Brazil, with a regulatory system that adopted the CPB regulatory model, presently is approving agricultural biotechnology traits for commercial release at a quicker pace and a lower cost than probably any other country in the world. In 2011, Brazil approved a transgenic food bean, not a soya bean, developed by the national agricultural research system (Embrapa) for poor farmers to grow and Brazilians, particularly poorer Brazilians, to eat as part of the national dish of beans and rice. Indeed, the trend for agricultural biotechnology adoption has been better and more favourable in developing countries, in comparison to industrialised countries, for the past several years. In light of the adoption rate in developing countries, one could conclude that there is a favourable climate for agricultural biotechnology in the developing world.

Unfortunately, the CPB and UNEP-GEF have also had substantial negative effects for agricultural biotechnology in the developing world. Many developing countries appear caught in an interminable debate about agricultural biotechnology. Many developing nations are uncertain and confused about whether or not to adopt agricultural biotechnology, particularly when the EU regulatory system threatens the developing nation's agricultural exports to Europe, and thus cannot reach legislative and executive agreement about an appropriate regulatory system for agricultural biotechnology. Three countries – Zambia, Kenya and Zimbabwe – have gone so far as to prohibit the import of transgenic grain, even when grown in South Africa, to provide food security for domestic populations facing famine. Thus, starving people did not gain access to agricultural commodities that hundreds of millions of people elsewhere eat routinely as part of their diets. Concurrently, India spent more than 7 years testing and considering a transgenic brinjal (eggplant) before the Genetic Engineering Approval Committee (GEAC) granted approval to commercial release. The GEAC decision was promptly overturned by a higher political authority in the Ministry of Environment that also changed the name of the GEAC to the Genetic Engineering Advisory Committee. India still grows 9.4 million hectares of transgenic cotton (more than 80% of total cotton hectares) but the Indian regulatory structure denied poor farmers and poor consumers transgenic brinjal with clearly established benefits in yield, reduced pesticide applications, farm safety, labour productivity, farm income, and product quality and product pesticide-residue safety.

Negative impacts of inappropriate and burdensome regulatory structures have been particularly significant for developing nations in three ways:

- First, developing nations could benefit especially from agricultural biotechnology generated as public goods through publicly funded or foundation-funded agricultural research. However, despite impressive public research results in laboratories and greenhouses, only a very limited number of public-goods agricultural biotechnology crops have reached farmers' fields from public research. Public research institutions do not receive funding to cover the costs of regulatory compliance – the millions of dollars required to take a transgenic product through the regulatory system. Public research institutions do not have the personnel to guide biotech products through the regulatory process with its documentary demands and time delays. Facing these seemingly insurmountable regulatory barriers, public research institutions have shied away from agricultural biotechnology despite the advantages that agricultural biotechnology offers for crops with a poverty focus.

- Second, developing nations could benefit from capacity building in the latest plant breeding techniques that modern biotechnology offers. Plant breeding itself is a difficult and complex endeavour without imposing upon the plant breeder a scientifically unjustifiable regulatory structure. Young scientists who might eagerly accept the challenges of plant breeding instead are discouraged from undertaking modern biotechnology, instead leaving the field of plant breeding entirely or adopting and using conventional techniques purposefully to avoid regulatory systems. Rather than asking what breeding technique and combination of techniques make the most scientific sense to address a particular crop problem, farmer need or consumer benefit, plant breeders now often ask first and foremost, what technique(s) avoid regulation? With each new breeding technique (e.g. cisgenic breeding, zinc finger targeted insertion of transgenes, transformation method transcription activator-like families of type III effectors (TALEs)), scientists must expend time and effort in consultation with lawyers and regulators to determine whether these new breeding techniques are within or without the burdensome regulatory systems. Scientific capacity building for plant breeding, especially for developing nations, thus suffers from reduced interest from plant biologists, reduced funding from public and private sources and greater uncertainty about the legal status of transformed plants in the various regulatory systems of individual legal systems around the world.
- Third, developing countries need agricultural biotechnology to address present and critical food security, food safety and food nutrition issues. Regulatory structures that impose excessive and unnecessary barriers to the adoption of agricultural biotechnology crops thus impose tremendous lost opportunity

costs upon developing nations. Lost opportunity costs encompass more than lost economic benefits; lost opportunity costs most sadly encompass the lives lost to hunger and malnutrition and the lives stunted by inadequate nutrition. No better example exists than Golden Rice – a rice developed as a public good for the benefit of subsistence farmers – biofortified to contain beta-carotene. By eating Golden Rice, subsistence farmers and poor consumers would attain, in their daily diet, adequate vitamin A, saving millions from death and millions more from blindness. Rice with beta-carotene cannot come from non-transgenic breeding techniques because the rice genome does not contain the genes that can produce beta-carotene. Despite its origin, its effectiveness and its benefits for the poor, regulatory structures have needlessly trapped Golden Rice for a decade; current hopes focus on a possible approval in the Philippines in 2013. The Golden Rice saga is so heart-wrenching that the mind responds with Michelangelo's painting of the *Last Judgment* in the Sistine Chapel – a painting with beauty and meaning appealing to believers and non-believers alike.

REGULATORY STRUCTURES: NEEDED CHANGES

Present regulatory systems for agricultural biotechnology embody the misconception that agricultural biotechnology is different in kind from other breeding techniques and presents unique and unprecedented risks. Present regulatory systems have not budged from this misconception despite 16 years of positive experience with transgenic crops grown on more than a billion hectares, and the pretence that the 'Precautionary Principle' will liberalise regulatory oversight as safe experience accumulates. As the Public Research and Regulation Initiative (PRRI, 2011), a worldwide initiative of public-sector scientists conducting biological research for the public good, stated in 2011, '*The substantial scientific evidence accumulated* [after 15 years of GM crops] *shows that there are no verifiable reports of any adverse effect to environment or human health.*'

How refreshing it would be if governments around the world would change their laws and regulations to remove this genomic misconception from their regulatory systems. If this were to happen, regulatory systems would regulate plant breeding by modern biotechnology just as they regulate any other breeding technique – based on the risk of the plant created by the breeding technique. At present, regulatory systems for all other techniques of plant breeding impose minimal regulatory oversight because plant breeding protocols have proven that plant breeders can produce safe, efficacious, socially beneficial and environmentally benign genetically improved varieties of microorganisms, plants and animals. By removing the genomic misconception, regulatory systems would treat various plant breeding techniques alike, would focus on the product rather than the process, and would regulate properly and proportionately according to scientifically identified risk. First and foremost, developing nations should avoid this genomic misconception in their regulatory systems.

Even if a particular nation, due to inertia or opposition, is not able to change its regulatory system to abandon the misconception, governments can adopt a risk analysis attitude towards agricultural biotechnology rather than a precautionary attitude. Governments should recognise that the precautionary attitude must not blind their regulatory systems from the 16 years of favourable experiences with agricultural biotechnology. Governments can build on these favourable experiences to create categorical exemptions from regulation for many agricultural biotechnology crops grown now for many years. Governments can greatly lessen the data requirements and streamline the decision-making process for new crops and varieties using genetic traits that have already received regulatory approval in other crops and varieties. Governments can harmonise their regulations with those of other governments and, more importantly, accept and honour the regulatory decisions made by competent regulatory agencies of fellow states. Regulatory systems should not require plant breeders to prove again and again what is already known from previous regulatory dossiers. Translating this shift to a risk analysis into practical legal consequences means that for countries to implement the CPB they need only

- identify transgenic crops that are recognised as not having adverse effects (Article 7.4)
- adopt simplified procedures as often, if not by default, appropriate (Article 13), and
- reach bilateral and multilateral agreements about the approved legal status of transgenic crops for which favorable real-world experiences exist (Article 14).

Developing nations should adopt a permissive, encouraging attitude towards agricultural biotechnology and their regulatory systems should reflect these favourable attitudes.

More fundamentally, regulatory structures in agriculture need to recognise, not only in words but in deeds, that science and innovation for agriculture are beneficial and desirable. Within the past year, scientists in Sweden, the United Kingdom and the United States have issued calls for regulatory systems to reflect science-based decisions and to lessen regulatory burdens upon agricultural biotechnology. Dr Thomas Lumpkin, Director of the International Wheat and Maize Improvement Center of El Batán, Mexico (better known by its Spanish initials, CIM-MYT) has pithily expressed this sentiment when he said, 'We need science to come back to farming.' Science needs to return to regulatory systems for agricultural biotechnology too. The scientific consensus is that agricultural biotechnology is safe, efficacious, beneficial and socially wise. Developing nations must adopt this scientific consensus and their regulatory systems must allow science and innovation for agriculture.

Developing nations must not trap themselves in underdevelopment by rejecting modern plant breeding and technological advances due to political and institutional interests that ignore or mischaracterise the positive scientific evidence and favourable accumulated experience of agricultural biotechnology. Farmers, consumers and society in general deserve better from their political and regulatory structures.

FOOD SECURITY FOR DEVELOPING NATIONS

The Food and Agricultural Organization (FAO) has published several recent reports indicating that world agricultural systems will have to produce about 70% greater amounts of food to meet the food security demands likely to exist in the year 2050. This significant increase in food production derives from several present and projected scenarios. At present, more than 900 million of the world's 7 billion people are hungry because they do not consume sufficient macronutrients (carbohydrates, fats and protein). Furthermore, about 1 billion people are malnourished because they do not consume sufficient micronutrients (vitamins and minerals) in their daily diets. While global society already faces significant hunger and malnutrition, projected growth in population indicates that by 2050 approximately 9 billion people (an additional China plus India) will inhabit the Earth. Moreover, this additional population will come almost exclusively in developing nations that even now are struggling to meet the food security needs of their people.

Production of adequate quantity and quality of food in the coming years must also account for changing dietary preferences and changing climates. As people hopefully become wealthier, people change their diets from subsistence foods to more protein-rich foods, historically with increased consumption of meat and animal products. While animal protein is an excellent source of food, increased animal production means increased land area devoted to growing animal feed, particularly maize and soya beans, or to grazing animals on quality grasses. As for climate change, farmers and pastoralists will be forced to adapt their crops and animals to thrive in the climate as it exists at the time the crops and animals are grown. While predicting what and where climate changes will occur with precision is not possible, many agricultural developmental specialists are particularly worried about the water resources and rainfall patterns that agricultural production requires.

As if hunger, malnutrition, population increases, dietary preference and climate change were not enough on the agricultural plate, agricultural production must also take into account environmental obligations, particularly preservation of wild lands (e.g. rainforests) and habitat to protect and preserve biodiversity. Agriculture must take into account a life-cycle analysis of its inputs and outputs and its interactions with other global systems (e.g. the hydrological cycle and the carbon emissions cycle). To use a common phrase, agriculture must produce more while simultaneously reducing its footprint on global lands, waters and air.

In a broad sense, agricultural production can increase only through either extensification or intensification. Extensification means the opening of additional land and resources to agricultural production. Russia, Central Asia, Africa and South America do have lands that could prove excellent for agricultural production. However, these additional lands are rather limited in comparison to lands already in agricultural production. Moreover, extensification presents major conflicts with the environmental goals of preserving wild lands and biodiversity. Some extensification may be justifiable but extensification by itself will not be sufficient to satisfy the demands for agricultural production in 2050 unless agriculture ignores environmental obligations.

By contrast, intensification means the increased production of additional foods, fibre, fuels and feeds on a per-hectare basis, producing metaphorically and literally two blades of grass where before only one grew. In fact, agriculture has increased production through intensification at an amazing pace since 1950. Agriculture has achieved this increased production through a combination of factors – mechanisation, chemical inputs (fertilisers, herbicides and pesticides), irrigation expansion and improved genetics for seeds and animals. Realistic hope exists that agriculture can continue to increase production through these factors, along with additional knowledge from ecology and climate sciences, to create a sustainable intensive agriculture that balances its necessary production goals with its resource demands for land, water, air, genes and human and animal labour. Sustainable intensive agriculture has become the agreed goal for agricultural development.

For developing nations, sustainable agricultural development has several important implications. If sustainable agricultural production can become the norm, developing nations will simultaneously increase both the quantity and quality of their agricultural products. Increased sustainable agricultural production for subsistence farmers directly means increased food security for themselves and their families, and (given adequate infrastructure) for sale at markets. Subsistence farmers are likely to have surplus agricultural products to sell and improved personal safety through reduced labour and chemical inputs. Reduced labour inputs especially allow for children and parents to improve their educational levels and agricultural skills. Thus, sustainable intensive agriculture indirectly means improved economic and social well-being for subsistence farmers.

For developing nations, the agricultural sector is a dominant sector in terms of GDP and in terms of employment. Sustainable intensive agriculture can thus become a driver of overall economic and social development for the nation. Food security for subsistence farmers means additional agricultural products in domestic and international markets, reducing food costs for urban consumers and earning additional trade income for the nation. Sustainable intensive agriculture can hopefully become a virtuous cycle of poverty reduction, affordable food and national economic growth while concurrently protecting the nation's environment and natural resources. Sustainable intensive agriculture can allow economic growth and environmental protection to be complementary, not conflicting, national goals.

Science and innovation in agriculture will be crucial drivers of sustainable intensive agriculture. Developing nations must create policies and attitudes that promote science and innovation in agriculture. Developing nations must be willing to devote additional national resources to agricultural research and development. Developing nations must also be willing to coordinate and to accept scientific knowledge and technology transfer from both the domestic and international public and private sectors. While sustainable intensive agriculture by itself cannot address all the developmental needs of developing nations (e.g. infrastructure, governance, functioning markets and international free trade), it is a necessary component of the social, economic and environmental welfare for development to occur and become sustaining.

Agricultural biotechnology will be one crucial component of sustainable intensive agriculture. The World Bank, in a 2008 report, estimated that as much as 50% of the increased yield in crop agriculture in the 1980s and 1990s came from genetic improvements in new varieties. In the coming years to 2050, genetic improvements are likely to add as much or more to increased yield to meet the food demands of global societies. Sustainable intensive agriculture will not occur if agricultural biotechnology – or other agricultural innovations such as cloning, synthetic biology and nanotechnology – are excluded a priori from consideration and adoption (Foresight, 2011). Sustainable intensive agriculture will not occur in time to meet the needs of the twenty-first century if developing nations formulate or implement non-scientific regulatory structures that impede, delay, disrupt or discourage scientific creativity, agricultural investment and farmer adoption of agricultural biotechnology.

In the face of the challenges of the twenty-first century, as well expressed in the United Nations Millennium Development Goals that will persist well beyond 2015 throughout the entire century, present regulatory systems that are hostile to agricultural biotechnology must change. Developing nations must set their own independent course for sustainable, intensive agricultural development regardless of the unsustainable attitudes and actions of any developed nations about science, innovation and agricultural biotechnology. As the Chinese proverb states, 'A person who has food has many problems. A person who has no food has only one problem.'

If developing nations look for an inspiring vision of agricultural development, look not to the descendants of Thomas Malthus who thought that population growth would preclude society's progress. Look rather to the descendants of Norman Borlaug, agronomist, humanitarian and father of the Green Revolution.

REFERENCES

Ammann, K. (2011). The GM crop risk–benefit debate: science and socio-economics. In R. A. Meyers (ed.) *Encyclopedia of Sustainability Science and Technology*. Springer, New York.
Foresight (2011). *The Future of Food and Farming*, Final Project Report. Government Office for Science, London.

National Academy of Sciences (1987). *Agricultural Biotechnology: Strategies for National Competitiveness*. NAS, Washington, DC.
Organisation for Economic Co-operation and Development (1986). *Recombinant DNA Safety Considerations*. [Also known as *The Blue Book*.] OECD, Paris.
Public Research and Regulation Initiative (2011). *Public Research in Modern Biotechnology*. http://greenbiotech.eu/
World Health Organization (1982). *Biotechnology: The First Laternal Review*. WHO, Geneva, Switzerland.

KEY FURTHER RESOURCES

DESCRIPTION OF THE REGULATION OF AGRICULTURAL BIOTECHNOLOGY

Cantley, M., and Lex, M. (2010). Genetically modified foods and crops. In J. B. Wiener, M. D. Rogers, J. K. Hammitt, and P. H. Sand (eds.) *The Reality of Precaution: Comparing Risk Regulation in the United States and Europe*, ch. 3. RFF Press, Washington, DC.

IMPACTS OF THE REGULATORY STRUCTURES UPON AGRICULTURAL BIOTECHNOLOGY

Brookes, G., and Barfoot, P. (2011). *GM Crops: Global Socio-Economic and Environmental Impact 1996–2009*. PG Economics Ltd, Dorchester, UK.
James, C. (2011). *Global Status of Commercialized Biotech/GM Crops 2010*, ISAAA Brief No. 42. International Service for the Acquisition of Agri-biotech Applications, Ithaca, NY.
Royal Swedish Academy of Agriculture and Forestry (2008). Golden Rice and other biofortified food crops for developing countries: challenges and potential. Report from the Bertebos Conference, Falkenberg, Sweden.
Strauss, S. H., Kershen, D. L., Bouton, J. H., et al. (2010). Far-reaching deleterious impacts of regulations on research and environmental studies of recombinant DNA-modified perennial biofuel crops in the United States. *BioSciences* 60, 729–741.

REGULATORY STRUCTURES: NEEDED CHANGES

Durham, T., Doucet, J., and Unruh-Snyer, L. (2011). Risk of regulation or regulation of risk? A *de minimis* framework for genetically modified crops. *AgBioForum* 14, 61–70.
Gressel, J. (2008). *Genetic Glass Ceilings: Transgenics for Crop Biodiversity*. Johns Hopkins University Press, Baltimore, MD.
Paarlberg, R. (2008). *Starved for Science: How Biotechnology Is Being Kept Out of Africa*. Harvard University Press, Cambridge, MA.

FOOD SECURITY FOR DEVELOPING NATIONS

Potrykus, I., and Ammann, K. (eds.) (2010). Transgenic plants for food security in the context of development, proceedings of a study week of the Pontifical Academy of Sciences. *New Biotechnology* **27**, 445–718.

World Bank (2008). *World Development Report 2008: Agriculture for Development.* World Bank, Washington, DC. www.worldbank.org/

ALFREDO AGUILAR, DANUTA CICHOCKA, JENS HÖGEL,
PIERO VENTURI AND IOANNIS ECONOMIDIS

17

Biotechnology research for innovation and sustainability in agriculture in the European Union

INTRODUCTION

By 2050 the global population is estimated to reach 9 billion. The demand for food, feed, fibre and fuel is expanding while the available land and natural resources are evidently finite. The fast economic growth of recent decades has led to overexploitation of the Earth's ecosystems and it is becoming ever more evident that, if we are to provide future generations with access to the same resources as we benefit from today, we urgently need to switch to a much more sustainable approach to global economic growth.

In response to this challenge the global political 'sustainability' agenda has emerged. The Rio Declaration 'Agenda 21' and the 'Rio conventions' (Convention on Biological Diversity (CBD), United Nations Convention to Combat Desertification (UNCCD), United Nations Framework Convention on Climate Change (UNFCCC)) have paved the way for the concept of a 'bio-based economy' which emerged from the Organisation for Economic Co-operation and Development (OECD) in 2002 (United Nations, 1992). In 2005 the concept was introduced to the European Union's political framework as the *European Knowledge-Based Bio-Economy* (KBBE). The bio-based economy links industrial development with sustainability and implies improving industrial performance in an economically viable, environmentally friendly and socially responsible manner (OECD, 2001; European Commission, 2002). However, in order to achieve eco-efficient growth Europe needs to invest significant

Successful Agricultural Innovation in Emerging Economies: New Genetic Technologies for Global Food Production, eds. David J. Bennett and Richard C. Jennings. Published by Cambridge University Press. © Cambridge University Press 2013.

funding in its knowledge base and fully realise the potential of the most advanced and innovative technologies. The bio-economy sectors of the European Union (EU) are worth €2 trillion in annual turnover and account for more than 22 million jobs, approximately 9% of the workforce. It is estimated that direct research funding associated with the bio-economy under the Horizon 2020 Framework Programme for Research and Innovation (2014–20) could generate about 130 000 jobs and €45 billion in value added in bio-economy sectors by 2025 (European Commission, 2012a).

The need to increase public funding for research and innovation in support of the bio-economy has been recognised under Horizon 2020. Almost €4.7 billion have been proposed for the Challenge *Food Security, Sustainable Agriculture, Marine and Maritime Research and the Bio-Economy*. Such funding will be complemented by other elements (particularly biotechnology) from other Challenges in Horizon 2020 and by research and innovation in the Programme Leadership in Enabling and Industrial Technologies (European Commission, 2011).

Biotechnology is one of the key technologies which may help close the gap between the growing world population, increased longevity, improved quality of life and depleted natural resources. It is a rapidly developing field which encompasses a very wide range of tools with huge potential for the more efficient use of bioresources, improved agricultural yields, energy saving and environmentally friendly industrial processes. Since the beginning of the European Union's 7th Framework Programme for Research, Technological Development and Demonstration Activities (FP7) in 2007, biotechnology has received substantial attention and has been placed among priority areas for European support. The FP7 Theme: *Food, Agriculture and Fisheries and Biotechnology*, more commonly referred to as the *Knowledge-Based Bio-Economy* (KBBE), with a total budget of just under €2 billion for the period 2007–13, covers agricultural production, fisheries and aquaculture, forestry, food quality, safety and processing, and a wide range of non-medical biotechnologies. Grouping biotechnology with food and agriculture underlines the links and interdependence of these three sectors.

The biotechnology part of KBBE includes:

(a) novel sources of biomass;
(b) marine and freshwater biotechnology;
(c) industrial biotechnology and biorefineries;

(d) environmental biotechnology for the clean-up
of contaminated environments and development of
environmentally friendly industrial processes; and

(e) new trends in biotechnologies (including bioinformatics,
systems biology, synthetic biology and nanobiotechnologies).

Although the biotechnology research supported under the KBBE programme focuses on non-food and non-agricultural applications, it
includes research that has implications across the agricultural sector,
such as applications in biosafety and the risk assessment of genetically
modified organisms (GMOs).

Nowadays, farmers are faced with the challenge of rapidly
increasing demand for crop production while remaining sensitive to
environmental concerns. In order to meet the socio-economic and political expectations for efficient and sustainable crop production, all available technologies need to be considered. The use of GMOs in agriculture,
but especially in the production of non-food biomass, can no longer be
neglected. The development of alternative raw materials for industrial
applications, combined with new industrial techniques that integrate
thermochemical processes with biological ones (such as in biorefineries), have become the subject and focus of research activities in
Europe and elsewhere in the world. The introduction of advanced
processes and materials based on biotechnology and GMOs may not
only enhance the quality of life of Europe's citizens while reducing
environmental footprints, but can also improve the competitiveness of
European industry. However, as with all new technologies, the potential risks and benefits of biotechnologies and GMOs must be identified
and quantified. Furthermore, before any product containing genetic
modifications, or in the production of which GMOs are used, reaches
the consumer or the environment, its safety must be ensured. Despite
the very long history of EU-funded research on GMO safety (European
Commission, 2010a), the topic still raises a lot of controversy among
European citizens. The public debate on the research, development,
release and commercialisation of GMOs is still ongoing. Since such
work began in the early 1990s the European Commission has been
trying to make sure that a broad range of public views is being
reflected in the relevant GMO legislation. At the same time, every
effort is being made to establish a comprehensive set of methodologies
for GMO risk assessment, with the ultimate goal of providing policy-
makers with the tools to make policy decisions based on factual and
scientific evidence.

This chapter reviews the history and development of EU GMO legislation, summarises the outcomes of over 25 years of EU-funded research on GMO safety and risk assessment, and puts ongoing EU-funded biotechnology research, especially in the areas relevant for emerging economies, into an international context. Finally, the chapter outlines future trends and perspectives in biotechnology with a view to the up-coming EU framework programme for research and innovation, Horizon 2020.

THE LEGAL CONTEXT OF GMOS IN THE EU

Legislation governing research and development and the commercialisation of GMOs in the EU has been in place since the early 1990s. It has evolved and been updated considerably over time. Such rules, and the procedures they define, are in compliance with those of international bodies, such as the World Trade Organization (WTO) and are also in line with the requirements of the Cartagena Protocol on Biosafety (Secretariat of the Convention on Biological Diversity, 2000). Today, the main legislative EU Acts and recommendations in the context of agriculture and food safety cover the deliberate release of GMOs into the environment (European Union, 2001), GM food and feed (European Union, 2003a, 2003b), genetically modified microorganisms (European Union, 2009) and coexistence of GMOs and non-GMOs (European Union, 2003c). In line with the legislation, and before commercialisation or release for research purposes, every GMO is scrutinised for its potential effects on the environment, and on consumer and animal health.

There are three clearly distinguishable steps in scrutinising the biosafety of GMOs: following a risk assessment, there is a risk management step and finally continuous monitoring. In the ideal case, results of the monitoring activities should feed back into risk assessment and management steps, to allow for the continuous update of policy and management measures taken based on empirical scientific analysis.

Risk assessment of GM food and feed

The European Food Safety Authority (EFSA) is the responsible body for risk assessments of GM food and feed at the EU level. The EFSA GMO panel conducting the assessments comprises highly qualified experts from a number of European countries. Panel members are appointed on the basis of proven scientific excellence in their fields, following periodic open calls for applications and rigorous and open selection procedures. The EFSA is committed to independence and transparency, requiring selected experts to sign declaration of interests which remain publicly available.

Figure 17.1 provides a general outline of the different steps involved in the risk assessment and management of GM food and feed. EFSA generally does not conduct scientific studies on its own, but uses three main sources of information during the assessment:

- information provided by the applicants as described in the legislation and applicable EFSA guidance;
- additional evidence from scientific peer-reviewed and publicly available studies related to the GMO in question;
- evidence from EU or nationally funded research projects on the biosafety of GMOs.

An essential part of the risk assessment concerns the formal and informal communication between EFSA, the applicant and EU Member State authorities, for example for addressing specific concerns, providing additional evidence or simply to minimise uncertainties or misunderstandings. Before finalising risk assessments, Member States and a wide range of stakeholders are requested to submit their comments on the draft EFSA opinion. These are considered while preparing the final opinion.

EFSA also generates and publishes guidance for GMO risk assessments. These documents are regularly updated in the light of new scientific know-how. The agency operates a very comprehensive website, the main tool for publishing EFSA opinions, guidance documents, and others (European Food Safety Authority, 2012).

Risk management of GM food and feed

Risk management of GMOs in the EU starts with a decision to authorise (or to not authorise) a given GMO into the common market. Following a positive EFSA opinion, the European Commission submits a draft decision to Member States for their vote, in line with the applicable procedures of a relevant committee (so-called 'comitology' procedures). To date, in almost all cases, no qualified majority for or against the authorisation of a GMO has been obtained. This stalemate has obliged the European Commission to forward the draft decisions to the Council of Ministers while informing the European Parliament. Voting in the Council of Ministers in certain cases has also failed to deliver the qualified majorities required for the adoption of decisions authorising (or not) the marketing of GMOs.

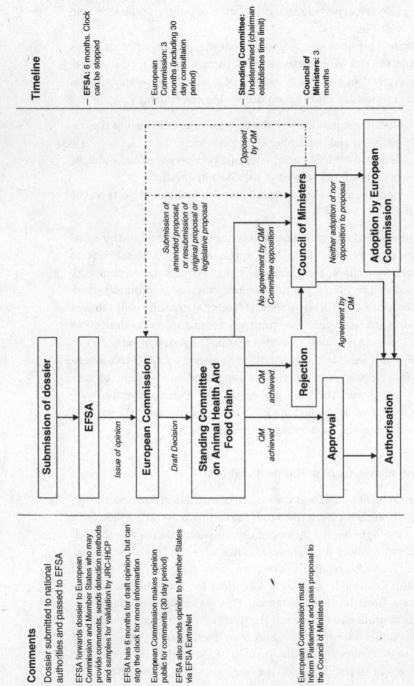

Timeline

- **EFSA:** 6 months. Clock can be stopped

- European Commission: 3 months (including 30 day consultaion period)

- **Standing Committee:** Undetermined (chairman establishes time limit)

- **Council of Ministers:** 3 months

Opposed by QM

Submission of amended proposal, or resubmission of original proposal or legislative proposal

No agreement by QM/ Committee opposition

Neither adoption of nor opposition to proposal

Agreement by QM

| Submission of dossier |

| EFSA |

Issue of opinion

| European Commission |

Draft Decision

| Standing Committee on Animal Health And Food Chain |

QM achieved

QM achieved

| Council of Ministers |

| Rejection |

| Approval |

| Authorisation |

| Adoption by European Commission |

Comments

- Dossier submitted to national authorities and passed to EFSA

- EFSA forwards dossier to European Commission and Member States who may provide comments, sends detection methods and samples for validation by JRC-IHCP

- EFSA has 6 months for draft opinion, but can stop the clock for more informartion

- European Commission makes opinion public for comments (30 day period)

- EFSA also sends opinion to Member States via EFSA ExrtraNet

- European Commission must Inform Parliament and pass proposal to the Council of Ministers

Figure 17.1 The authorisation process for GMOs in the EU. EFSA, European Food Safety Authority; QM, qualified majority. (Food Chain Evaluation Consortium, 2010)

Although there is a comparable situation with regard to decisions on the cultivation of GMO in the EU, some Member States objecting to the authorisation of a GMO (a variety of maize MON810) for cultivation sought to prohibit cultivation of this particular GMO on their territory by invoking a 'safeguard clause' in the legislation. The possibility of invoking safeguard clauses in case of evidence coming to light of dele-terious effects of GMOs is a general principle in many areas of food and environmental law. Using safeguard clauses, six EU Member States have currently banned cultivation of MON810 in their territories.

Monitoring

Monitoring the potential effects of GMO cultivation and/or consumption is a legal requirement in the EU. Such monitoring can be conducted at all stages of the lifetime of a GMO, from initial field trials to commer-cialisation. Monitoring is generally to be divided into general surveil-lance and case-specific monitoring.

- General surveillance is conducted to verify or refute the results of the initial risk assessment where no specific monitoring requirements were requested as a result of the risk assessment.
- Case-specific monitoring could include a range of actions such as surveying the development of resistance of insects against specific pesticides produced by a GMO.

Depending on the GMO and the results of the risk assessment, either one or both types of monitoring might be required. Applicants who have successfully achieved an authorisation for a GMO are required to report annually to the European Commission upon any observations made. As such, general surveillance and case-specific monitoring are important tools which should be used by decision- and policy-makers to adjust decisions where necessary. Active and constructive use of this tool would contribute to close the gap between the initial risk assessment and practical experience gained and could help to avoid unnecessarily forfeit-ing potential benefits of new products or technological developments.

THREE DECADES OF RESEARCH ON SAFETY OF GMOS

On the basis of the Precautionary Principle, the EU has developed research programmes and practices to evaluate the risks and benefits of GMOs to animal or public health and the environment. Other international organisations such as the OECD follow the same principles, not only to ensure consumer safety, but also to harmonise risk assessment approaches and facilitate the international trade of agricultural commodities and industrial products.

The results of this research from the European Union Framework Programmes were published and reviewed in two volumes (Kessler and Economidis, 2001; European Commission, 2010a). These demonstrate different scientific approaches to the issue of safety of GMOs. The inclusion of projects on the safety of GMOs in the research programmes started more than 25 years ago as a response to policy-makers' and public concerns regarding the safety of the technology. Projects of recent Framework Programmes dealing with the development of new products and processes based on GMO technology fully integrate safety assessments into their conception, experimentation and development.

In 2001 the European Commission published the first overview of the accumulated results of *EC-Sponsored Research on Safety of Genetically Modified Organisms (GMOs)* (Kessler and Economidis, 2001). This publication included work supported over the preceding 15 years from the 1st to the 5th Framework Programmes for research, technological development and demonstration activities. It featured 81 projects, involving over 400 laboratories, and the results covered a range of subjects: horizontal gene transfer; environmental impact of transgenic plants; plant–microbe interactions; transgenic fish; recombinant vaccines; food safety; and other issues. This publication attracted the attention of the scientific community but also of regulators, public services, non-governmental organisations and other stakeholders. In 2010 a second volume followed presenting the outcomes and conclusions of studies supported in subsequent Framework Programmes (European Commission, 2010a). This publication coincided with the development of the concept of the bio-economy in which the issue of GMOs plays a focused but significant role. In this case the development, use and safety of GMOs, mainly GM plants for the production of agricultural and industrial commodities, attracted interest for the potential responses to challenges for which there are currently no solutions available.

The second publication presents the results of 50 projects, involving more than 400 European research groups with financial support of around €200 million. This figure brings the total European Commission funding of research on GMO safety to more than €300 million since its inception in 1982 in the Biomolecular Engineering Programme.

The main conclusion to be drawn from the efforts of more than 130 research projects, covering a period of more than 25 years of research, and involving more than 500 independent research groups, is that GMOs are not per se more risky than conventional plant breeding technologies. The research on biosafety has produced a large amount of scientific evidence for these so-called 'negative' results: data that fail to demonstrate any specific hazard linked to the GM technology. In addition, the different experimental settings and observations have increased our awareness and understanding in molecular terms of many ecological processes and interactions, and have helped in developing the notion of an environmental baseline from which changes due to the use of GM, or any other technology, can be measured and any impact assessed.

Another important conclusion is that today's biotechnological research and applications are much more diverse than they were 25 years ago, which is also reflected by the current seventh Framework Programme and by the ideas embedded in the forthcoming Horizon 2020. This diversity ensures that biotechnology remains at the core of the Knowledge-Based Bio-Economy, a concept applicable from primary production through to industrial and pharmaceutical applications. Modern biological know-how is used to address major societal challenges, including food and feed security and safety, the development of renewable resource platforms for the production of biomaterials and bio-energy, and pharmaceuticals, while improving environmental sustainability.

The results and experience gained in biosafety research, together with the current research approach to the development of new products and processes based on GMO technology, aim to fully integrate safety assessments in their conception and implementation. This approach aligns with the principles laid down in the *Europe 2020* Strategy, where building the bio-economy is one of its main targets aiming to '*re-focus R&D and innovation policy on the challenges facing our society, such as climate change, energy and resource efficiency, health and demographic change. Every link should be strengthened in the innovation chain, from "blue sky" research to commercialisation*' (European Commission, 2010c). These principles will help Europe to thrive in an even more competitive and resource-limited global economy, providing education, knowledge, health support and, above all, job opportunities for generations to come.

THE INTERNATIONAL SCIENTIFIC DIMENSION

Cooperation between the European Commission and Member States to develop a global dimension of the European Research Area

Biotechnology for agriculture is a major tool for tackling global challenges such as food security and energy supply. Member States and the European Commission make significant efforts to coordinate and contribute to major European policies (Common Agricultural Policy (CAP), environmental, energy, etc.) and international objectives such as Millennium Development Goals of the United Nations.

Since the beginning of 7th Framework Programme, the activities of the European Commission have been integrated with the actions taken by Member States and other countries in the area of plant biotechnology and specific attention has been given to international cooperation. During the same period, several Member States have defined roadmaps aiming to identify the political strategies for development of national and regional approaches to the bio-economy. At the European level some examples are *A Natural Resource Strategy for Finland: Using Natural Resources Intelligently* (Sitra, The Finnish Innovation Fund, 2009) and German *National Research Strategy BioEconomy 2030: Our Route towards a Biobased Economy* (Federal Ministry of Education and Research, 2011).

The Member States are particularly active in their cooperation with many countries around the world for historical, economic and political reasons. The Strategic Forum on International science and technology Cooperation (SFIC) was created to coordinate these efforts (Council of the European Union, 2012). The activities of SFIC are driven by a European Partnership in the field of international scientific and technological cooperation, based on consultation and sharing of information. This Partnership aims to identify common priorities which could give rise to coordinated or joint initiatives and positions vis-à-vis non-European countries and within international forums. A major tool for cooperation between EU Member States and other countries is the Partnering Initiative which foresees a more systematic cooperation at the level of R&D programmes. The overarching aim of a Partnering Initiative is to better tackle challenges common to both partners, the EU and the third country/region, through more coordinated and, hence, more efficient research efforts. Partnering Initiatives on biomass and biowastes with India and on fibre crops and genetic crop improvement with China have been established.

Cooperation between the European Commission and other countries

The European Commission Communication *Enhancing and Focusing EU International Cooperation in Research and Innovation: A strategic Approach* is the current reference for international cooperation in European research and development (European Commission, 2012d). The framework programmes are characterised by being open to the participation of partners from any country in the world. The European Commission is actively cooperating with industrialised countries such as the USA, Canada, New Zealand and Australia on plant biotechnology research. It is also cooperating with emerging economies such as Brazil, Russia, India and China. Plant biotechnology is also addressed in international forums such as the EU–US Task Force on Biotechnology Research (Aguilar *et al.*, 2008; European Commission, 2012c) and the International Knowledge-Based Bio-Economy Forum (European Commission, 2012b).

An overview of the main international cooperation activities in research carried out by the European Commission is outlined in Table 17.1. Significant funding has been provided for cooperation with emerging countries through coordinated or joint calls in specific areas such as plant vaccines (with Russia) and on wastewater treatments and reuse in agricultural systems (with India). Work on fibre crops has been identified as a Partnering Initiative with China. Work on sweet sorghum and *Jatropha* in agricultural systems in the tropics has also been supported. In Work Programme 2013 for the Theme Food, Agriculture and Fisheries and Biotechnology a call aimed at cooperation with Latin America on sustainable biodiversity in agriculture was published.

Finally, one example of specific projects aimed to integrate activities with international programmes in the same area is the *Project 3 to 4* supported by the EU that aims at introducing the characteristics of C4 carbon fixation into C3 carbon fixation crops with increased productivity and reduced inputs associated with the C4 pathways. This project integrates partners and experiences of the 4CRice funded by the Bill and Melinda Gates Foundation.

FUTURE PERSPECTIVES: TOWARDS HORIZON 2020

The Commission proposal for Horizon 2020 is at the heart of the *Europe 2020* strategy to promote smart, sustainable and inclusive growth and it is designed to help to deliver jobs, prosperity and a better quality of life (European Commission, 2010b, 2011). Horizon 2020 should be seen as a

Table 17.1 *Overview of the research activities between third countries and the European Commission in plant biotechnology*

Countries		Tools
Emerging countries	Russia	Twinning EU–Russia Working Groups Agro-bio-food, Partnering Initiative on microbes and plant biodiversity
	Chile, Mexico, Argentina, Latin American countries	Partnering Initiative (biodiversity in agriculture)
	Argentina, China	Partnering Initiative (fibre crops, genetic crops), networking at programme level (EC–China Task Force on Food, Agriculture and Biotechnologies)
Industrialised countries	Canada	Twinning of projects (biomass, biorefinery), networking at programme level (KBBE form)
	USA	Networking at programme level (EU–US Task Force on Biotechnology)
	Australia, New Zealand	Networking at programme level (KBBE form)

policy initiative making a major investment in growth and jobs across the EU through innovation and research. Horizon 2020 activities will run from 2014 till 2020 with a proposed budget of €80 billion.

Horizon 2020 will incorporate innovation elements aimed at addressing the main European Social Challenges in a much more visible way than did previous programmes. It will also enhance the competitiveness and leadership of European industries and agriculture. Recently, an outline of the EU biotechnology programme in FP7, summarising the main ongoing projects, has been published (Cichocka *et al.*, 2011).

Biotechnology has been explicitly identified as the engine of the bio-economy in Europe under the recent German and Belgian Presidencies of the European Union (Aguilar *et al.*, 2009). The bio-economy encompasses the sustainable production of renewable biological resources and their conversion into food, feed, bio-based products, biofuels and bioenergy. Its sectors and industries have a strong innovation potential due to their use of a wide range of sciences, enabling technologies and available knowledge in Europe. The bio-economy's cross-cutting nature offers a unique opportunity to address interconnected societal challenges – such as food security, natural resource scarcity, fossil resource dependence and climate change – in a comprehensive manner while achieving sustainable economic growth.

The European Commission has recently adopted a Communication on the Strategy for *Innovating for Sustainable Growth: A Bio-Economy for Europe* (European Commission, 2012a). The Bio-Economy Strategy, through its Action Plan, will aim to pave the way to a lower emission, resource efficient and competitive society that reconciles food security with the sustainable use of renewable resources for industrial purposes and environmental protection. It has three main goals:

(a) to create a more coherent policy environment and governance mechanisms to better coordinate national, EU and global bio-economy policies and engage in public dialogue

(b) to drive research and innovation in primary production and processing sectors, improving their productivity, resource efficiency and overall sustainability and develop new industrial concepts, infrastructures and business models

(c) to assist policy-makers by providing scientific evidence for policy development and implementation.

ACKNOWLEDGEMENTS

We thank John Claxton for fruitful discussions and critical reading of the manuscript.

DISCLAIMER

This publication expresses the views of the authors and should not be regarded as a statement of the official position of the European Commission nor of its Directorate-General for Research and Innovation.

REFERENCES

Aguilar, A., Bochereau, L., and Matthiessen, L. (2008). Biotechnology and sustainability: the role of transatlantic cooperation in research and innovation. *Trends in Biotechnology* **26**, 163–165.
Aguilar, A., Bochereau, L., and Matthiessen, L. (2009). Biotechnology as the engine for the knowledge-based bio-economy. *Biotechnology and Genetic Engineering Reviews* **26**, 383–400.

Cichocka, D., Claxton, J., Economidis, I., *et al.* (2011). European Union research and innovation perspectives on biotechnology. *Journal of Biotechnology* 156, 382–391.

Council of the European Union (2012). Strategic Forum for International S&T Cooperation website. www.consilium.europa.eu/policies/era/sfic?lang=en

European Commission (2002). *Life Sciences and Biotechnology: A Strategy for Europe*, COM(2002)27. Communication from the Commission to the European Parliament, the Council, the Economic and Social Committee and the Committee of the Regions. Office for Official Publications of the European Communities, Luxemburg. http://ec.europa.eu/biotechnology/pdf/com2002-27_en.pdf

European Commission (2010a). *A Decade of EU-Funded GMO Research (2001–2010)*, EUR 24473. Publications Office of the European Union, Luxemburg.

European Commission (2010b). *Europe 2020 Flagship Initiative Innovation Union*, COM(2010)546 final. Communication from the Commission to the European Parliament, the Council, the European Economic and Social Committee and the Committee of the Regions. European Commission, Brussels.

European Commission (2010c). *Europe 2020: A Strategy for Smart, Sustainable and Inclusive growth*, COM(2010)2020 final. Communication from the Commission. European Commission, Brussels.

European Commission (2011). *Proposal for a Regulation of the European Parliament and of the Council establishing Horizon 2020: The Framework Programme for Research and Innovation (2014–2020)*, COM(2011)809 final. European Commission, Brussels.

European Commission (2012a). *Innovating for Sustainable Growth: A Bio-Economy for Europe*, COM(2012)60 final. Communication from the Commission to the European Parliament, the Council, the Economic and Social Committee and the Committee of the Regions. European Commission, Brussels.

European Commission (2012b). *International Knowledge-Based Bio-Economy Forum*. http://ec.europa.eu/research/bioeconomy/international-cooperation/forum/index_en.htm

European Commission (2012c). EU–US Task Force on Biotechnology Research website. http://ec.europa.eu/research/biotechnology/eu-us-task-force/index_en.cfm

European Commission (2012d). *Enhancing and Focusing EU International Cooperation in Research and Innovation: A strategic Approach*. European Commission, Brussels.

European Food Safety Authority – EFSA (2012). EFSA website. www.efsa.europa.eu/

European Union (2001). Directive 2001/18/EC of the European Parliament and of the Council of 12 March 2001 on the deliberate release into the environment of genetically modified organisms and repealing Council Directive 90/220/EEC. *Official Journal of the European Union* L106, 17.04.2001.

European Union (2003a). Regulation (EC) No. 1829/2003 of the European Parliament and of the Council of 22 September 2003 on genetically modified food and feed. *Official Journal of the European Union* L268, 18.10.2003.

European Union (2003b). Regulation (EC) No. 1830/2003 of the European Parliament and of the Council of 22 September 2003 concerning the traceability and labelling of genetically modified organisms and the traceability of food and feed products produced from genetically modified organisms and amending Directive 2001/18/EC. *Official Journal of the European Union* L268, 18.10.2003.

European Union (2003c). Commission Recommendation of 23 July 2003 on guidelines for the development of national strategies and best practices to ensure the coexistence of genetically modified crops with conventional and organic farming (notified under document number C(2003) 2624). *Official Journal of the European Union* L189, 29.07.2003.

European Union (2006). Decision No. 1982/2006/EC of the European Parliament and of the Council of 18 December 2006 concerning the Seventh Framework Programme of the European Community for Research, Technological Development and Demonstration Activities (2007–2013). *Official Journal of the European Union* L412, 30.12.2006.

European Union (2009). Directive 2009/41/EC of the European Parliament and of the Council of 6 May 2009 on the contained use of genetically modified micro-organisms. *Official Journal of the European Union* L125, 21.05.2009.

Federal Ministry of Education and Research – BMBF (2011). *National Research Strategy BioEconomy 2030: Our Route towards a Biobased Economy.* Bundesministerium für Bildung und Forschung/Federal Ministry of Education and Research – BMBF, Bonn. www.bmbf.de/pub/bioeconomy_2030.pdf

Food Chain Evaluation Consortium (2010). *Evaluation of the EU Legislative Framework in the Field of GM Food and Feed,* Final report. http://ec.europa.eu/food/food/biotechnology/evaluation/docs/evaluation_gm_report_en.pdf

Kessler, C., and Economidis, I. (eds.) (2001). *EC-Sponsored Research on Safety of Genetically Modified Organisms,* EUR 19884. Office for Official Publications of the European Communities, Luxemburg.

OECD (2001). *The Application of Biotechnology to Industrial Sustainability: A Primer.* OECD Publishing, Paris. www.oecd.org/dataoecd/61/13/1947629.pdf

Secretariat of the Convention on Biological Diversity (2000). *Cartagena Protocol on Biosafety to the Convention on Biological Diversity: Text and Annexes.* Secretariat of the Convention on Biological Diversity, Montreal, Canada.

Sitra, The Finnish Innovation Fund (2009). *A Natural Resource Strategy for Finland: Using Natural Resources Intelligently.* Sitra, Helsinki. www.sitra.fi/julkaisut/muut/A%20Natural%20Resource%20Strategy%20for%20Finland.pdf

United Nations (1992). *United Nations Conference on Environment and Development.* UN, Rio de Janeiro, Brazil.

18

Europe, GM crops and food: understanding the past, looking to the future

INTRODUCTION

This chapter seeks to give a brief commentary on the situation for genetically modified (GM) crops in the European Union (EU), focusing on the way in which EU decision-making has operated in order to understand how the policy thinking, legal framework and regulatory decisions have developed as they have – and what that means for how things might develop as we look towards the future.

The first section is an overview of the background to the development of EU GM policy, at a global level, but also Europe-wide. This looks at the way the arguments around GM crops have played out over the past 20 years or so in Europe as a whole, and the resultant outcomes we have seen with regard to the way in which they are currently regulated. This section therefore focuses on the issues at stake and the parties outside of government who have been involved in the discussions, both publicly and more privately.

The chapter then looks at and seeks to understand the different forces that affect how policy, legislation and regulatory decisions are taken, and see how they interconnect – often in ways in which the designers of the processes might not have intended, at least when it comes to decisions involving GM crops and food. In this section, the focus is on the way that the political system has digested all of the combination of scientific, commercial, social and national issues involved, in an effort to understand how the checks and balances have produced the results they have.

Successful Agricultural Innovation in Emerging Economies: New Genetic Technologies for Global Food Production, eds. David J. Bennett and Richard C. Jennings. Published by Cambridge University Press. © Cambridge University Press 2013.

Against this backdrop, the chapter then concludes by reviewing the latest developments in the EU, and the efforts to resolve the deadlock and controversy on this issue, and offers some reflections on how they may evolve, and what that means for the future of GM crop technology in Europe, and the implications that has on other parts of the world as well as for Europe.

BACKGROUND

There can be few issues, at least in the EU, that have attracted as much sustained controversy as GM crops and food. There are many other food 'crises' that one can recall, which have been high profile but relatively short-lived – think contamination of wines with illegal antifreeze or vegetables from southern Europe being found in northern European supermarkets to contain higher than allowed pesticide residues. And then there are long-running problems from the 1980s which have faded now, such as the 'Mad Cow' scares around bovine spongiform encephalopathy (BSE). This took many years to be tackled and involved extremely heated arguments, pitting some countries against others in ways in which one had the impression that a crucial national interest was threatened and the very future of the countries most affected was at stake – but ultimately these were solved too, as public attention on the issue waned as a result of effective political compromises thrashed out in many difficult EU discussions. Of course, outside the realm of agri-food policy, many other topics are controversial in the EU and have divided Member States to a degree where the very idea of a cohesive Union was challenged – the Iraq war polarised EU Member States in a profound way. Also, for example, financial challenges facing the Eurozone. Such challenges are obviously the result of deep differences in the way economies have performed despite intentions and efforts that they converge. These problems could genuinely affect the future of European integration. But the controversy around GM certainly does not affect the entire economy or its military security, so the fact that it remains in 2012 almost as intractable as it was in the 1990s when it surfaced, is indeed remarkable. So why is this?

The context in which the initial furore over 'Frankenstein foods' (as they were originally called in a *Daily Mail* UK newspaper front page headline) developed was, of course, not favourable to them from the outset. There had been food scares, such as BSE in the late 1980s and early 1990s, which meant that there was a sensitive political context on any issue concerning food safety, and a public sensitised to the topic and ready to hold their politicians to account. Moreover, an active bureaucracy, at least at EU level, that could seek to fill a vacuum in an area of 'EU integration' had not yet developed., Also, activist organisations sought issues by which to gain publicity and political influence, and a voracious media (albeit with slightly different appetites in different countries within the EU and elsewhere in the world) was hungry for stories that would build on this.

However, on its own that was perhaps necessary, but not sufficient to generate the controversy over GM food which, by the end of the 1990s, had polarised in such a way that the underlying scientific debate (which was itself not resolved) became secondary to a wider political set of issues. Or at least, if not secondary, it was interpreted through a very public, political lens rather than a relatively discrete, scientific academic one.

From the point of view of industrial players seeking to commercialise the new technology in which they had invested very significant amounts of money over previous years, and held associated patent rights, there is a general recognition that they, and in particular the most active and ambitious company, Monsanto, were insufficiently attentive to the context in which they sought to market their new products. They failed to properly research and understand public opinion and consumer sentiment on food-related issues in Europe, and consequently sought to promote the benefits of the technology in ways which manifestly failed to resonate.

Indeed, the lack of emphasis on what they could present as real benefits to their consumers was clearly why, when they confronted opposition or even simply scepticism, there was no basis of goodwill, let alone enthusiasm, for their new technology. In other words, they had lost half the battle even before it had begun. Compare this with other new technologies, notably in the consumer electronics sector, and, in Europe, the difference is stark. This is less so in other parts of the world, notably the USA, where the technology was first marketed to Mid-West farmers in terms of benefits that they could appreciate, and thus the technology did not encounter the same problems as encountered in the EU. But then, these parts of the world had not experienced the same food safety 'crises' as had been the case in the EU. Had the proponents of the technology considered this fully when seeking to pursue the same approach on the 'other' side of the Atlantic, the myriad of other cultural explanations that can be adduced to account for the difference in levels of acceptance of GM crops and food might have remained squarely an academic discussion, rather than one which has featured so clearly in the media and public mind.

But a marketing failure alone is not an adequate explanation of why the controversy took hold and has been sustained since, and so the question remains, what is? Of course, as is the case even with subjects like climate change, the science is never fully 'settled', even if there is a broad consensus amongst that scientific community on the risks involved and how they might be managed. We know that very well and can see that a minority view can easily become an apparently mainstream alternative in the public mind if there is a sufficiently powerful light shone upon it. There is a broad scientific consensus that has emerged around 'green' agricultural biotechnology in general (rather than on individual GM 'events', to use the EU's jargon for specific plant applications). The most eminent of scientists, from the independent academic to those, for example, holding the role of Government Chief Scientist in the UK, now publicly make the case that the risk–benefit analysis is positive and favourable for GM crops and food, given considerations around the challenge of climate change, food security and population growth. As a result, though this has taken time to emerge, most natural scientists now argue we should embrace and not reject GM crops and food.

No, the controversy over GM food is not fundamentally due to a scientific disagreement, even if that has existed and there will always remain areas of uncertainty. The underlying reason that the issue of GM food became so controversial is that it was identified early on by a number of far-sighted, active and ingenious non-governmental organisations (NGOs), notably Greenpeace and Friends of the Earth, as being a lightning rod for a much wider set of concerns that they had and which they wanted to advance and popularise. Though not wholly uniform themselves, these concerns were a series of growing environmental problems and the way in which capitalism at a general level and 'big business' in a more vernacular sense was responsible for them and indeed perpetuating them. These concerns were the late twentieth-century continuation of the ideas initiated by, for example, Rachel Carson in the 1960s in *Silent Spring*.

With a world-view and accompanying mission that identified global business and a 'captured' government (in the USA and Europe at least) as the champions of what they depicted to be the opposite ideal, the scene was set for a great fight between the forces of 'good' and 'evil' – and the early marketing mistakes of the companies trying to commercialise the technology as fast as they could provided the missing ingredient. It was helped too by the fact that for 'green' biotechnology (as opposed to 'white', industrial, or 'red', medical, biotechnology) there

Figure 18.1 Triangulating for sustainability the EU way.

were a small number of players involved, and the control they might exert through their patent portfolios added to the worries that food supply was becoming dominated by too few, private rather than public sector organisations. So this then helped to turn a possibly arcane scientific question or a short-lived marketing failure into a surrogate for a discussion about what 'sustainable development' really means, for developed and developing worlds alike, whether the current globalised 'Anglo-American' model of political economy is 'fit for purpose' or whether with each individual decision we take in the same framework, we are moving closer to an environmental and social breakdown.

These broader 'sustainability' issues are without doubt completely central to the human condition in the early twenty-first century, and developing wider agreement about the nature of the broader issues faced, how interconnected they are and what we must collectively do about them, are as urgent as they are important. But with arguments and evidence on both sides about how each proposed approach is in fact a better one from the point of view of these bigger issues, it is not the purpose of this chapter to explore the merits of each perspective. Rather, it is to look at how the European political system processes these issues. This process takes place within its own evolving interpretation of sustainability, with the trade-off and tensions inherent within that. See Figure 18.1 for a graphic indication of how the 'Europe 2020 strategy' does this – and what the consequences are of that.

Against this backdrop, it is perhaps not surprising that the current EU legislative framework and policy thinking is criticised by most environmental NGOs as being much too weak and by most business groupings as being unnecessarily strict – there was possibly no way that the EU could produce a 'win–win' situation to keep all sides happy, so the result has been something that is essentially keeping all sides equally unhappy – though there are caveats to that.

Certainly, compared to other parts of the world, the EU's regulatory regime is held up by the European Commission, and considered by most governments in the EU as well as overseas, to be the most rigorous in the world, and presented as being the guarantor of consumer safety and environmental protection. Whether it is with respect to

- the conditions attached to proving a single GM event is safe to be grown at all,
- the precautions necessary to have when actually growing it,
- the obligations on any organisation marketing seeds within the EU,
- the provisions to avoid 'cross-contamination' for segregation of non-GM from GM imports, whether for human or animal food,
- right up to the marketing and labelling provisions for any product sold in shops that contains any GM at all,

the legal and regulatory framework is undoubtedly an onerous one on any company wanting to operate in this area.

All of these provisions have been agreed already – but have singularly failed to remove the controversy around GM crops and food. Each candidate for approval is contested throughout the process by NGOs opposed to them generally, and any that does reach approval is invariably criticised as being achieved as a result of heavy business lobbying and a distortion of 'the science'. On the other hand, the industry associations will frequently state that the numbers of authorised GM events for cultivation is not only lower than in the USA and most other parts of the world, but the speed with which new events are being approved under accepted processes is so slow that it undermines the system's credibility and indeed the commercial incentive to invest in the approval process at all. The process of approval, involving scientific consideration by the European Food Safety Authority (EFSA), followed by expert discussions between representatives of national governments and the European Commission, and finally an approval that requires support from the European Parliament, is undoubtedly complex (Figure 18.2). But despite operating within the context of agreed laws, it is also highly controversial – so why is that, when the whole objective of having

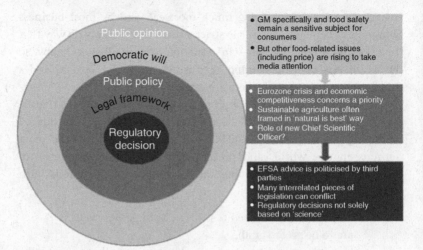

Figure 18.2 Layers of influence and the role of science in decisions.

legislation to protect the environment or consumers should reassure all parties that the outcome should reflect this intention?

UNDERSTANDING EU POLICY-MAKING ON GM CROPS AND FOOD

As indicated above, the way in which the EU takes decisions can be understood in relation to three levels of decision, which applies generally as much as specifically in this case. Starting with the most detailed, a specific regulatory decision about the approval (or otherwise) of a GM crop event is taken within the context of the relevant legislation that has been previously agreed. Whether that legal framework is controversial or not, it is the result of an agreed, standard process, albeit one that has several variants depending on the subject being considered, and which has undergone various changes as the EU's size and complexity has grown over the years, and new treaties have been agreed to reflect that and ensure decision-making still operates well.

Until such time as the policy thinking around a given topic changes, therefore, the legislation is likely to remain the same. Of course, that means that EU policy-making is key to determining not just the shape of any new legislation, but also whether there should be any change to the status quo at all. In this sense, changes of government in a particular country can often be a moment where such policy thinking changes, though it can also change due to external events.

So a final, umbrella level of influence, obvious as it may be, is that of public opinion, which is often transmitted via the media (which is itself undergoing something of a revolution due to changes in technology and the rise of social networks) as well as through the periodical expression of political preference through elections – at least as we have in representative democracies such as those in the EU.

There is therefore a clear interrelationship between these different levels of influence on decisions. Understanding how they work together is important to be able to understand the specific case of GM crop decision-making within the EU. Figure 18.2 outlines this relationship and gives specific instances of how it operates in the case of GM crops and food, and where scientific advice comes into play as well.

In the case of the EU's GM legislation, there are different legal bases taken for different pieces of legislation, a mix of those based mainly on environment provisions, those based on laws and those based mainly on internal market provisions.

Although the processes for agreement of these between the main EU institutions varies, they result from a series of checks and balances which ensures that all perspectives are taken into account, from those of each country's national government, through to the directly elected and accountable politicians in the European Parliament, and finally the 'European perspective'. The 'European perspective' is provided by the Commission, which has a *modus operandi* which is deliberately very consultative and involving of third-party interests from all the different perspectives an issue such as this can throw up.

As Figure 18.3 indicates, each level of decision-making – regulatory, legislative and public opinion – has a specific set of characteristics which define the way decisions are supposed to be taken, the outcome that is desired and the mindset of participants.'

In the case of public opinion, each individual participates in shaping any development, and this is obviously highly subjective, working according to an individual belief system that has the media as its main transmission mechanism, outside of elections. Given the variety of historical experiences, cultural contexts and basic socio-economic situations within the EU, it is no great surprise that there is little that is truly uniform about European public opinion – and the best expression of this is perhaps in the EU's own unofficial motto 'United in Diversity'!

On the other hand, in the case of regulatory decisions that are taken on highly specific issues, the process could not be more different

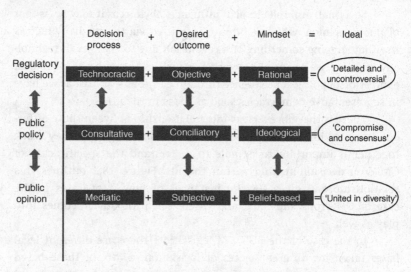

Figure 18.3 Characteristics of different decision-making levels.

in character, at least in theory. Here, decisions are shaped through a very technocratic process commonly known as 'comitology', where the aim of all participants is usually for the decision that results to be the most objectively justified possible, and therefore their mindset in approaching this process would normally be very rational.

The ideal outcomes are there characterised as detailed and uncontroversial. If one considers the many thousands of such decisions that are taken by the EU each year and scarcely reach the pages of even specialist trade or scientific publications, let alone more widely read or seen media, this is indeed how this system tends to work. So why, therefore, is it the case that such decisions when sought on GM crops and food issues are so out of keeping with this more typical or desired character?

To understand this, one must consider the middle range of decision-making in the EU around the public policy that shapes legislation, where the most seasoned observers will frequently draw attention to the aim of having a 'good' compromise and consensus on any issue. As indicated in Figure 18.3, for the EU and its institutions, this means that decision-making on policy frameworks is characterised by a high degree of consultation with third parties or 'stakeholders', a process which is exemplified by the role of the European Commission.

Indeed, when such a picture is painted on an issue such as this, it is not difficult to have a degree of sympathy with the European

Commission officials charged with identifying the 'European' interest amidst the clamour of voices on all the different angles, as it will clearly be challenging to do this, even where the issues are not so controversial.

The role of the European Parliament in such decisions has grown over time and is now very significant as well. And for that institution, being the sole forum for the direct representation of democratic opinion at EU level, the role of party politics typified by ideologies of the left and right, greens or others, is of course central. Figure 18.4 illustrates the range of groups, numbers of MEPs and number of elections at national level at a period in the late 2000s by way of example.

And last, but perhaps most significantly, the need to reconcile the various interests, notably national, at play is exemplified by the Council of Ministers, which seeks on every issue to avoid putting Member States in a minority of one and to achieve a genuine consensus if at all possible. Of course, in the case of GM crops and food, national positions have been very divergent indeed, and this degree of diversity is captured in Figure 18.5, which is a snapshot of the views present in 2007, by way of example.

What has occurred in the case of GM crops and food decisions, however, is that the boundaries between the different levels of decision-making, which are never totally firm, have crumbled almost entirely, as a result of the highly effective efforts on the part of a number of NGOs to work the issue as an exemplar of a wider problem. They represent it as the 'thin end of the wedge', and deliberately campaign to ensure that public concerns are raised by any development away from their desired outcome. Such developments are represented as the result of corporations who do not have a public interest in mind at all, and are in effect set against it. And with views so polarised between Member States on the subject, the envisaged role of the European Commission in the process of comitology, whereby it is able to force a conclusion in the absence of consensus, has not been played out either, essentially because of the concern bordering on fear that each such decision would be portrayed as illegitimate imposition by an unelected bureaucracy, whether the case be made by the 'losing' Member States or the NGOs opposed to the outcome. The result, as indicated earlier, is the near deadlock in decisions on individual events for approval, whether for cultivation in the EU or for import, for example as feed.

In such a situation, the consideration of specific scientific evidence on an individual GM event was invariably contested in a way

EPP-ED (centre right)	227
PES (centre left)	218
ALDE (liberal democrat)	105
GREENS-EFA	42
GUE-NGL	40
ID	23
UEN	44
ITS	20
Non-Affiliated	14

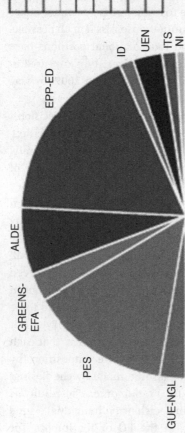

Parliamentarians

Germany	99		Austria	18
France	78		Bulgaria	18
UK	78		Slovakia	14
Italy	78		Finland	14
Spain	54		Denmark	14
Poland	54		Iceland	13
Romania	35		Lithuania	13
Netherlands	27		Latvia	9
Portugal	24		Slovenia	7
Belgium	24		Luxemburg	6
Czech R	24		Cyprus	6
Hungary	24		Estonia	6
Greece	24		Malta	5
Sweden	19		**Total**	**785**

Elections

Estonia	Mar-07		Lithuania	Oct-08
Finland	Mar-07		Slovenia	Oct-08
France	May-07		Romania	Nov-08
Iceland	Jun-07		Denmark	Feb-09
France	Jun-07		Malta	Mar-09
Latvia	Jun-07		Slovakia	Apr-09
Belgium	Jun-07		Europe	Jun-09
Iceland	Jul-07		Lithuania	Jun-09
Slovenia	Nov-07		Luxem.	Jun-09
Czech R	Jan-08		Bulgaria	Jun-09
Cyprus	Jan-08		Poland	Sep-09
Spain	Mar-08		Germany	Sep-09
Greece	Mar-08		Romania	Dec-09
Malta	Apr-08			

Figure 18.4 Party politics: what's the big idea?

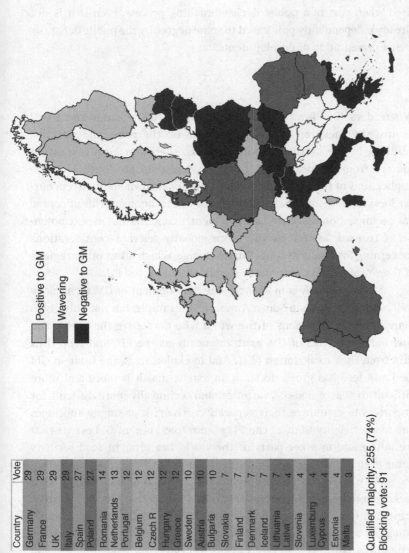

Country	Vote
Germany	29
France	29
UK	29
Italy	29
Spain	27
Poland	27
Romania	14
Netherlands	13
Portugal	12
Belgium	12
Czech R	12
Hungary	12
Greece	12
Sweden	10
Austria	10
Bulgaria	10
Slovakia	7
Finland	7
Denmark	7
Iceland	7
Lithuania	7
Lativa	4
Slovenia	4
Luxemburg	4
Cyprus	4
Estonia	4
Malta	3

Qualified majority: 255 (74%)
Blocking vote: 91

Positive to GM
Wavering
Negative to GM

Figure 18.5 Power politics: national interests.

that the process was never intended to deal with. This is not to pass judgement on the rights and wrongs of the case each stakeholder had in the discussion, all of which have legitimacy in the widest public sense of the term, but it does highlight how the scientific process is itself politicised when part of a public decision-making process, even if it is not already independently politicised to some degree by the public debate or media presentation of developments.

LOOKING AHEAD

Where does that leave the EU with respect to GM crops and food? Against the backdrop of the polarisation on the issue, and effective deadlock in the decision-making process, there has been a growing concern from a wider range of stakeholders than previously that the implications of the EU's stance is negative for its own ability to encourage local investment and development of GM technology. This of course has an impact on associated employment, economic and export potential if realised, as well as the more globally relevant considerations concerning food security and climate change, which affect other regions of the world perhaps even more significantly that the EU.

Perhaps equally importantly, the development of GM technology well beyond the USA, in South America for example, has not only given more examples of regions of the world who do not see the health and environmental risks of GM as dramatically as the EU, and who thus gain from their exploitation of it. And in exploiting it, the trade in GM food and feed has grown to such an extent that it is more and more difficult to source non-GM varieties, and technically more difficult for importers to guarantee their separation. The risk of supply shortages due to an inability to meet the EU's 'zero tolerance' of GM events that are authorised in other parts of the world has brought food security home even to the EU.

With respect to the latter issue, the Commission has in effect been seeking and achieving authorisation of GM events that are allowed to be present at low levels in imports of non-GM feed, and is moving towards doing the same in food, without provoking a degree of public discussion around the issue. This in effect seeks to protect EU food supplies, and keep prices lower than they would otherwise be, without making that case in a highly public way. Whether that approach can and will be maintained when it concerns food for human consumption instead of feed for animal consumption is an interesting question – and at the time of writing still not known.

At the same time as moving as indicated above, in a bid to cut the Gordian knot locking together countries totally opposed to the cultivation and use of GM with those who are supportive of their development, the European Commission proposed in 2010 a new approach to the issue, whereby, essentially, the scientific-focused risk assessment and consequent approval for cultivation of GM events would remain an EU-level task, whilst the decision about whether to allow cultivation of that event would be taken at national level. The thinking behind this approach was that it might enable a new legal framework to be established which enabled GM events to be approved in a scientific process whilst providing for countries to invoke other reasons for not authorising them in their territory.

However, whilst the 'renationalisation' proposal may have some attractions in this respect, it has proven fraught with problems, amongst them the precedent it sets for other products, for invoking non-scientific issues as a reason for withholding approval for marketing an otherwise 'safe' product – which runs counter to the intentions of the European single market, a flagship project for Europe still today, as well as World Trade Organization (WTO) provisions. The resulting arguments over the proposal in both the European Parliament and the Council of Ministers have therefore once again come up against familiar coalitions for and against GM in general. Even the most creative efforts at forging a compromise – for example agreeing a mechanism whereby companies wanting to cultivate an approved GM crop Europe-wide would not do so in those countries which requested they did not – have failed to break the deadlock.

If one returns to the three levels of decision-making that characterise the EU processes in general, then one must conclude that the role of public opinion is the crucial element of the discussion. Only when public opinion is resolved will it be possible to find a solution at the policy level and at the regulatory level. Indeed, this is recognised clearly by all involved, whether for or against, and they have invested significant efforts in this regard. In some parts of the EU, one can now discern some change, with other issues coming to the fore and countering the specific focus on GM as the main preoccupation. Opinion polls indicate a slight shift in favour of GM in countries such as the UK, where the government stance has been favourable for some time, but where media coverage is often still highly charged (Eurobarometer, 2011). However, in other countries, whose governments are opposed to GM, there is little sign that opinion is shifting. The imperfect situation looks set to continue for some time to come, therefore, and it might be that until a crisis that defines the issue either much more clearly as 'unsafe' for human consumption or essential to maintain security of supply for either the EU or indeed developing countries whose food security problems are of a different dimension to those of the EU, that we will just have to accept it as such.

REFERENCES

Carson, R. (1962). *Silent Spring*. Houghton Mifflin, Boston, MA.
Eurobarometer (2011). *Biotechnology*. TNS Opinion and Social, Brussels.

19

US international engagement in agricultural research and trade

INTRODUCTION

By 2050, the global population will surpass 9 billion and require 70% more food to meet its needs. At the same time, the world's arable land will be increasingly challenged by water scarcity and climate change, raising the risk of production shortfalls in a world where approximately 1 billion people already go to bed hungry every night (FAO, 2011).

Increasing food production alone will not be enough. Growing awareness of the need to produce food more sustainably means that we must use less water, land, fertiliser and pesticides than we do today, even as we increase yields. These challenges can only be met through the appropriate use of agricultural science and technology and it will necessitate not only the development of innovative technologies, but the widespread adoption of appropriate agricultural practices and policies, including trade policies.

In 2011, farmers/growers produced genetically modified (GM) crops in 29 countries in North and South America, Europe, Asia, Africa and Australia (James, 2012). While the United States remained the single largest producer of GM crops, GM technology provided benefits to both large-scale farmers and smallholder farmers in developing countries. Ninety per cent of farmers growing GM crops live in the developing world, including India and China, and the West African nation of Burkina Faso experienced the second fastest adoption of GM seeds in 2011 (James, 2012).

Despite a proven record of safety, sustained growth of 10% or more a year for the last 15 years and income gains of nearly 20% for

Successful Agricultural Innovation in Emerging Economies: New Genetic Technologies for Global Food Production, eds. David J. Bennett and Richard C. Jennings. Published by Cambridge University Press. © Cambridge University Press 2013.

poor farmers in Africa, opposition to the technology continues to limit its impact. Such opposition has led to heightened regulatory and labelling requirements in some countries, and delayed or blocked adoption of the technology altogether in others.

In order to address food security challenges globally, the United States is actively engaged in international agricultural research and promoting transparent, predictable and science-based regulatory systems that will foster innovation and facilitate technology adoption and trade of GM crops and products. GM technology is one tool among many that the United States is pursuing to address global challenges and tackle the problems of hunger and poverty. But GM technology is essential if farmers are to produce more food on less land with fewer inputs. GM products are also a critical component of US agriculture and play an important role in US agricultural exports.

This chapter will consider some of the ways the US government is working to promote research and trade in GM products through bilateral, multilateral and regional engagement.

AGRICULTURAL RESEARCH AND SUPPORT FOR MODERN PLANT IMPROVEMENT AND BIOTECHNOLOGY

The US government, largely through the work of the US Department of Agriculture (USDA), has long supported research in the food and agriculture sector, beginning in 1862 with implementation of the Morrill Act establishing the Land Grant research colleges and universities. This commitment has continued uninterrupted for the past 150 years, impacting the food and agriculture community through the research, education and economics (REE) mission area. The REE is the home of the Agriculture Research Service (ARS) and the National Institute of Food and Agriculture (NIFA) and provides funds for intramural research and extramural research, respectively. Most, but not all, of the programmes that support research and capacity building and training in agriculture and agricultural technology in developing countries reside in these agencies.

The focus of research conducted under the auspices of the USDA is largely to build the success of agriculture in the United States and includes fundamental and translational research, agriculture extension, growth and stabilisation of rural economies, education and training. There is an ongoing commitment to the support of research and training, including in agriculture biotechnology.

The US Department of Agriculture supports international agriculture research through several of its Mission Areas, including:

(1) Animal and Plant Health Inspection Service (APHIS) supports research in plant and animal health and provides information related to food safety, including on GM crops. It is also the agency that ensures the safety of GM crop varieties and crop products that are imported into the United States. APHIS has offices around the world and works with Foreign Agriculture Service personnel to provide opportunities for training scientists in country and through fellowships in the United States with a particular focus on issues related to food safety and the import of agriculture products, including of products of agricultural biotechnology.

(2) The Foreign Agriculture Service (FAS) administers and awards fellowships for training of scientists and students under different types of programmes, including the Cochran Fellows Program and the Borlaug Fellows Programs. These fellowships can be used for a variety of training activities, including for research in agriculture biotechnology and for training in biosafety and regulatory issues, among other programmes. These opportunities are generally brief, of less than 6 months duration, but can be extended under specific circumstances.

(3) The Agriculture Research Service conducts research in several facilities outside of the United States and hosts scientists from other nations in their research facilities and programmes. These programmes are long term in nature and provide opportunities for training in fundamental sciences as well as in translational/applied biology, including in biotechnology.

(4) Through its Center for International Programs, the National Institute of Food and Agriculture administers funds provided by the US Agency for International Development (USAID), FAS, APHIS and other USDA and non-USDA agencies, for research and training in programmes that are designed for country-led initiatives.

Many of the USDA programmes are considered to be 'translational research' in the broadest sense and include fundamental and targeted research and training in molecular genomics, the use of advanced breeding tools in plant and animal improvement, genetic engineering of plants, greenhouse and field evaluation of crops and animals, and training in natural resource and water management, agriculture statistics, agriculture economics and rural development.

The National Science Foundation (NSF) encourages and facilitates collaborations between scientists in US universities and research institutions in a number of areas of science, including in plant science. The NSF does not have a programme that targets agriculture research per se: nevertheless it supports basic research that is expected to impact agriculture as fundamental data gathered through NSF-funded grants are translated to applied projects. The NSF, however, partners with other agencies and organisations to sponsor research collaborations that engage university scientists in developing economies in collaborations for research and training with public-sector scientists in the United States. Some of the areas of support include research in plant genomics, genetics and genome evolution, cell biology, biochemistry and other disciplines; projects funded by the NSF may involve the use of plant biotechnology and genetic engineering.

In addition to funds provided through funding of the NSF, two programmes administered through the NSF involve funds from non-NSF resources.

(1) BREAD (Basic Research to Enable Agriculture Development) receives its funding from the Bill and Melinda Gates Foundation and is administered through the National Science Foundation. 'The objective of the BREAD program is to support innovative basic scientific research designed to address key constraints to smallholder agriculture in the developing world' (www.nsf.gov). This programme supports research that is fundamental in nature and was not established for application of such research. Nevertheless, these funds are used to gather information that can be applied, including in agriculture, through advanced and molecular breeding. The programme is highly competitive and provides an opportunity for US scientists to collaborate in discovery research with scientists from less privileged countries.

(2) PEER (Partnerships for Enhanced Engagement in Research) is funded by USAID (see below) and is co-administered by the NSF. The programme encourages scientists from developing countries to identify an NSF-funded researcher with like interests and to establish a scientific collaboration that builds upon existing NSF-funded projects, extending them to include the project of the collaborator. USAID funds are used to

support research in the collaborating laboratory. The first PEER projects were funded in 2012 and additional competitions are anticipated in coming years. USAID has a long history of support for development of agriculture-based economies. It has supported the work of the Consultative Group for International Agriculture Research (CGIAR) (www.cgiar.org) from its inception, providing support for core programmes as well as for specific projects. Amongst the longest-running are the Collaborative Research Support Program (CRSP) projects, which are managed by one or more of the Land Grant colleges and universities for crop-specific and region-specific research to improve crop productivity. These CRSP projects have been an important mechanism for training students and other scientists from less developed countries in US institutions, including in advanced breeding and GM technologies, extension, food safety and others.

USAID also funds research and training programmes in biotechnology through awards made to universities, including Cornell University and Michigan State University, which in turn support research and training projects in biotechnology at universities and research institutes in the developing world. Some of these projects provide training in science and technology as well as in product biosafety and regulatory sciences and oversight. In other cases, USAID makes direct awards for specific projects in US and non-US laboratories to advance agriculture sciences and technologies in developing countries. The agency supports activities to advance training in biosafety programmes and relevant policies that enable local policy-makers to establish regulatory frameworks and concordant legislation to oversee the testing and commercialisation of GM products. One such programme is the Program for Biosafety Systems (PBS) administered through the International Food Policy Research Institute in Washington, DC, one of the CGIAR institutes. USAID also provides funds for research programmes that are administered through other US agencies that support research, including research in food and agriculture. These programmes are included in descriptions of the activities of other agencies.

The most recent initiative of USAID that targets food security is the Feed the Future (FtF) initiative. Through Feed the Future the agency works with ministries in developing countries to develop country-led plans to achieve food security and build agriculture economies. In some cases FtF projects support adaptation of long-used approaches to improving productivity and marketing, and in other cases it supports new technologies such as biotechnology and information technologies to achieve the goals. USAID recently embarked on a new programme to identify Grand Challenges for Development, enlisting scientists and technologists from across the many disciplines in US universities and research institutes to stimulate new solutions to both old and new problems in developing countries. Some projects supported by the Grand Challenges for Development will be conducted in collaboration with scientists in developing country institutions.

In addition to support by agencies of the US government, there has long been support for research and training in this sector by foundations, private-sector companies, and individual donors. For example, the Ford Foundation and Rockefeller Foundation provided support for research and training and capacity building in institutes affiliated with the CGIAR, thereby expanding the support of governments for agriculture research in developing countries. In the last two decades the Bill and Melinda Gates Foundation has become a primary donor/supporter for research and training programmes in developing countries. Some foundations, including the Rockefeller Foundation and the Bill and Melinda Gates Foundation, have provided support for research in advanced plant breeding technologies, biotechnology, and other technologies with the goal of increasing food production.

Some of the projects funded by the Bill and Melinda Gates Foundation are conducted in collaboration with private-sector companies or company foundations. Such collaborations are either funded directly or through donations of intellectual property (IP), technical knowledge, and services in kind. For example, the development of Golden Rice began as a collaboration between scientists in Switzerland, led by Professor Ingo Potrykus, and in the Syngenta Company, led by Dr Peter Beyer. Golden Rice required application of more than 40 different patents to achieve the goal of a variety of rice with high levels of provitamin A. Many of the patents used in the project were donated by a variety of sources; funds for laboratory, greenhouse and field research and product development were provided by many foundations as well as government funding agencies, including USAID. In other cases public–private partnerships in agriculture and agricultural biotechnology are funded by private company sources to support public-sector researchers in advanced laboratories and/or laboratories in developing countries. In some cases, bilateral partnerships develop into multilateral relationships with a mix of direct funding and in-kind support.

There are many examples of funding of agriculture biotechnology and genetically engineered plants through such projects over the past 20 years. To date, however, there has not been an example of biosafety approval and commercial release of an agricultural product developed through biotechnology supported by these projects, largely because of (1) the lack of established biosafety regulations in developing countries for which products are being developed, (2) lack of familiarity of public-sector scientists with the process of de-regulating products of agriculture biotechnology, and (3) the complexity of the process of deregulation and the associated costs.

REGULATORY SYSTEMS AND TRADE

During the food price crisis of 2007–8, a number of countries, including Egypt, Indonesia and Vietnam, imposed export bans on food to dampen domestic food price increases. The result was higher prices on the global market and food shortages in other parts of the world, which led to protests and indirectly to violence in some countries. The spike in food prices, and the riots that followed, helped put agriculture back on the global agenda of priority issues facing governments and highlighted the role that trade can play in either promoting or undermining global food security.

The continued high food prices experienced since 2008 have been a boon to farmers in some parts of the world, including the United States. In the fiscal year ending 30 September 2011, the United States exported a record US$137.4 billion in farm products, more than US$20 billion more than the previous year. High food prices have had a negative impact on global food security, however, pushing ever more people into poverty and expanding the ranks of the hungry.

High food prices present both a challenge and an opportunity to the United States with respect to GM foods. Nearly half of US agricultural exports consist of, or contain, biotechnology commodities or ingredients, including most of the soya, maize, cotton and canola as well as processed foods including these ingredients. As a result, the United States has a particular interest in ensuring that countries institute transparent, predictable and science-based regulatory systems that promote free and fair trade in agricultural products.

The US government supports regulatory development and policy analysis for GM crops in the context of broader economic, food security, environment and trade issues. Promoting trade in GM products is not enough to ensure that developing countries will be able to harness the technology to meet their own food security needs. In order for developing countries to access and utilise GM crop technology they must first establish biosafety systems or regulatory frameworks that make it possible to import and produce such materials.

The US government works with developing countries to establish approval mechanisms for the import and production of GM crops primarily through partnerships between USAID and organisations such as the International Food Policy Research Institute's (IFPRI) Program on Biosafety Systems, the Center for Environmental Risk Assessment (CERA), the South Asia Biosafety Program (SABP), the Agricultural Biotechnology Support Program (ABSP II), the African Agricultural Technology Foundation (AATF) and the Donald Danforth Plant Science Center. Such partnerships aim to strengthen environmental safety and food safety policies, and to build local capacity for the development and implementation of science-based regulations related to GM crops. Activities include enhancing the capacity of regulatory staff in these countries to conduct science-based risk assessments and monitor compliance with biosafety regulations. The US government views local capacity as an important part of establishing fully functional regulatory frameworks in countries that may want to commercialise GM crops or import GM commodities.

The United States is engaging both bilaterally and multilaterally to promote trade in and cultivation of GM crops. Regional policy cooperation promotes the exchange of experience and technical expertise among neighbouring countries, harmonisation of regulations, facilitation of trade, and technology transfer.

USAID supports African regional organisations, such as the Common Market for Eastern and Southern Africa (COMESA) and the Economic Community of West African States (ECOWAS), as they work to develop regional approaches to biotechnology policy. Other US agencies, such as the USDA and the US Department of State, have also conducted outreach activities targeting these regional organisations.

Similarly, the US government supports cooperative policy dialogues and capacity building with regional Asian forums such as the Asia–Pacific Economic Cooperation (APEC) and the Association of Southeast Asia Nations (ASEAN). APEC has been a focus of attention for various US agencies. In particular, the United States has chaired APEC's High Level Policy Dialogue on Agricultural Biotechnology (HLPDAB). According to the APEC web portal (www.apec.org), the HLPDAB was established in 2002 in *'recognition of the importance APEC Ministers and Leaders place on member economies' work on the safe introduction of biotechnology products, and on obtaining public acceptance of these products. Policy makers use the HLPDAB to develop regulatory frameworks, facilitate technology transfer, encourage investment and strengthen public confidence regarding biotechnology in order to increase agricultural productivity and protect the environment, with the ultimate objective of promoting food security.*' In 2012 the United States passed the chairmanship of the HLPDAB to Russia, which hosted the 2012 APEC forum. In addition to the HLPDAB meeting itself, APEC economies host several workshops and meetings each year under the auspices of the HLPDAB with participation by the US government. Past meetings have focused on the low-level presence of unapproved GM products in the food supply and on preparation for meetings of the Convention on Biological Diversity's Cartagena Protocol on Biosafety, which will be discussed later.

The US government also promotes bilateral and trilateral policy dialogues with a number of countries that promote harmonisation of regulations and/or the sharing of information on policies, legislation and regulations affecting GM products.

The US government's most significant trilateral dialogue is the North American Biotechnology Initiative (NABI). The USDA co-chairs the NABI meeting each year with Canada and Mexico, along with counterparts from each country's Ministry of Agriculture. The officials share information on regulatory developments related to GM crops in their countries and discuss areas of mutual interest with respect to international developments.

The US government engages a number of countries in bilateral discussions on GM technology with respect to research, trade and regulations. For example, several dialogues have taken place between the US government and the European Commission since 1990; the EU–US Task Force on Biotechnology Research has been '*coordinating transatlantic efforts to promote research on biotechnology and its applications for the benefit of society. The Task Force was established in June 1990 by the European Commission and the White House Office of Science and Technology*' (EC, no date). This task force focuses on research, science and other technical issues and not trade. The Office of the US Trade Representative leads US government discussions with the European Commission related to the World Trade Organization biotechnology case, which is discussed later.

Other important bilateral dialogues include the US–China High-Level Biotechnology Working Group (BWG) and the US–China Technical Working Group (TWG), which focus on policy and technical discussions respectively. Other broader dialogues with China in which GM issues sometimes arise include the US–China Joint Committee on Cooperation in Agriculture (JCCA) and the US–China Joint Commission on Commerce and Trade (JCCT).

The US government also engages with a number of countries on a regular, though not necessarily formal, basis, including Brazil, Australia and New Zealand, which share some common interests in supporting transparent, science-based regulatory systems.

The US government actively participates in a number of multi-lateral organisations where GM technology, regulations and policies are discussed. Some of the most active discussions have occurred in the Cartagena Protocol on Biosafety to the Convention on Biological Diversity, the Codex Alimentarius Commission and the World Trade Organization (WTO).

The Cartagena Protocol on Biosafety (CPB) to the Convention on Biological Diversity (CBD) is an international agreement which regulates the safe handling, transport and use of living modified organisms (LMOs), such as GM seeds, microorganisms and animals that are able to replicate in the environment. The CPB was adopted on 29 January 2000 and entered into force on 11 September 2003. While the United States is not a party to the agreement, US companies shipping LMOs to countries that are parties to the agreement still need to comply with laws implementing the treaty in the country of import. As a result, the US government was active in the negotiating of the treaty and has remained an active participant in subsequent meetings of the CPB as an observer.

The US government has been an active participant in meetings of the Codex Alimentarius Commission and its committees, including its work related to GM products, which was established by the Food and Agriculture Organization (FAO) and the World Health Organization (WHO) in 1963. The Commission develops harmonised international food standards, guidelines and codes of practice to protect the health of consumers and ensure fair trade practices in the food trade. While the standards established by the Commission are voluntary, the character of these standards was given heightened importance as a consequence of the special status given to these standards by the WTO. These issues have become more important as free trade agreements expand and tariff barriers fall, leaving food safety standards as one of the few ways of protecting domestic markets from foreign exports. Given the trade implications of decisions taken by the Commission, it is no surprise that the Commission has been a regular focus of attention for discussions on GM products.

The Commission has taken up the questions of GM standards in a number of its committees over the last decade and the United States has been an active participant in these discussions. The Commission has established two biotechnology task forces that have successfully established principles and guidelines for risk assessment of GM plants, including nutritionally enhanced plants, animals and microorganisms, as well as guidelines for risk assessments in situations of low-level presence of unapproved GM products. These guidelines were adopted by consensus of the Commission, which now numbers approximately 180 members. The Commission's work on GM products in other committees has sometimes been quite contentious. The Codex Committee on Food Labelling (CCFL), for example, spent nearly 20 years on the question of if, when and how to label foods derived from modern biotechnology. The CCFL completed its work in 2011 without a clear answer to these questions.

The WTO is, perhaps, the organisation in which the differences in various countries' approaches to GM products have been most clearly laid out. On 7 August 2003, the United States, Canada and Argentina announced their intention to seek a Dispute Settlement Panel to address the European Union's (EU) moratorium on approving GM products and over EU Member State bans of previously approved products for cultivation. On 29 September 2006, the WTO issued the final report in the case. The WTO found that the EU measures were in breach of the EU's obligations under the WTO Agreement on the Application of Sanitary and Phytosanitary Measures (WTO, 2012).

On 21 November 2006, the WTO Dispute Settlement Body (DSB) adopted recommendations and rulings calling for the EU to bring its measures into compliance with WTO obligations. The panel noted that the moratorium and EU Member State bans were inconsistent with the scientific findings of the EU's own safety assessments. While Canada and Argentina have since settled their narrower cases with the EU, the United States maintains that the EU has yet to comply with the DSB ruling and has continued to hold regular meetings with the EU urging the regional body to remove barriers in the approval system for GM products. In addition, the US government continues to raise the issue of the EU's non-compliance during meetings of the WTO's Committee on Sanitary and Phytosanitary Measures.

CONCLUSION

US international engagement in research and trade on agricultural biotechnology reflects recognition of the challenge of producing sufficient food, fibre and energy to meet the needs of a growing world in facing environmental degradation and climate change. For the scientists who work on GM crops, agriculture biotechnology is merely a tool of translational biology and which is used, as appropriate, to achieve specific goals in agriculture. Public-sector research institutions as well as private-sector interests establish the goals.

GM crops are an integral part of the agriculture system in the United States and are anticipated to play a key role to increase crop productivity around the world in the face of variable climate and weather patterns. The research infrastructure in the United States was developed to provide fundamental knowledge, and translational studies (including via biotechnology) to adapt research results to usable goods and to transmit the information to producers and consumers. In this respect, the United States has much to offer to emerging economies in terms of experience, capital and capacity building to facilitate technology adoption and implementation. With these objectives in mind, the US government has established a variety of programmes to provide training in research and in biotechnology for scientists from developing countries. It also provides training to develop transparent and science-based methods to evaluate the safety of products developed by biotechnology, with the goal of encouraging the sharing of information related to food production practices to ensure safety of all foods and feeds.

Because of the urgency of meeting global food needs and building food security worldwide, the US government strives to reduce barriers to trade that reduce access to food and lead to increased food cost without a concomitant increase in safety, and to encourage science-based systems that will lay the foundation for more robust trade policies that promote access to safe and affordable food products.

DISCLAIMER

Jack Bobo serves as the Senior Advisor for Biotechnology and the Chief of the Biotechnology and Textile Trade Policy Division in the Bureau of Economic and Business Affairs at the US Department of State. The views in this chapter are his alone and do not necessarily reflect the policies or views of the government of the United States.

REFERENCES

FAO (2011). *The State of the World's Land and Water Resources for Food and Agriculture (SOLAW): Managing Systems at Risk*. Food and Agriculture Organization of the United Nations, Rome and Earthscan, London.

EC (no date). EU–US Task Force. http://ec.europa.eu/research/biotechnology/eu-us-task-force/index_en.cfm

James, C. (2012). *Global Status of Commercialized Biotech/GM Crops: 2011*, ISAAA Brief No. 43. International Service for the Acquisition of Agri-biotech Applications. Ithaca, NY. www.isaaa.org/resources/publications/briefs/43/executivesummary/default.asp

WTO (2012). www.wto.org/english/tratop_e/sps_e/spsagr_e.htm

Part 4 Social, legal, ethical and political issues

Introduction

DAVID J. BENNETT

The hard-learned – and oft-forgotten – lesson from many science-technology/public/political debates – from the Luddite riots in early nineteenth-century Britain against the changes of the Industrial Revolution and the Locomotive Act in 1865 (the 'red flag Act') which limited these new-fangled fire- and smoke-breathing monsters to 10 mph, 4 mph in rural areas and 2 mph in towns to, more recently, the chemical and nuclear industries, genetic modification (GM) and climate change – is that they are 'hearts and minds' debates. People are interested first in what affects them, their families and their friends and colleagues directly – then in other things if they are entertaining or stimulating – and then scientific and technological matters if they are directly relevant to them (having a cystic fibrosis child, astronomy as a hobby, whatever) when they become experts. Just think of one's own interests – that is unless you are a scientist when science is most probably both your child and your hobby!

Yet, whether we recognise it or not, science and the technologies which derive from its discoveries totally underpin every aspect of life for one-fifth of the world's population. In the developed, so-called 'post-industrial', societies people count on science for their health, their wealth, their welfare, their standard of living and lifestyles, and their

Successful Agricultural Innovation in Emerging Economies: New Genetic Technologies for Global Food Production, eds. David J. Bennett and Richard C. Jennings. Published by Cambridge University Press. © Cambridge University Press 2013.

very environment. This will also soon be so for a further one-third living in China and India. For the remaining just under one-half in the emerging economies, forecast to increase in population by far the most by 2050, it is their aim rapidly on the way to being achieved. Science has been one of the most powerful, if not arguably the most powerful, of human enterprises. In developed societies we have come to depend totally upon it and all indications show that this will be ever-increasingly the situation throughout the whole world.

It is almost a paradox, and quite contrary to what many scientists believe and are often astounded to hear, that they routinely come top of the list in European public opinion polls as those most trusted to explain the impact of science on society. This is so whether they work in university, government or industry laboratories as the Europe-wide Eurobarometer survey shows in Figure 1 (European Commission, 2010). In the UK, for example, a series of surveys carried out between 2000 and 2011 show that, overall, public attitudes to science are positive and interest in science has increased (Department of Business Innovation and Skills, 2011)). Here four-fifths (82%) agree that 'Science is such a big part of our lives that we should all take an interest', a similar proportion (79%) that 'On the whole, science will make our lives easier' while over half (54%) agree that 'The benefits of science are greater than any harmful effect.'

Yet in the 'hearts and minds' debates about new applications of science and technology of which we are speaking, history teaches us that it is when their advantages can be clearly recognised and their disadvantages are outweighed, unrecognised or ignored that they succeed. But what is seen as an advantage by one person, or nation, because it is seen as desirable may be seen as just the opposite by other people and in different parts of the world in different circumstances. And this of course is the case with GM and may become so with recent allied non-GM techniques.

George Gaskell and Sally Stares begin this Part 4 by asking 'Have GM crops and food a future in Europe?' in their Chapter 20. They have been centrally involved at the London School of Economics in a series of large Europe-wide surveys, the Eurobarometer, and chart what they call the 'rollercoaster ride' in the European public's attitudes to biotechnology from the early enthusiasm of 1991 to the present-day ambivalence. They discuss three potential explanations for Europe's resistance to GM food – the perceived 'unnatural' image of genetic modification, the perceived absence of benefits, and food in its associations with culture. They conclude that for the first generation of commodity crops

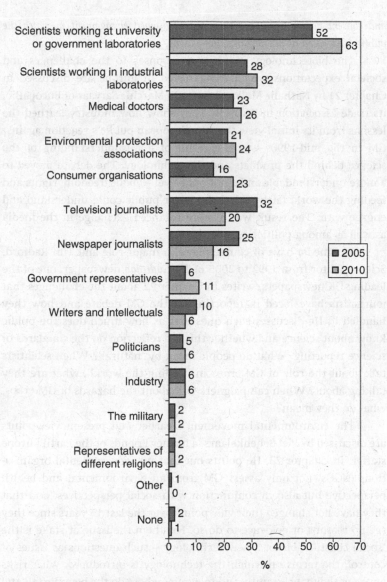

Figure 1 Best qualified to explain the impact of scientific and technological developments on society. (*Science and Technology: Special Eurobarometer* 340, 2010)

'*the outlook seems bleak*', but if the second generation achieves the predicted consumer benefits of reduction in pesticide residues on food and delivers environmental benefits with reduced chemical spraying supporting biodiversity, then the public may warm to them. But public support should not be taken for granted; it will need cultivating: '*Trust and consent has to be earned through enlightened engagement in which the public are treated as participants, not pawns.'*

The biotechnology industry's responses to the challenges and societal expectations which it has faced in Europe are discussed in Chapter 21 by Nathalie Moll and Carel du Marchie Sarvaas of EuropaBio, its trade association in Brussels. They show how industry learned the lessons from its initial view that the European public's reaction against GM in the mid-1990s was the result of a misunderstanding of the science behind the products and the technology. The debate moved to a more understandable and emotional level – about freedom, rights and feeding the world fairly – that the wider public could understand and engage with. The result was a more balanced exchange in the media as well as among politicians.

The media have of course played a major role and Tim Radford, Science Editor from 1992 to 2005 of *The Guardian* newspaper, one of the leading UK newspapers, writes in Chapter 22 about the challenges that journalists have faced in reporting on the GM debate and how they handled it. He discusses such questions as how much does the public know about science and whether this is a reflection on the standards of science reporting. What do people mean by 'natural'? When scientists talk about the role of GM crops in 'feeding the world', what are they talking about? When campaigners talk about the hazards of GM crops, what do they mean?

The environmental movement's earlier and present viewpoints are discussed by Piet Schenkelaars, a former Friends of the Earth Europe staffer, in Chapter 23. He points out that non-governmental organisations (NGOs) not only assess GM from an environmental and health perspective but also in conjunction with social perspectives, and that they have not changed their viewpoints over the last 15 years since they see no reasons or evidence to do so. For them the issue at stake is the 'social constitution' of a new technology – such questions as issues of control, the terms on which the technology is introduced, what risks apply, with what certainty, and to whom, where do the benefits fall and whether they fall to the same people, and who takes responsibility for resulting problems?

Richard Jennings takes a philosopher's view of the GM debate in Chapter 24. He focuses on two of the principal social and ethical issues that are contested from the standpoints of both the pro- and anti-GM advocates – the debates over the social economic implications of GM technology and whether GM foods affect human health. He suggests that they are effectively living in different worlds – looking at different evidence and listening to different voices. Trenchantly he ends by concluding 'Perhaps through a sympathetic discussion of all points of view we can achieve a closer engagement between these two worlds of discourse – perhaps we can realise the best of both worlds. It is imperative that we find a way to listen to each other and resolve these differences because, after all, we do live in the same world, a world that faces significant problems.'

And finally, in conclusion, Klaus Ammann, former Director of the Bern Botanical Garden and Professor at the University of Bern in Switzerland, argues for advancing the cause of successful agricultural innovation in emerging economies. He discusses the premises for a realistic debate on successful bio-economical approaches, the science behind modern crop breeding, the high costs of regulation as a consequence of what he terms the 'genomic misconception', the concepts of sustainability from agro-ecology to bio-economy, and closes with three success stories of modern agriculture in emerging countries.

REFERENCES

Department of Business Innovation and Skills (2011). *Public Attitudes to Science 2011*. Ipsos MORI, London. www.ipsos-mori.com/researchpublications/researcharchive/2764/Public-attitudes-to-science-2011.aspx

European Commission (2010). *Europeans, Science and Technology*, Eurobarometer 340. European Commission, Brussels. http://ec.europa.eu/public_opinion/archives/ebs/ebs_340_en.pdf

GEORGE GASKELL, SALLY STARES AND CLAUDE FISCHLER

20

Have GM crops and food a future in Europe?

Intensive mono-functional agriculture, typical across many European Member States, is designed to increase the efficiency and productivity of the agricultural sector. This is accompanied by frequent spraying of crops with chemicals for protection against pests and diseases. While the health and environmental impacts of pesticides and their residues are debated among scientific experts, in the minds of European citizens they constitute the most significant food risk. The Eurobarometer public opinion survey on *Food-Related Risks* commissioned by the European Food Safety Authority (EFSA) in 2010 looked at risk perceptions in two ways. First, an open-ended question which invited respondents to say '*what comes to mind when they think about possible problems and risks associated with food and eating*', and second a closed question asked respondents to rate the extent to which they worry about 17 food-related risks, including pesticides in fruit and vegetables (EFSA, 2010).

The most frequent response to the open question was '*chemicals and pesticides*', mentioned spontaneously by 17% of Europeans – 7% mentioned genetically modified organisms (GMOs). In the closed question 74% of Europeans say they are fairly or very worried about pesticides in fruit and vegetables – the highest percentage of worry across 17 food risks. Sixty-seven per cent say they are fairly or very worried about GMOs. The 2010 survey replicated a number of questions from EFSA's first *Risk Issues* Eurobarometer in 2005 (EFSA, 2005). The findings on worry about pesticides in fruit and vegetables show an increase of 4% over the period – 12 countries show a 4% or more increase in worry,

Successful Agricultural Innovation in Emerging Economies: New Genetic Technologies for Global Food Production, eds. David J. Bennett and Richard C. Jennings. Published by Cambridge University Press. © Cambridge University Press 2013.

including Austria, Belgium, Denmark, France, Germany, Hungary, The Netherlands and Sweden. Overall, it appears that pesticides in fruit and vegetables are not only the top concern among Europeans, but also an increasing concern.

In the context of public concerns about pesticides, two potentially significant developments in genetically modified (GM) crops are noteworthy. In the British countryside in Hertfordshire, a field trial of a new GM wheat is under way at the time of writing (Rothamsted Institute, 2012). The GM wheat contains an added synthetic gene that causes the plant to exude an insect pheromone – farnescene – which occurs naturally in many plants including the peppermint plant – to deter cereal aphids. If the trial provides proof of this concept, this GM wheat will require less chemical insecticide spraying, reduce pesticide residues and support biodiversity.

A second field trial on the Fortuna potato was initiated by BASF. Current potato production in Europe can involve spraying a crop some 12–15 times to prevent potato blight (the cause of the Irish famine in the mid nineteenth century). The Fortuna potato has two additional genes taken from a wild potato variety found in the mountains of Mexico that is blight-resistant, holding out the prospect of a new potato strain with consumer and environmental benefits. However, the immediate future of the Fortuna potato is in doubt following BASF's decision to terminate agricultural biotechnology research in Europe on account of public resistance. A spokesman for the company, whose Agbio research group will be relocated to the United States, said 'there is still a lack of acceptance for this technology in many parts of Europe – from the majority of consumers, farmers and politicians – it does not make business sense to continue investing in products exclusively for cultivation in this market' (BASF, 2012).

The commitment to develop eco-friendly strains may well herald the arrival of the long-awaited second generation of GM crops bringing tangible consumer benefits and contributing to environmental sustainability. How will such GM crops be viewed by the European public?

In the Eurobarometer on *Biotechnology and the Life Sciences 2010* we asked respondents about a second-generation GM apple (Gaskell *et al.*, 2010, 2011). Commercial apple growers spray crops with pesticides and fungicides on a frequent basis – in some locations 20 times a year – in order to prevent diseases such as canker, scab and mildew. However the more or less inedible crab apple, a closely related species which can cross naturally with modern apples, has genes that provide resistance to the common apple diseases. While classical breeding to introduce such

Table 20.1 *Perceptions of safety, environmental impacts and naturalness of GM food and transgenic apples, EU27 (excluding 'Don't knows')*

Responses (%)	GM food	Transgenic apples	Cisgenic apples
Safe/not risky	27	37	53
Not harmful for the environment	30	55	63
Unnatural	76	78	57
Support	27	33	55

genes into modern varietals would be a painstakingly slow process, it can be achieved by the technique of genetic modification. This process, called cisgenics, involves adding genes from the same species or from plants that are crossable with the recipient plant in conventional breeding programmes. This contrasts with transgenics, in which genes are taken from other species or bacteria that are taxonomically very different from the gene recipient and transferred into plants to promote resistance to herbicides or to insect pests – the latter via the incorporation of a gene that codes for *Bacillus thuringiensis* (Bt) toxin, for example.

How do the public respond to cisgenics? Is the transfer of genes within a genus ('life form') more acceptable than transfers of genes across the genus? The species combined in a genus are generally perceived to be phenotypically equivalent, and genetic transfers may therefore be imagined as more natural.

In the Eurobarometer survey, GM apples were described with pictures to illustrate both the transgenic and cisgenic methods. Subsequent questions allowed for two main comparisons to be made. The first is between transgenics and cisgenics in apple production: is genetic modification within a species more acceptable to the public than modifications which cross the species barrier? Second, we can also compare public perceptions of transgenic apples with perceptions of traditional GM food. In principle there should be no difference with transgenic apples as the process is identical. Table 20.1 shows the contrasting perceptions of the safety, environmental impacts and 'naturalness' of GM food, transgenic apples and cisgenic apples.

While both GM food and transgenic apples are seen to be 'unnatural' by three out of four respondents, transgenic apples are perceived as safer and less likely to harm the environment than GM food. This suggests that the preamble describing how transgenics as a technique would *'limit use of pesticides, and so pesticide residues on the apples would be minimal'* may have suggested a benefit both to the environment and to food safety. But only one in three Europeans say that they support transgenic apples – 6% more than the support for GM food. It may be concluded that the additional benefits offered by transgenic apples are not sufficient to offset the concerns about the unnaturalness of crossing the species barrier – the introduction of 'foreign' genes.

By contrast, there are some striking differences between perception of transgenic and cisgenic apples. Cisgenic apples are seen to be safer, less harmful to the environment and crucially less unnatural. More than one in two Europeans support cisgenics. In contrast to transgenics, it seems that people may see cisgenic apples as not transcending the 'life form' barrier separating living beings. Cisgenics appears to be more natural, perhaps comparable to hybridisation in 'natural' horticulture.

FORTY YEARS OF GM FOOD

Europeans' perceptions of GM food have been documented in a series of social surveys commissioned by the European Commission's Research Directorate-General, DG12, now Directorate-General for Research and Innovation. The Eurobarometer surveys involve face-to-face interviews with a statistically representative sample of 1000 adults in each Member State. The main series of Eurobarometers started in 1991, but earlier assessments of public opinion flagged up concerns about GM food.

In the late 1980s, when biotechnology was still confined to the laboratory, the importance of public perceptions was recognised. Mark Cantley noted that the *'public and political opinion was learning to see gene technology, genetic engineering, biotechnology and so on as a single, vague and disquieting phenomenon'* (Cantley, 1992). A Eurobarometer survey in 1979 found that while only 23% of Europeans thought that *'the development of research on synthetic food'* was worthwhile, 49% thought it was an unacceptable risk. Cantley recommended to the European Federation of Biotechnology Task Group on Public Perceptions of Biotechnology that while education and information were needed to overcome irrational public fears, above all, efforts should be made to build trust, through scientific, financial, political and environmental accountability. However, such warnings had little impact. The new life sciences were seen by scientists, industry and in politics as the twentieth-century equivalent of the Industrial Revolution, bringing a step-change in the health, agricultural and environmental sectors.

To complete the early years of food biotechnology, the 1990s saw the arrival of consumer applications – the first of which was so-called

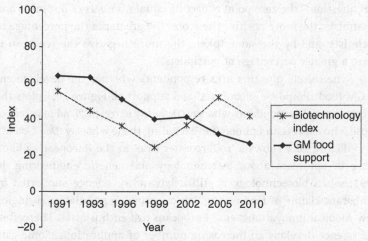

Figure 20.1 Optimism index on biotechnology and percentage support for GM food.

vegetarian cheese. Instead of the traditional animal rennet, the enzyme chymosin, a biotechnological product, was used. Next came the genetically modified tomato. Amid considerable publicity, Calgene launched the Flavr Savr™ tomato in the USA. In the UK, Zeneca's GM tomato puree arrived on the shelves of supermarkets in 1996. It was described as 'Californian tomato puree: made with genetically modified tomatoes'; and on the back of the can, it was noted that 'the benefits of using genetically modified tomatoes for this product are less waste and reduced energy in processing'. For the same price as a can of conventional puree, the GM puree offered 20% more weight.

While this sold well, for technical reasons the product was withdrawn. Notwithstanding headlines in the UK national press about 'Frankenfoods', there was little public debate or consumer opposition in the UK, which may have led both industry and government to assume that GM products as a whole would meet with consumer acceptance.

What was the state of public opinion around this time, and how has it evolved over the following 20 years? Figure 20.1 contrasts two questions put to respondents in the Eurobarometer surveys from 1991 to 2010.

The first question concerns whether people are optimistic or pessimistic about the contribution of technologies to everyday life. Respondents were asked 'Do you think biotechnology/genetic engineering will improve our way of life in the next 20 years, it will have no effect, or it will make things worse?' An index of optimism was derived from the responses to this question – the zero point reflecting equal numbers of optimists and pessimists; the more positive the score, the greater is the percentage of optimists, and by the same token, the more negative the score, so is there a greater percentage of pessimists.

The second question asks respondents whether the development of GM food should be encouraged and supported. Figure 20.1 plots the percentages of respondents who say 'Definitely agree' or 'Tend to agree', of those who express an opinion (i.e. excluding those who say 'Don't know').

Figure 20.1 shows a 'rollercoaster ride' in the European public's sense of optimism about biotechnology and genetic engineering. In 1991, when biotechnology is still a laboratory science supported by exuberant claims of feeding the world's growing population, bringing new medical innovations, etc., Europeans are enthusiastic. Thereafter, the science develops an increasing number of applications. Some gain public support, for example, genetic testing and GM-based medicines and vaccines. Others are not supported, for example GM crops and foods, laboratory animals for research, and cloning – Dolly the Sheep. In this period optimism about biotechnology and genetic engineering declines. In 2000 the human genome is sequenced, followed by developments in gene therapy, pharmacogenetics, research on human embryonic stem cells, and industrial biotechnologies. During this phase, European optimism in biotechnology increases, returning to the level seen in 1991. Finally, between 2005 and 2010 there is a small decline.

The timeline for GM foods runs in parallel to optimism about biotechnology over the period 1991 to 1999, with both showing a decline. The years 1996 to 1999 are dominated by three issues which in different ways set the agenda for policy-making, dominate media coverage, raise the public profile of biotechnology and set the tone for public perceptions. In 1997, the cloning of Dolly the Sheep turns science fiction into reality, leading to fears about what this science would do next. While nothing to do with biotechnology directly, the BSE crisis in 1996 shows the limitations of scientific expertise and introduces the public to industrialised agriculture. Finally, there is the long-running GM food debate. Starting in 1996, the importing of the first GM soya into Europe leads to consumer and environmental non-governmental organisations (NGOs) mobilising opposition, hostile coverage in sections of the media, supermarket boycotts, and eventually, in 1999, a European-wide de facto moratorium on the planting of GM crops in Europe.

With the moratorium, GM crops and food lose prominence as a public issue, although at a policy level, new regulations on deliberate releases of GM crops, labelling of foods above a specified threshold of

GM content, and public consultation are enshrined in European legislation. The concept of local decisions on the coexistence of conventional, organic and GM agriculture is also introduced.

Yet, all these policy responses to the crisis times of 1996–9 do not apparently allay public anxieties over GM food. The *Eurobarometer* in 2002 shows no uplift in support, and in 2005 and 2010 a downward trend in support is in evidence.

THE BASES OF PUBLIC CONCERN

The search for a single defining cause behind the declining support in Europe for GM food and crops is an attractive proposition, but unlikely to be successful. Any consideration of what lies behind public perceptions of GM food must take into account the diversity of social concerns and political opinions across Europe. This is witnessed in the plurality of movements that engaged in the GM food debate. The issue attracted environmentalists and those promoting consumer rights, anti-globalisation, and organic and 'natural foods', to name but a few. In different countries and at different times, all of these movements campaigned against GM foods. Sometimes, particular campaigns took on a broader political dimension when they were taken up by prominent public figures or were debated in national parliaments and the European Parliament.

Equally, identification of the causes needs to take account of some interesting consistencies across the European countries, as shown in Table 20.2, which shows the level of support for GM foods over time and across the increasing number of European Member States. From 1996 to 2010 the UK has been one of the most supportive of GM food. Over the same period, Denmark, Norway, Sweden and Austria are also consistent in showing relatively low levels of support. In most of the other European countries, support has been cut by one half, and in some countries, for example France, by considerably more. This suggests that while national events may play a role, the causes may be pan-European.

What pan-European issues might be at stake? We will consider three possible causes: the public's imagination of genetic modification; the absence of consumer benefits; and food and culture.

THE PUBLIC'S IMAGINATION OF GM

Lying behind public perceptions of technologies is a deeper level of representations or images. These bring together hopes, fears and expectations that come to people's minds as they think, read and talk

Table 20.2 *Trends in support for GM food (excluding 'Don't knows')*

	Percentage of respondents who agree or totally agree that GM food should be encouraged				
	1996	1999	2002	2005	2010
United Kingdom	52	37	46	35	44
Ireland	57	45	57	43	37
Portugal	63	47	56	56	37
Spain	66	58	61	53	35
Denmark	33	33	35	31	32
Netherlands	59	53	52	27	30
Norway	37	30			30
Finland	65	57	56	38	30
Belgium	57	40	39	28	28
Sweden	35	33	41	24	28
Italy	51	42	35	42	24
Austria	22	26	33	24	23
Germany	47	42	40	22	22
Switzerland	34				20
Luxembourg	44	29	26	16	19
France	43	28	28	23	16
Greece	49	21	26	14	10
Czech Republic				57	41
Slovakia				38	38
Malta				51	32
Hungary				29	32
Poland				28	30
Estonia				25	28
Slovenia				23	21
Latvia				19	14
Lithuania				42	11
Cyprus				19	10
Iceland					39
Romania					16
Bulgaria					13
Croatia					13
Turkey					7

about innovations such as biotechnology. The importance of these imaginations lies in their role of moving beyond a given reality. They look backwards – anchoring or giving meaning to the novel and unfamiliar by reference to past events, objects and experiences. They also look forward – anticipating whether a proposed technological

Table 20.3 *Dystopian imaginations*

Question	Agree 1996 (%)	Agree 1999 (%)
Ordinary tomatoes don't have genes but genetically modified ones do	36	37
By eating a genetically modified fruit, a person's genes could also become modified	49	42
Genetically modified animals are always bigger than ordinary ones	36	35

innovation suggests a negative or a positive future. Such judgements about the past and future are informed by social values – normative positions on what should and should not be done.

In parallel to the development of the 1999 *Eurobarometer* survey, a series of focus group discussions were conducted in 11 European countries (Wagner *et al.*, 2001). The findings highlighted a number of key themes which, to a greater or lesser extent, were shared across Europe. First, the public are not anti-science. Although many admit with some regret that they are poorly informed, they appreciate the role of scientific research in the systematic testing of new developments, and express confidence in science as a system of knowledge acquisition.

Yet, considerable ambivalence about biotechnology was evident. On the one hand, the medical and pharmaceutical applications directed towards the detection and curing of diseases were supported with enthusiasm. By contrast, applications such as the cloning of animals and GM foods raised a number of related concerns. Expressions such as tampering, meddling, fiddling and interfering with nature were commonplace.

Nature is discussed in two ways. First, in the traditional and spiritual form carrying the imperative of veneration; second, in a secular form in which nature is seen as a complex system with interventions running the risk of unknown and even unknowable consequences. The moral objection to certain biotechnological applications seems to be primarily due to the image of nature-as-a-spiritual-force. This image carries powerful sacred connotations that are valid not only for the expressly religious, but also for the more secular public.

In particular, the focus groups revealed some striking imaginations about food biotechnologies that were translated into questions in the 1996 *Eurobarometer* survey (Table 20.3).

(1) 'Ordinary tomatoes don't have genes but genetically modified ones do'
(2) 'By eating a genetically modified fruit, a person's genes could also become modified'
(3) 'Genetically modified animals are always bigger than ordinary ones'.

More than one in three Europeans agree with the proposition that food biotechnology is associated with fears about adulteration, infection and monstrosities (Wagner et al., 2006). Such concerns echo the idea of magical thinking, described by the anthropologist Fraser (1930) and more recently by Rozin. For example, the consumption of chicken slumped for a time when people heard that bird flu had reached Europe, even though one cannot catch bird flu from properly cooked chicken. This is an example of the law of contagion, the transfer of essences, 'once in contact, always in contact' (Rozin and Nemeroff, 1990).

It is not claimed that people actually held such views before being asked the question in the interview. In all probability, many would not have come across the issue before. But when the question is posed, people try to make sense of it and a combination of their unease about the technology, anxieties about food and magical thinking lead them to imagine the worst.

GM FOOD: WHO BENEFITS?

Returning to the focus group discussions, we heard a strong current of critical opinion about some biotechnologies including GM food. This concerned the absence of perceived benefits and the possibility of non-GM alternatives to achieve similar ends. People would question the point of genetic modification of food: was it necessary when there is plenty of food in the shops? Why change the character of food when it is already good and wholesome? Questions of a similar nature were raised around xenotransplantion: would it not be easier to encourage people to carry organ donation cards rather than developing transgenic pigs? Arising from such views, people wondered why society should take the risks that might be involved when the claimed benefits appear to be non-existent, or the ends achievable by other 'tried and tested' means.

To investigate this in more detail, the 2002 Eurobarometer respondents' data were divided the into four groups according to their answers to questions about whether GM foods 'will bring benefits to many people', and second, whether they 'pose risks for future generations' (Gaskell et al., 2004).

Table 20.4 *Risks and benefits of GM food*

		GM food poses RISKS for future generations			
		Agree		Disagree	
		'Trade-off'		**'Relaxed'**	
		Useful and risky		Useful and not risky	
	Agree	Of total sample:	18%	Of total sample:	14%
GM food		Percentage of these		Percentage of these	
will bring		who encourage:	52%	who encourage:	81%
BENEFITS		**'Sceptical'**		**'Uninterested'**	
to many		Not useful and risky		Not useful and not risky	
people	Disagree	Of total sample:	62%	Of total sample:	6%
		Percentage of these		Percentage of these	
		who encourage:	17%	who encourage:	27%

Those who thought GM foods were beneficial and without risk were labelled 'relaxed'; beneficial and risky 'trade-off'; not beneficial and risky 'sceptical' and not beneficial and not risky 'uninterested'. Table 20.4 shows the percentage of respondents in each of the four groups as a whole, as well as the percentages of each group that agree to the statement '*GM food should be encouraged.*'

It is notable that there is a sizeable number of respondents in the group labelled 'sceptical'. Fully 62% of the sample believes that GM foods offer no benefits and carry risks – only 17% of them think GM food should be encouraged. By contrast, 81% of the 'relaxed' group (useful and not risky) support GM food, but this group is a mere 14% of the European population.

One frequently held explanation for European's resistance to GM food is that the public simply misperceive the risks on account of misleading media coverage. But our analyses point to the absence of benefits as a key driver of public opinion.

Comparing the 'trade-off', 'sceptic' and 'relaxed' groups suggests that each group may use a different decision strategy to arrive at a view about support for GM foods. The 'trade-off' group weighs up the benefits and costs following the tenets of rational choice. By contrast, for the sceptics, it may be that the absence of perceived benefits acts to truncate their deliberation on the issue; the attribute of risk is deemed irrelevant, and accordingly has less influence on the final judgement of encouragement. Here, the implied decision model is based on a single dominant attribute: in this case, the absence of benefit.

For the 'relaxed' group, the implied decision rule is less clear. Their perception of benefits may lead them to ignore the risks, or they may deliberate on the risks, judge them to be minimal, and combine the two attributes according to the rational choice criteria.

For the sceptics, by far the largest group, the perception of the absence of benefits associated with GM foods means that GM food fails to meet the key criterion of an innovation, an improvement on the status quo, and that acts as a dominant attribute: a non-conditional prerequisite of any level of support.

FOOD AND CULTURE

Food and genetic modification bring together the new and the old. Genetic modification has a short history, dating back only to the 1970s. By contrast, the production, preparation and consumption of food are as old as human society. Over the millennia and across the world, food and eating, while arguably the most basic biological function, have evolved to be a central feature of culture – shaping social organisation, the division of labour, and demarcating religions, races, communities, classes and genders (Zwart, 2000).

Food is constitutive of both cultural and individual identity through the process of 'incorporation' – the crossing of the barrier between the 'outside world' and the 'inside world' of the body. Food intake is not merely physical, the intake of calories or fuel. With the food we also absorb beliefs and collective representations – as suggested by the age-old aphorism *'you are what you eat'* (Fischler and Masson, 2008).

In the Western world, the last half-century has witnessed a change from food shortages to surpluses. Anxieties about having enough to eat have been replaced with concerns related to the ever-increasing distance between 'the farm and the fork'. The modern eater, according to Fischler (1990), is an increasingly anxious consumer, torn between the appeal of cheap, convenient and palatable processed food, and the repulsion or menace of factory farming and pesticides, and of additives to replace natural ingredients. The perceived and, to some extent, real consequences are new and subtle dangers: less visible, understood or controllable. Herein possibly lie the reasons for the popularity of 'organic' and 'natural' foods discussed later.

While culture gave earlier generations principles about what and what not to eat, the clues of texture, flavour and even intuition fail to protect us from the perceived hazards of eating in modern times. The consequent psychological, political and ethical distress resonates in worries about being 'at risk' from pesticides, residues, pollutants and additives. The most common complaint about contemporary processed foods typically is that *one does not know what one is eating any more*. It encapsulates the contemporary food consumer's dilemma: *'I am what I eat; I don't know what I eat; thus I don't know what/who I am.'*

With the decline of tradition and culture we see the emergence of individual choice – for some desirable, for others the cause of anxiety, bewilderment and the state of 'gastro-anomy'. Individuals are often at a loss as to how to make choices in the general nutritional cacophony – conflicting norms (or normlessness), prescriptions and proscriptions about food. Anxieties about food are evidenced in food scares, pathological eating disorders and a normlessness that some social scientists link to the trend of rising obesity. All in all, in modern societies food is as likely to be seen as a source of stress as one of pleasure.

To 're-identify' with food, to re-appropriate it (*'knowing what they eat'*), and to introduce a new logic into everyday eating, people search for new strategies, as seen in the demand for food labelling, legal protection against the use of chemicals and biotechnology, the adoption of individual alternative diets, ranging from more or less rigid vegetarian, to organic, low-calorie or low carbohydrate, etc.

Food anxieties have led to new 'strategies of confidence', including the development of repertoires of trusted food, for example organic food, fair trade, vegetarian or local food; and brand loyalty – with brands standing for familiarity, and reassurance about safety and quality.

Finally, there is the issue of naturalness, which we have already seen is a crucial determinant of greater public support for cisgenic over transgenic apples. In many, if not all cultures, but at the very least in the United States, France, Britain, Switzerland, Italy and Germany, the adjective 'natural' associated with food is considered positive. As such, it is a common theme in the advertising of food products. The perception of naturalness is affected more by addition, even infinitesimal, than it is by subtraction; by process rather than by content; by chemical rather than physical transformation; and by contaminants (Rozin *et al.*, 2004; Rozin, 2005).

Thus, while medical biotechnologies promise new ways to alleviate diseases and appeal to the ethic of the 'duty of care', food biotechnologies may confront cultural norms and preferences. Perhaps the most extreme example is cloning animals for food products, which in the Eurobarometer 2010 was supported by only 18% of Europeans, yet beef and milk from cloned cattle are routinely eaten and drunk in the United States. Why is there such opposition? Of all the food products digested by human omnivores, meat is both the most adored and the most abhorred. On the one hand, almost all of the food taboos (as well as the most violent aversions) centre on meat or animal foods, some by religious fiat, others reflecting no less constraining cultural preferences – *'would you care for a real dog hot dog?'* – yet eating dogs and many other animals, birds and insects which are frequently eaten in other cultures is abhorrent to most Europeans. On the other hand, meat is culturally focal – the English Sunday roast, the daily *filet de boeuf* in France. With animal cloning for food products, the elision of food as an aspect of culture, sensitivities around meat and genetic modification combine to make the idea very unpalatable.

Three potential explanations for Europe's resistance to GM food have been discussed – imagination of genetic modification, the perceived absence of benefits, and food and culture. These are probably not an exhaustive account. Equally, they may appear to be complementary, with one highlighting the absence of benefits, while the other two, in different ways, highlight risk – in terms of process, consumption and culture. In reality, each may play a role in different countries and in different sections of the public to different extents.

In closing, we look again at cisgenics, and investigate the extent to which 'naturalness' and benefits contribute to public support. This is explored using a latent class model that allows us to describe different classes of orientations towards transgenic and cisgenic forms of GM. Essentially, this method segments the public into a number of types or groupings in which people take similar positions. This analysis suggests that five classes or groups capture the key orientations. Figure 20.2 summarises the results, with pattern and shading to denote the most likely response for each item, class by class. Responses for transgenic apples and cisgenic apples are aligned side by side for ease of comparison. Along the top of the table are the percentages of Europeans expected to fall into each class, followed by the interpretative label attached to the class.

To take the first class, for example, the typical respondents in this group strongly agree that transgenic apples are promising; tend to agree that they are safe and should be encouraged; strongly disagree that they are bad for the environment and that they feel uneasy about such a technology, and tend to disagree that transgenic apples are unnatural. They strongly agree that cisgenic apples are useful, are not risky and should be encouraged; and they strongly disagree with all the negatively worded statements about cisgenic apples. So this class is labelled as one

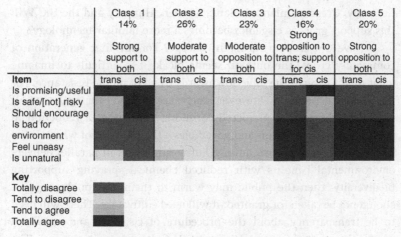

Figure 20.2 Segmenting the European public.

of strong support for both types of GM. Note that among these 14% of the European public, enthusiasm seems to be a little more certain for cisgenic than for transgenic apples.

The second class (covering 26% of Europeans) represents moderate support for transgenic and cisgenic apples, with 'tend to agree' responses typical for the positively phrased items and 'tend to disagree' for the negatively phrased – albeit that transgenic apples are still most likely to be judged unnatural.

The third and fifth classes represent moderate and strong opposition to both types of GM, and together cover 43% of the European population.

It is the fourth class which is the most intriguing, and whose composition echoes the findings of the previous survey analyses and focus groups. This 16% of the population typically express strongly negative views on transgenic apples alongside enthusiasm for cisgenic apples. They strongly agree that the latter are promising, tend to agree that they are not risky and to be encouraged; strongly disagree that they are bad for the environment and that they feel uneasy about them; and strongly disagree that they are unnatural.

Overall, this analysis shows about 40% of Europeans supporting transgenic and cisgenic apple production. A further 16%, while against transgenics, are persuaded by the claimed benefits of cisgenics. Looking at all those respondents who are opposed to transgenics (classes 3, 4 and 5), the percentage who are attracted by cisgenics is at or above 30% in 20 countries (of the 32 countries in the dataset), including Denmark,

Germany, Greece, Finland, Ireland, Sweden, Hungary and the UK. Will this support grow as cisgenics becomes a more familiar technology?

Have GM crops a future in Europe? For the first generation of commodity crops, the outlook seems bleak; it is difficult to imagine what might reverse the long downwards trend of public resistance. For the second generation, however, there are grounds for optimism. If cisgenics achieves the predicted consumer benefits of a reduction in pesticide residues on fruit and vegetables – the presence of which is the leading food risk concern among Europeans – and in parallel delivers environmental benefits with reduced chemical spraying supporting biodiversity, then the public may warm to them. But public support should not be taken for granted; it will need cultivating. There will need to be transparency about the procedure of cisgenics and equally a recognition that this may raise concerns about 'unnaturalness'. The benefits should be independently verified, and not overstated. The lessons of the past should not be ignored. The developers of the first generation of GM crops blandly assumed that any public resistance was irrational and would evaporate once GM products were on the market. Trust and consent has to be earned through enlightened engagement in which the public are treated as participants, not pawns.

REFERENCES

BASF (2012). www.basf.com/group/pressrelease/P-12–109.
Cantley, M. (1992). Public perception, public policy, the public interest and public information: the evolution of policy for biotechnology in the European Community, 1982–92. In J. Durant (ed.) *Biotechnology in Public*, pp. 18–27. Science Museum, London.
European Food Safety Authority (2005). *Risk Issues: Special Eurobarometer* 238. www.efsa.europa.eu/en/riskperception/docs/riskperceptionreport.pdf
European Food Safety Authority (2010). *Food Related Risks: Special Eurobarometer* 353. www.efsa.europa.eu/en/riskcommunication/riskperception.htm
Fischler, C. (1990). *L'Homnivore*. Odile Jacob, Paris.
Fischler, C., and Masson, E. (2008). *Manger: Francais, Europeens et Americains face à l'alimentation*. Odile Jacob, Paris.
Fraser, J. G. (1930). *The Golden Bough: Studies in Magic and Religion*. Macmillan, London.
Gaskell, G., Allum, N., Wagner, W., *et al.* (2004). GM foods and the misperception of risk perception. *Risk Analysis* 24, 183–192.
Gaskell, G., Stares, S., Allansdottir, A., *et al.* (2010). *Europeans and Biotechnology in 2010: Winds of Change?* European Commission Publications, Brussels.
Gaskell, G., Allansdottir, A., Allum, N., *et al.* (2011). The 2010 Eurobarometer on the life sciences. *Nature Biotechnology* 29, 113–114.
Rothamsted Institute (2012). Rothamsted Wheat Trial: second-generation GM technology to emulate natural plant defence mechanisms. www.rothamsted.ac.uk/Content.php?Section=AphidWheat.

Rozin, P. (2005). The meaning of 'natural': process more important than content. *Psychological Science* 16, 652–658.

Rozin, P., and Nemeroff, C. (1990). The laws of sympathetic magic. In J. Stigler, R. Sweder, and G. Herdt (eds.) *Cultural Psychology: Essays on Comparative Human Development*, pp. 205–232. Cambridge University Press.

Rozin, P., Spranca, M., Krieger, Z., *et al.* (2004). Preference for natural: instrumental and ideational/moral motivations, and the contrast between foods and medicines. *Appetite* 43, 147–154.

Wagner, W., Kronberger, N., Gaskell, G., *et al.* (2001). Nature in disorder: the troubled public of biotechnology. In G. Gaskell and M. W. Bauer (eds.) *Biotechnology 1996–2000: The Years of Controversy*, pp. 80–95. Science Museum, London.

Wagner, W., Kronberger, N., Berg, S. F., and Torgersen, H. (2006). The monster in the public imagination. In G. Gaskell and M. W. Bauer (eds.) *Genomics and Society: Legal, Ethical and Social Dimensions*, pp. 150–168. Earthscan, London.

Zwart, H. (2000). A short history of food ethics. *Journal of Agricultural and Environmental Ethics* 12, 318–335.

21

Dealing with challenges and societal expectations: the industry's response

INTRODUCTION

This chapter sets out the main challenges the biotechnology industry has had to deal with over the last few years. After setting out the challenges it explains how the industry has dealt with them in Europe, and concludes with a series of lessons learned that could be useful for other regions to consider.

The focus is on Europe because this is where most challenges have occurred and because this is the geographical area in which the authors operate. Nonetheless we have tried to draw lessons from the European experiences that are relevant for emerging economies.

The contrast in uptake and cultivation of genetic modification (GM) technologies is very strong between geographical areas for different reasons. Northern and southern American countries have witnessed staggering adoption rates concerning planting of large-scale commodity crops over the last 10 years. Regarding cotton, the same can be said for other selected countries such as India, Burkina Faso, Australia and others (James, 2011).

Agricultural biotechnology is the fastest-growing crop technology in the history of agriculture. Globally, it is clear that since the first commercial planting in 1996, biotechnology crops have become an integral part of farming, and developing countries are now leading the way. According to the International Service for the Acquisition of Agri-biotech Applications (ISAAA), in 2011, developing countries adopted biotechnology crops at twice the rate of developed countries. Moreover, approximately 50% of biotechnology crops are now grown in developing countries.

Successful Agricultural Innovation in Emerging Economies: New Genetic Technologies for Global Food Production, eds. David J. Bennett and Richard C. Jennings. Published by Cambridge University Press. © Cambridge University Press 2013.

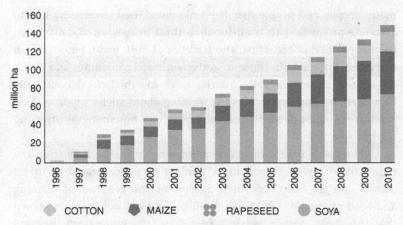

Figure 21.1 Agricultural biotechnology is being adopted at record speed around the world: global planting of GM crops per year since 1996. (LIS Consult, 2011)

In 2011, 16.7 million farmers planted 160 million hectares of biotechnology crops in 29 countries, up by 12 million hectares (8%) and 1.3 million farmers (8%) from 2010, when there were 15.4 million farmers planting biotechnology crops on 148 million hectares. In Europe, the number of hectares of the only GM maize permitted to be cultivated increased from 91 643 hectares to 114 607 hectares, an increase of over 20%. Europe is not keeping pace with its global competitors, who have now been growing a wide array of biotechnology crops for years (James, 2011).

THE MAIN POLITICAL, REGULATORY AND PUBLIC ACCEPTANCE CHALLENGES

There are a range of political, legal and regulatory challenges related to the understanding, acceptance, regulation and adoption of biotechnology crops around the world. These challenges differ between countries and regions depending on a variety of factors specific to them. Such factors include agronomic development, sophistication of political systems, culturally determined attitudes and economic development which are considered here.

A new technology in an emotive area: food

Though the basic tenets of plant breeding are thousands of years old, the use of biotechnological techniques is relatively new. The concept of using technology to improve the result – food or feed – is alien to many people, and people fear the unknown. Food advertisers like to associate products with traditional cultural images: the family farm, the happy child at breakfast, the traditional milk maid, natural fresh ingredients ripening in the sun, and so on. Food companies and retailers do not attract attention to the ways in which food is actually harvested, produced or processed. Talking about such aspects of food production might detract from the image of wholesome, healthy and happy foods.

The result is that in many countries – and especially so in the richer nations where consumers tend to be concentrated in urban areas – people have very little notion of the realities of farming and food production. Public opinion polling has consistently shown that the closer people feel to agriculture, the more open they are to consider innovations to improve production. In addition, cultural traditions about food greatly influence attitudes to biotechnology. Polling shows there to be great differences in attitudes between northern and southern European countries towards innovation in agriculture. Northern Europeans tend to be more open to new technologies or ways of doing things than those in Mediterranean countries.[1]

To illustrate, consider the cultural contrasts between the world's second and third largest agricultural exporters by value: France and The Netherlands. France has vast and rich agriculture lands, a long agricultural history and many desirable products. French products are overwhelmingly marketed as 'produced in the traditional way' or 'natural'. In contrast, The Netherlands is a highly densely populated country with little agricultural land and thus embraces new techniques in agriculture, like hothouses, modern animal breeding, hydroponic agriculture growing crops in water with nutrients, computer-controlled feeding systems, etc. In France there is widespread aversion to biotechnology among the public, though interestingly not among farmers, while in The Netherlands, there is much more support. The reason for the different attitudes lies in different cultural traditions – the French culture places more value on the traditional, the artisanal, the natural aspects of food, while the Dutch culture places more value (by geographical necessity) on efficiency, innovation and economic return.

---

[1] Private polling available upon request.

Lack of knowledge and understanding among politicians

One of the major political challenges for biotechnology acceptance and adoption around the world is the lack of knowledge and understanding by politicians and policy-makers about the basic science underlying the technology. Only a handful has a scientific background and public administrations are also not encouraged to attract civil servants with scientific training. As a result, these politicians and administrators are challenged when in charge of managing the rules governing a technology such as biotechnology. In addition, there remain many politicians around the world whose lack of understanding of biotechnology results in them being uncomfortable in speaking publicly about the technology, let alone its benefits.

Coupled closely to politicians' lack of understanding is their political acceptance of the technology. In Europe biotechnology first came into the public arena at a time when consumer faith in authorities' honesty concerning scientific issues and capacity to protect public health was deeply damaged as a result of the 1994 crisis over BSE (bovine spongiform encephalopathy, commonly known as 'mad cow disease'). Politicians and scientists alike had reassured the public about the safety of bonemeal feed to animals in the early 1990s only to be shown up by evidence of BSE infection a few years later. The harm this single event did to the public perception of science and administrations in Europe should not be underestimated. The introduction of biotechnology crops came just a few months after this episode.

Some politicians have gained political advantage out of opposing biotechnology. Consider the political rejection in 2002 of food aid that contained GM maize, during terrible droughts and famines, by the political leaders of some sub-Saharan African countries. Three of the four African states that initially opposed eventually accepted milled food aid but the Zambian president stated that Zambians would rather starve than eat GM 'poison' and continued to reject the food aid, effectively making his statement a sad reality (BBC News, 2002). Many former and current European ministers still openly attack the safety of the technology despite the thousands of safety assessments and almost two decades of safe use of biotechnology crops without any safety incidents.

Interests opposed or uncomfortable with biotechnology adoption

A challenge for the acceptance and adoption of the technology are some vested interests that openly (and not so openly) oppose biotechnology (Figure 21.2). Consider the following:

Figure 21.2 Vested interests in Europe protect their patch. (Greenhouse Communications)

Attitudes of **food manufacturers and retailers** vary widely from region to region. In most of the Americas there is widespread acceptance and use of GM foods. In Europe, there is reluctance to adopt. This reluctance could be due to public acceptance issues and/or fear of brands being targeted by anti-GM campaign groups as has occurred on occasion.

The **main international traders of commodity products** are not opposed to biotechnology production, but the need to separate international commodity supply chains into GM and non-GM streams complicates their operations. In some cases this may be beneficial to business generating premiums and margins but it can also cause supply and liability problems, particularly when some GM varieties are approved in some jurisdictions, but not (yet) in others.

Many **organic farming movements** oppose GM production for ideological and economic reasons. The ideological rationale is that biotechnology is 'unnatural' and therefore conflicts with the principles of organic production. The official economic rationale of the organic movement is that organic production methods are incompatible with GM production methods due to possible cross-fertilisation of crops in bordering fields. A plausible concept not often cited could be that GM production often allows for more targeted and therefore reduced pesticide use and this type of marketing poses a competitive threat to the more costly brand value of organic production. Interestingly, the head of the French farm union FNSEA believes that the future of organic production is GM cultivation (*Libération*, 2011).

Some, but not all, international **development groups** oppose GM production methods, despite ample evidence from leading international experts, as well as examples in countries such as South Africa, India and Burkina Faso of the important role the technology can play in farmers' development and in more predictable sustainable food production. There is large-scale research under way in the national research centres of many developing countries to develop biotechnology varieties for staple crops such as cassava, bananas, sorghum and maize. Moreover, some leading charities, such as the Bill and Melinda Gates Foundation (2011), support the use of GM crops and invest in their development for the developing world.

Some **countries with large production** compete with other regions cultivating biotechnology products. A good example of this competition is the trend in the Americas to accelerate the time it takes to approve new GM varieties. Brazil currently has the fastest system. In 2011, the Argentinian government announced it was revamping its approval system to be in line with Brazil so that Argentinian farmers had access to the latest GM varieties at the same time as Brazilian farmers. The US government announced in February 2012 that it too intended to increase the efficiency of its approval process with the aim of halving the time for product approval. The rationale provided by the USA was to grant access to new varieties to US farmers at the same time as global competitors.

Europe relies heavily on land outside its borders to meet its food needs. Its net imports are equivalent to outsourcing an area of arable land almost as big as the area of Germany – and growing every year. On average, Europe imports 30–40 million tons of protein every year (AERI, 2010; Witzke, 2010). In 2011, European Union (EU) officials announced their intention to require European farmers to leave fallow (set-aside) 7% of their land. **Farming groups in countries that export to the EU** probably welcome the move seeing that, because EU demand for raw materials would continue to increase but domestic EU produc-tion would be more limited, they could export more to the EU.

Non-existent, dysfunctional or politicised decision-making systems

A major challenge for the biotechnology industry is non-existent, dysfunctional or decision-making systems where scientific analysis is overruled in order to meet political needs. In some countries there are simply no legal and technical processes to assess and approve GM products, or allow field trials. A number of developing countries fall into this category. These were, or are, being set up, but they require financial resources and trained personnel to maintain. In some countries there are legal and technical processes, but they are dysfunctional.

Maybe most problematic are the politicised decision-making systems. The major example of this is the system in the EU, and particularly the approach to approving GM products. The EU authorisation system for new GM products was designed to meet both political and scientific needs and focuses on the process by which a product is produced rather than the end product itself. In theory the EU system is science-based; the reality is different – it has three design flaws.

First, it sets the independent scientific judgement regarding safety made by the European Food Safety Authority (EFSA) against the political views of national politicians who vote on that safety assessment to approve GM products. The result is a constant calling into question of the scientists' assessments about safety by some politicians and a consequent damaging of public trust in the science, the procedure and the products. It has also led to constant tinkering – to meet political needs – of the scientific requirements to be met by GM applications. The scientific assessment remains science-based, but the approval process is politicised as is demonstrated by the voting behaviour of those countries that consistently vote against the scientific opinion of EFSA scientists.

Second, the EU system separates the approval for a product for import from that for cultivation. The result is that products for import are regularly approved in order to prevent trade pressure because they are grown in major commodity-growing countries that export to the EU. However, they are not approved for cultivation. This has created the strange situation that Europeans are importing GM produce but are not allowed to benefit from growing the same materials on European soil. Put simply, it is permissible to import a GM variety, to eat it, to feed it to animals, but not to grow it.

Third, the EU chose a labelling threshold level, above which products that have GM content must be labelled. However, Europeans indirectly consume (through animal feed) upwards of 30 million tonnes of GM protein every year in the form of meat, eggs, milk and other animal products whether they recognise it or not (Figure 21.3). That is the equivalent of 60 kg per year for every one of the 500 million European consumers – man, woman and child (AERI, 2010; Witzke, 2010). Europeans are consuming animal-derived produce, and every

Figure 21.3 Food labels: confusing or helping consumers?

time they travel to North and South America and parts of Asia they are also eating directly derived plant GM produce.

Media, misinformation and public acceptance

A lot has been written about the role of interest groups opposed to GM. It should be noted that these groups have significantly influenced policy-making in agriculture and GM public acceptance in some regions of the world, yet failed to make any impact whatsoever in other regions.

Some of these interest groups have been particularly successful in injecting incorrect or false information into the public domain. These myths persist and are difficult to counter, especially amongst those people who prefer, and can afford to, let their ideology and not facts guide their lifestyle.

All kinds of wild claims have been made about biotechnology crops over the years since their introduction, but time has shown many of the negative claims about GM crops to be untrue. None of the scare stories has ever materialised. Twenty years ago, before we had experience with cultivating GM crops, it may have been understandable to be wary about the technology. However, biotechnology crops are now part

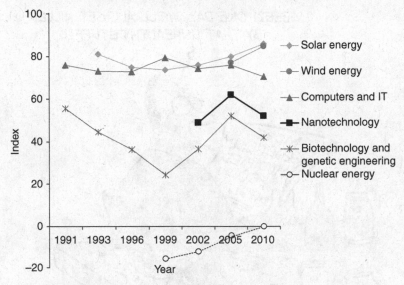

Figure 21.4 Index of optimism abour six technologies. (Eurobarometer: Europeans and Biotechnology, 2010)

of the food landscape and the record is clear: scientific studies by leading scientists and over 2 trillion GM food meals eaten have shown no negative effects on health or the environment (European Commission, 2010).

Anti-GM groups claim that Europeans are overwhelmingly opposed to GM food and crops, but often they base these claims on incorrect readings of public opinion polls. Some polling results and questions are misleading. For example, some polls asked people to rank their levels of concern and asked them to agree or disagree with statements like 'GM food is unnatural', 'makes you feel uneasy' or 'GM food is not good for you.' Questions that ask people to quantify 'how worried they are' obviously record high levels of concern. Reliable opinion pollsters do not use such methods; instead, they ask people to rank their concerns instead of prompting them with suggestions of what those concerns might be.

The EU's polling division Eurobarometer did this in 2010, asking 26691 Europeans: '... in your own words, what are all the things that come to your mind when thinking about possible problems or risks associated with food and eating? Just say out loud whatever comes to mind and I will write it down.' Only 8% of Europeans spontaneously say they are worried about GM in food. In fact, when asked how optimistic they are about various technologies and possibly contra-intuitively to many people, after a low in 1999 of only 20% saying they were optimistic about biotechnology, this rose to some 40% to 50% from 2002 onwards (Figure 21.4).

Many mainstream media have admitted that they may have made mistakes in their coverage of GM issues in the past. Time was when the media were quick to publicise claims by anti-GM groups, but many have learned that they had often been duped into giving credence to inaccurate and/or junk science. Part of this is due to a structural lack of people with scientific training in the mainstream media. The supposed David versus Goliath conflict between *'NGOs protecting consumers'* and *'domineering corporations taking control of our food chain'* was, and for many media, still is, a fracas too attractive to resist. In 2011, the BBC Trust released a review of impartiality and accuracy of the BBC's coverage of science, noting that: *'Equality of voice calls for a match of scientists not with politicians or activists, but with those qualified to take a knowledgeable, albeit perhaps divergent, view of research. Attempts to give a place to anyone, however unqualified, who claims interest, can make for false balance...'* (BBC Trust, 2011).

Some governments have also actively participated in injecting questionable data into the public arena. The Ministry for Health of one particular European government joined with Greenpeace and organised a press conference to present new research that claimed possible health damage from ingesting GM materials. Embarrassingly, the research later turned out not to meet basic scientific standards, was not peer-reviewed, could not be reproduced and was rejected by mainstream scientists. The Ministry was later forced, in front of EU government representatives, to admit the research was flawed and officially retract it.

Cost of regulatory compliance

Considerable resources are used to develop GM crops. A survey report produced in September 2011, *'The cost and time involved in the discovery, development and authorisation of a new plant biotechnology derived trait'*, sets this out for biotechnology traits in major commodity crops that had received cultivation approval in two countries and import approvals from at least five countries (CropLife International, 2012). Main conclusions are that the cost of discovery, development and authorisation of a new plant biotechnology trait introduced between 2008 and 2012 is €140 million. The time from discovery to commercial launch of a product is 13.1 years on average. This does not include the time to develop and obtain regulatory approval for stacked trait varieties with many incorporated genes.

There is pressure between different global regulatory systems. Argentina announced in 2011 it would reduce the time it takes to approve a GM product. The USA announced in February 2012 it would implement new efficiencies which would halve the time needed. The cost and complexity of the EU regulatory system for GM crops, along with the lack of evidence of harm to people or the environment, should lead to a more risk-based regulatory system.

HOW HAS INDUSTRY DEALT WITH THESE CHALLENGES?

Industry has deployed a range of techniques to address the challenges identified above. These include proactive communications, helping governments understand or create regulatory approaches and, of course, meeting all the regulatory requirements set by governments.

Communications

Industry initially viewed the European public's uproar against GM in the mid-1990s as resulting from a misunderstanding of the science behind the products and the technology, rather than a more complex combination of lack of knowledge, lack of trust in science and policy-makers, and romantic misconceptions of food production. The public concern was addressed on a scientific level when it was mostly being expressed with feelings and emotions. Most of the initial communication focused on the safety of the GM products and how they were the same as existing products while failing to address the core point of the debate in Europe – fear of science and manipulation of nature. Industry learned that safety is too much of a relative and subjective value to be used as the focus of a communication effort.

The communication work on the GM dossier is one of history's case studies in terms of how to build effective local communication activities to address specific local communication challenges. What worked in one region of the world, for example in the USA where GMOs were first launched, did not work in Europe where GMOs arrived just after the 'mad cow disease' debacle when politicians had claimed safety and then had had to retract their statements.

Anti-GM groups were very quick to use their own scientific research outcomes and claim the upper hand by coining words or phrases that still speak volumes today, such as 'Frankenfood', and were able to instil a fear and mistrust that no amount of scientific information was able to dispel. The ensuing fight between more or less renowned scientists just helped inflame the debate and left European citizens even more confused and afraid than they were at the beginning.

The turn of the twenty-first century found European citizens deeply opposed to the introduction of GM, albeit not more knowledgeable about the products or the processes than they had been 5 years earlier. By then, industry had understood the need to address the discussions locally. However, the malfunctioning approval system in Europe, with an enforced de facto moratorium, yielded no products on which to have a concrete discussion with the public and mistrust in products on which government had had to impose a moratorium. The scientific discussion was based on products existing elsewhere and theoretical benefits for Europe, given that the products were not allowed onto the European market. This was not enough to regain the trust of the estranged citizens of Europe who were caught in a 'luxury/full-stomach discussion' about whether yet another type of tomato was really what they wanted.

From trying to communicate about science in a battlefield of pro and con to a population who likely had neither the time nor interest to delve deeper, the biotechnology debate has only recently moved towards 'global responsibilities' and 'freedom of choice'. The discussion has shifted from a focus on the fact that products on the market had to be approved as safe in order to get onto the market to a focus on where, when, why and by whom the products were being used (rather than focus on their safety). This change in the dialogue surrounding the GM debate brought the products and the processes back into their original *raison d'être*. These products were invented to produce more and better food, and with fewer inputs, to meet the demands of a growing world population.

The debate and the communication therein finally moved towards a more understandable and emotional level, that of freedom, rights and feeding the world fairly, which the general public could understand and engage in and resulted in a more balanced exchange of views in the media as well as among politicians.

Dedicated public affairs

Public affairs describes an organisation's relationship with stakeholders in order to explain the organisation's policies, provide statistical and factual information and represent issues that could impact upon the organisation's ability to operate successfully. In the case of GMOs, all products that reach the market have to be approved by a regulatory body following a series of scientific assessments. The nature of those scientific assessments as well as the regulatory pathway to approval is specific region by region and often country by country, and has been developed mostly in consultation with the various interested parties, including industry and non-governmental organisations (NGOs).

Once again, it was quickly understood in the late 1980s that a generalist approach to public affairs, on a global scale, is not possible in the area of GM. Each region has decided to develop different rules to govern the approval and marketing of GM products with some regions preferring a product-based assessment such as North America and other regions such as Europe opting for a process-based approach. As the industry producing GM products, and therefore having to implement the developed legislation, the biotechnology industry certainly was and remains an interested party.

Public affairs activities to respond to the consultations on developing and implementing legislation around the world has had to be dedicated often on a country-by-country basis in order to ensure that the correct considerations and knowledge were shared at the required moment in a given place. At the same time, because GM production often results in agricultural commodities that are traded on a global scale, it has always been important to ensure that the various regulatory systems around the world did not impede free movement of goods and respected World Trade Organization (WTO) rules.

In the case of GM, public affairs followed a similar route to communications in that it began trying to unravel and settle the scientific discussions on the basis of approval systems around the world and then moved on to ensure that global trade and other factors such as influence of one region on another were taken into account.

LESSONS LEARNED BY INDUSTRY

Industry has learned many lessons over the past years on dealing with challenges. The most important of these are:

(1) Many countries, including those in the EU, have yet to develop a coherent approach that allows them to reap the benefits of biotechnology crops. Thorough science-based assessment is necessary, and it is legitimate to consider new facts. However, any tendency to introduce new risk assessment requirements for political reasons undermines the credibility of science-based systems. Measures produced by politicians without a scientific basis do not contribute to improved safety assessment or increased public trust.

(2) Groups with vested interests against GM have been extremely influential in decision-making processes. It is unfortunate that the regulatory systems in the EU and in other countries – instead of farmer and public need or scientific progress – will determine the rate of uptake of biotechnology.

(3) The biotechnology community was ill-prepared for the aggressive media campaigns against GMOs, and in the initial stages failed to counter them effectively. As a result the technology providers were on the defensive from the start. The lesson is to counter all misleading claims immediately and very precisely, and to encourage others to do the same. However fast one is, though, false myths are difficult to eradicate. Public engagement is important, but no amount of surveys, focus groups, citizen panels and national debates will change opinions. Discussions about existing products and about consistent safety and benefits will.

(4) Discussions that focus on science will not help the public debate. Discussions need to focus on the real and the tangible. A product, a benefit, a challenge to be met.

(5) For five reasons, many Europeans are reconsidering biotechnology crops and foods. Many have realised that:
- biotechnology crops are part of the solution to the greatest challenge of our generation – hunger and malnutrition – and Europeans have a responsibility in tackling that challenge,
- the scaremongering claims by anti-GM groups have turned out to be untrue,
- the world is adopting biotechnology crops at an incredible pace, leaving Europeans on the sidelines at their economic cost,
- biotechnology crops play an important role in the developing world to help raise production and help rural communities,
- biotechnology crops can help to provide solutions to environmental problems.

(6) Europe's consumers are part of a global food market and rely heavily on land outside its borders to meet their food needs. The EU's net imports are equivalent to outsourcing an area of arable land almost as big as the area of Germany. However, as time goes by, the countries from which Europe imports are becoming progressively less and less sensitive to the

non-GM preferences of some Europeans, and are acting accordingly.

(7) The differences in attitudes from different governments in the EU show that where government systems are mature, they are more balanced and there is better decision-making. It is noticeable that most of the countries in the EU that oppose GM cultivation come from political cultures where innovation is not held in the highest regard, and science is slow to be embraced. These are also the countries that rank lower on most innovation, competitiveness and productivity indexes.

(8) Farmers want to be able to choose to cultivate GM crops because they derive socio-economic benefits from their use. Wherever they are allowed to, millions of farmers choose to cultivate GM crops. In 2011 worldwide 16.7 million farmers cultivated GM crops – twice the number of farms in the EU. Most of these farmers stick to the technology once they have tried it.

(9) Europe's reluctance about GM crops has often inhibited the uptake of agricultural innovation in developing countries. Consider the views of Gilbert Arap Bor, a Kenyan farmer who wrote an editorial in the *Wall Street Journal* called 'Africa can feed the world'. Bor wrote that African *'governments currently follow the woefully misguided example of European countries that refuse to accept biotechnology, including genetically modified crops ...'* He quotes a fellow Kenyan, Dr M'mobyi, who said that the influence of the West is *'denying many in the developing world access to such technologies which could lead to a more plentiful food supply... This kind of hypocrisy and arrogance comes with the luxury of a full stomach'* (Arap Bor, 2011).

Disclaimer

The views expressed are those of the authors and should not necessarily be construed as representing the views of EuropaBio and/or its members.

REFERENCES

Agricultural Economics Research Institute (2010). *Study on the Implications of Asynchronous GMO Approvals for EU Imports of Animal Feed Products*. European Commission, DG for Agriculture and Rural Development, Brussels.

Arap Bor, G. (2011). Africa can feed the world. *Wall Street Journal*, 11 November. http://online.wsj.com

BBC News (2002). Famine-hit Zambia rejects GM food aid, 29 October. http://news.bbc.co.uk/2/hi/africa/2371675.stm

BBC Trust (2011). *Review of Impartiality and Accuracy of BBC's Coverage of Science*. BBC Trust, London. www.bbc.co.uk/bbctrust/assets/files/pdf/science_impartiality

Bill and Melinda Gates Foundation (2011). *Why We Fund Research in Crop Biotechnology*. www.gatesfoundation/agriculturaldevelopment/Pages

CropLife International (2012). *The Phillips McDougall Study: Cost of Bringing a Biotech Crop to Market*. CropLife International, Washington, DC. www.croplife.org/PhillipsMcDougallStudy

European Commission (2010). *A Decade of EU-Funded GMO Research*. European Commission, DG for Research and Innovation, Brussels. http://ec.europa.eu/research/biosociety

Libération (2011). Portrait of Xavier Beulin, 17 May.

Witzke, H. von (2012). The EU's 'virtual land grab': why agricultural innovation matters. http://seedfeedfood.eu/the-eus-virtual-land-grab-why-agricultural-innovation-matters

KEY FURTHER RESOURCES

GENERAL INFORMATION: AGRICULTURAL BIOTECHNOLOGY

Agricultural Biotechnology www.europabio.org/agricultural
Croplife International www.croplife.org/
GMO Compass www.gmo-compass.org/eng/home/
ISAAA www.isaaa.org

SCIENTIFIC BLOGS AND WEBSITES

Biofortified www.biofortified.org/
CropGen www.cropgen.org/
Harvard Belfer Center – Biotechnology. http://belfercenter.ksg.harvard.edu/topic/71/biotechnology.html
PRRI www.pubresreg.org/

SUSTAINABLE AGRICULTURE

Agriculture for impact http://www3.imperial.ac.uk/africanagriculturaldevelopment
CGIAR Research Programme on Agriculture, Food Security and Climate Change http://ccafs.cgiar.org/
Farming First www.farmingfirst.org/
Foresight report: *The Future of Food and Farming* www.bis.gov.uk/assets/bispartners/foresight/docs/food-and-farming/11–546-future-of-food-and-farming-report.pdf
Overview of recent global and European reports http://seedfeedfood.eu/report-and-factsheets
Seed.Feed.Food www.seedfeedfood.eu

22
_____

Media and GM: a journalist's challenge

The public rejection of GM technologies in Britain and in Europe took many journalists by surprise, and not just science journalists. When the Calgene Flavr Savr tomato turned up in the early 1990s its arrival was greeted with a mixture of mild cynicism and mild approval. (Who wants squishy tomatoes? But could they not have genetically engineered some tomato flavour into the supermarket tomato at the same time?) But people in Britain never saw it: they had to wait for Zeneca to deliver a genetically modified tomato paste, and the arrival of this packaged entity was greeted with something approaching watery enthusiasm: at least one newsdesk despatched young reporters in search of media-friendly chefs who could be challenged to confect a pasta sauce with the stuff. The chefs' verdicts were if anything mildly approving, but what they thought does not in this context matter. At that time the brand-new, brave-new-world GM tomato paste served as little more than an excuse for an impromptu media stunt on a slow news day.

That is, it was not seen as a threat. Nor did the public seem to regard it as a threat. It sold in the shops for at least two years, and reportedly sold well. It was not alone: also available was a cheese made with a rennet equivalent delivered by genetically modified bacteria, rather than the rennet traditionally taken from the stomachs of freshly slaughtered and not-yet-weaned calves. This product was actively welcomed by a section of the community that did not care for some of the practices of the dairy industry, but did not want to go without cheese. Genetic modification was not, in itself, a problem at the time.

Successful Agricultural Innovation in Emerging Economies: New Genetic Technologies for Global Food Production, eds. David J. Bennett and Richard C. Jennings. Published by Cambridge University Press. © Cambridge University Press 2013.

The technology could certainly make a big difference in the medical world, and genetically modified mice were already in hundreds of research laboratories. Diabetics once depended on supplies of purified insulin extracted from dead animals: for more than a generation, pure human insulin had been guaranteed by genetically modified microbes. The new understanding of modern biology – that living things could be adjusted, augmented or rearranged, like bits of Meccano – had begun to spread into public debate. Reporters cheerfully reported some of the more enjoyable ideas floated in journals and at conferences about uses for genetic modification. Could a jellyfish gene be inserted into cereal plants programmed to 'switch on' and glow when stressed by sudden infestation? If so, farmers could spray just that bit of the field that shone faintly in the dark and spare the consumer any risk from unnecessary pesticides. Could tubers or other crops be genetically reprogrammed to deliver not surplus grain or wasteful fibre but industrial cellulose that could be harvested and used to make biodegradable plastics? Could the pharmaceutical industry devise ultra-nutritious superfoods, or develop fruit that carried within it a simple vaccine to save lives in the developing world?

Reporting at this level tended to reduce science to a procession of gee-whizz stories, and to present science as a kind of comic-strip adventure, rather like the children's television serial *Thunderbirds*, in which whatever Lady Penelope needed was provided within the same episode by someone with big glasses called Brains. There was a second problem: many reporters (me among them) had trouble understanding the science we were supposed to be reporting. We should have been warned of the dangers to come. I note that from the trial of Colin Pitchfork for murder in 1988 – Pitchfork was the first person to be convicted on the basis of forensic DNA identification – to the completion of the first draft of the human genome in 2000, I invariably coupled the acronym DNA with some rubric such as '*the four-letter alphabet of life carried in the chromosomes of almost every living cell*'. And whenever there was a big story in which DNA played a significant role (the O.J. Simpson trial springs immediately to mind) I and colleagues on other papers, and in the BBC and ITV, would be required to produce a little simple 'explainer' which addressed both the nature of DNA and the probabilities that any two people shared the same DNA. I was never convinced that anybody who read these briefings ever quite appreciated, or retained, what they were told (and once again, the O.J. Simpson verdict springs to mind). This is not to say that we necessarily explained things badly, or that newspaper readers could not understand what we wrote. People read newspapers very selectively and lose interest very quickly, especially if the stories they read contain almost-meaningless acronyms and unfamiliar words such as 'mitochondrion', or 'genotype'. I also think that molecular biology is quite difficult to understand even at the simplest level: most of us are not accustomed to thinking of ourselves as composites of 100 trillion cells of 200 to 300 distinct types, organised by 3.8 billion years of natural selection, each cell being a mysterious entity in which biology happens at a molecular level. Then there is a problem with the lexicon of biology. By the 1990s, according to a passing reference in *Nature*, biology had added more than 60000 words and meanings to the language.[1] Shakespeare required only 30000 words for the entire Avon catalogue. So anyone reporting biology had to marshal ideas that were difficult to comprehend, couched in words that were increasingly meaningless. That is an explanation of the difficulty that reporters faced, not an excuse for any inadequate reporting.

And then in 1996 along came Dolly the Sheep. A public which had begun to feel faintly uneasy about a number of developments – among them the seemingly arbitrary patenting of genetic information, a growing sense that insurance companies might disqualify clients on the grounds that they might be doomed to contract this or that inherited disease, and a continuing low-level anxiety about chemical and biological warfare nurtured during the Cold War – really did get bothered. Dolly, conjured up in a dish from a speck of flesh taken from a dead sheep, sounded like an ethically alarming development. Paradoxically, although the science was at the time difficult to grasp, the basic idea was quite simple: science fiction had already prepared the world for Dolly. Specifically a mad scientist had already cloned Adolf Hitler in the book *The Boys from Brazil*, filmed in 1978 with Gregory Peck as the mad scientist, and Laurence Olivier as the Nazi-hunter on his trail. But science fiction – most obviously Mary Shelley's *Frankenstein* and H.G. Wells's *The Island of Dr Moreau* – had long ago provided the world with the lurid imagery of disturbed scientists 'interfering with nature'. In the same year, the public was presented with another reason to distrust agricultural science: from 1987 to 1996, government ministers had repeatedly claimed that they had it on the best scientific advice that BSE (bovine spongiform encephalopathy) – the so-called 'mad cow disease' then endemic in British herds, and linked to the grisly practice

---

[1] Galloway, J. (2009). The hidden language of cells. *Nature* **458**, 972 (review of Wolpert, L. (2009) *How We Live and Why We Die: The Secret Lives of Cells*. Faber & Faber, London).

of feeding grazing animals with a protein supplement based on the toasted remains of slaughtered cattle – could not spread to humans. In 1996, the government, in a dramatic about-face, announced that a human variant of the disease could have spread to an unknown number of humans. This announcement naturally provoked considerable public alarm. It also did nothing for the public image of biological science. There had been no genetic engineering involved, but the BSE episode once again invoked the spectre of scientists 'interfering with nature'. This widespread unease was not, in itself, anti-scientific. But it served as a kind of priming for the alarm, confusion, outrage, protest and posturing that characterised the by now well-described events of 1998 and early 1999, when – seemingly unexpectedly – Britain and other countries in Europe suddenly began to reject both the notion of GM food, and the practice of research into GM crops. This rejection was a surprise to scientists, and even to the campaign groups that organised protest, and to the journalists who had been covering events. Various episodes in this public rejection alarmed sections of the science community, catapulted Greenpeace and other campaigning voices onto prime-time television news bulletins, and provided a series of good stories for the media.

Although science journalists were taken by surprise, they were not taken aback. The public had begun to take a vociferous interest in the science we had been reporting: we thought that was a wonderful opportunity. We exploited this opportunity in different ways: some of us to attempt to explain the science a little more explicitly; some of us to increase the levels of public alarm (my favourite was a *Daily Mail* headline that said '*Scientists warn of GM crops link to meningitis*'. The same paper coined the label 'Frankenfood'). Some scientists came out fighting, and acquired honourable bruises in the public debate. Some retired hurt, or kept silent. It was around this time, at some forum in which the relationship between media and science became a subject of debate, that I framed a proposition that still seems to me to be true: in a democracy, we all have an obligation to explain what we do. But in a democracy, there is no corresponding obligation to listen. So anyone who wishes to be heard must speak in words that will make people listen. As quite a few British and European plant biologists will concede, the public did not listen.

That was then. This is now. What should a journalist (or a scientist who has to deal with the media) learn from the episode? And what questions should a journalist ask in trying to report the next round of the GM debate, because, of course, the story will not go away. A very large area of land in various parts of the world is already devoted to GM crops: there will be more. The more researchers learn about the molecular biology of plants, the more likely it is that they will find useful genetic material that they will wish to incorporate in modern strains of wheat, or maize, or rice, and for very good reasons. But people who have made it their business to challenge GM plant technology will also wish to go on resisting change, partly because many of them see the advances as needless or valueless, some because they see genetic manipulation as inherently dangerous, and not a few because they see science in the service not of the hungry but of global capitalism, and they don't care for the spectacle. In all this, the journalist has several responsibilities: they are not in conflict, but they do represent obligations that have to be considered.

Let us start with a piece of very old advice. A journalist should try to remain impartial. A journalist has a duty to report all science accurately and fairly. This is because science is an activity financed either by the taxpayer or the consumer, and both of these are also likely to be listeners, viewers or readers of the media. But journalists also have an obligation to reflect the anxieties and preoccupations of the society around them. It does not help to subscribe to those anxieties and preoccupations but it does help to listen to them, and to try to understand the basis for them. This is very general advice: now for some specifics. Many people – scientists, civil servants, journalists, even campaigners – were surprised by the strength, noisiness and apparent urgency of public reaction to the GM issue. But some useful lessons can be drawn, and I frame them as five huge, difficult-to-answer questions that, perhaps, I should have been asking myself all along. They provide a wider context to a debate that was always much more than just an argument about transferring genes from one strain or species to another.

HOW MUCH DOES THE PUBLIC KNOW ABOUT ANY SCIENCE? IS THIS A REFLECTION ON THE STANDARDS OF SCIENCE REPORTING?

The answers are: not much, and yes, probably. In 1989 John Durant of the Science Museum and other researchers famously presented a series of questions to a sample of Britons who claimed to be interested in science. One was: does the Earth go round the sun, or does the sun go round the Earth? How long does it take? And so on. One in three failed the first question; two out of three failed the second. Does this matter? Locked in any random collection of people is a considerable amount of unevenly distributed expertise, and also a certain capacity for sudden confusion when confronted with an unexpected question involving precise general knowledge. The lesson for journalists should be simple: assume the public knows nothing. Start from scratch. But never assume the public is stupid. Most people are quite good at assessing the immediate personal importance of scientific information. If it makes no difference to their daily lives, they feel free to discard it, or forget it. Interestingly, even if this information is clearly going to make a difference, most people will forget it, ignore it, or reason that though it could be serious for other people it will not affect them. Health chiefs, medical advisors and science reporters began warning about the hazards of smoking in the 1960s. Another 30 years elapsed before people in Britain began to ban smoking in public places. Government scientists, campaigners and environmental reporters began warning of the hazards of climate change in 1988. Twenty-five years on, most governments had failed to respond with the necessary vehemence, and most people had failed to radically reduce their dependence on fossil fuels. Such things take time: experienced reporters know that essentially the same story has to be told again and again, in different ways, and with different examples, before – gradually – public attitudes change. No reporter now needs to couple the acronym DNA with a general explanation. Along with tectonic shifts and quantum leaps, DNA is now part of popular metaphor. People may not be able to define it, but they know that it in some way defines us. The lesson is: be prepared for a war of attrition. Modern science has moved at astonishing speed. Public understanding also advances, but at a glacial pace.

IS AN APPARENTLY INSTINCTIVE PUBLIC REACTION OF SUSPICION OR DISGUST LIKELY TO BE SUSTAINED?

Not necessarily. Reporters of a certain age can remember flurries of scandalised or anxious public reaction to the first organ transplants; to the first 'open heart' surgery; to the first 'test tube' babies; to birth control pills; to vasectomy; to hormone replacement therapy; and so on. Suspicion ebbed as the new therapies began to deliver treatments that people valued.

WHAT DO PEOPLE MEAN BY 'NATURAL'? HOW MUCH DOES MODERN AGRICULTURE DEPEND ON SCIENCE ALREADY?

Everybody should be careful of the word 'natural'. Almost all farmed food – perhaps all farmed food – is very different from the wild ancestors first domesticated around 10000 years ago. No modern staple even remotely resembles its wild or 'natural' ancestor. Modern wheat, potatoes and maize, rice and soya have changed beyond recognition, and the most dramatic changes so far have occurred in the Green Revolution of the last five decades, in which harvest yields actually kept ahead of population growth. These yields depended to a great extent not just on new strains of cereals and pulse, but also on reliable supplies of nitrogen and phosphates, on increasingly sophisticated use of herbicides, pesticides and fungicides, on better control of irrigation, on more careful planting schedules, and of course on cheap fuel to drive greater mechanisation and keep labour costs down. Modern crops are likely to go on changing, as agricultural researchers absorb the new lessons to be gained from increasingly intensive studies of the DNA – and therefore of the genetic inheritance and basic biochemistry – of plants.

WHEN SCIENTISTS TALK ABOUT THE ROLE OF GM CROPS IN 'FEEDING THE WORLD' WHAT ARE THEY TALKING ABOUT? IS THERE A PROBLEM FEEDING THE WORLD? AND IF SO, WILL GM CROPS REALLY MAKE MUCH OF A DIFFERENCE?

There really are problems ahead, and it is not obvious that GM crops can or will provide a solution. That is because the problems are huge. In the first place, only about 11% of the planet's land surface is suitable for any kind of farming, and a significant proportion of this has already been degraded by overgrazing, erosion, mismanagement or the build-up of salts in the soil through constant irrigation. But in the next 40 years, another 2 billion souls will join the 7 billion here already. Each of these new citizens will need living space, water, wood, gravel, cement, fuel and building stone or brick clay from what is now farmland, and will sterilise an additional area of potentially fertile soil by covering it with pitch, pavement and brick, or by discarding waste into it. As even the poorest nations develop economically, demand for meat will grow: beef, lamb and pork will command land that often would be far more productive planted with cereals. So to feed the newly wealthy, the poorest must face ever more meagre rations. The farmer's old enemies – blights, rusts, mildews, locusts, aphids and worms – will remain as resourceful, determined and opportunistic as ever, but supplies of minerals, fertilisers and safe pesticides may not be guaranteed. The oil that powers the tractors that deliver the harvests is also likely to be available on an increasingly precarious basis at ever-greater cost. Because the oil will eventually run out, many farmers will see a commercial opportunity to grow biofuels – crops that can be turned to fuel rather than food. And climate change – itself the product of population growth, explosive fossil fuel burning and catastrophic changes of land use within the last 100 years – is likely to reduce the land available for agriculture still further, as sea levels rise, and traditional grain belts become more arid. It is quite possible to believe that future GM crops could help alleviate conditions in some places. There could be new strains that will survive sustained drought, and recover with the rains; or flourish in saline soil; or deliver higher levels of nutrition with each kilo. But it is not obvious that GM technology on its own could resolve the problems ahead. These demand political, social and economic willpower, and cooperation at an international level.

WHEN CAMPAIGNERS TALK ABOUT HAZARDS INHERENT IN GM CROPS, WHAT DO THEY MEAN?

They mean that some synthetic growths – that is, new strains confected in a laboratory rather than painstakingly crossed by traditional husbandry over several generations in a greenhouse or a farm plot – could deliver unintended consequences. There could be some unidentified hazard inherent in the technique used to splice an alien gene into a farmland species; there could be unexpected consequences from the introduction of a trait that natural selection would have prohibited, or eliminated wherever it first appeared, and which can only survive because of human intervention and sustained protection. These new, synthetic crops could represent danger in several ways. Their traits could cross species barriers and spread into the surrounding environment, altering the local ecosystem and putting endemic species at risk. The alien gene, apparently benign, could turn out to be in some unexpected way hazardous to the health of some consumers. A malicious or crazed scientist could wilfully devise an organism that represented real danger to everybody; an innocent scientist experimenting with all the options could do the same by accident. These are the possibilities that generate journalism's most sorry response: the cliché. In such scenarios, scientists are playing God, playing with fire, making a Faustian bargain, opening Pandora's box, creating Frankenstein's monster, embarking on a slippery slope and introducing the thin end of the wedge.

For every one of these questions, there is a response. Most plant scientists would point out that most modern cereal crops survive only because they are tended, weeded, sprayed, harvested and then sown or planted again in freshly cultivated land: in the wild, they would not survive at all, or they would revert. They would expect GM crops to perish as swiftly as traditionally developed species. They would point out that crops are most likely to be genetically modified to reduce pesticide or fertiliser use, and therefore might even benefit the immediate ecosystem. They would point out that GM crops have been harvested for nearly two decades, during which overall life expectancy has steadily increased: if GM crops are in some way dangerous to human health, where are the bodies? They would point out that the real hazard to biodiversity is not modern agricultural science, but population growth, intensive farming, climate change, pollution and the destruction of habitat. Alongside these, any additional consequences from the introduction of genetically modified crops would be small. They might also point out that most journalistic clichés are not just lazy, or inappropriate, but downright silly. Why would anyone introduce the thick end of a wedge?

All these responses have merit: they are not, however, responses that should settle an argument or necessarily placate the fearful. That is because they represent statements about the future, and the great thing about the future is that we really do not know what will happen, and are usually bad at predicting it accurately. Past experience suggests that some examples of GM introduction might turn out to be bad ideas; some seemingly innocent developments could contain unexpected dangers; and some advanced GM science may be misused in the service of terrorism, nihilism or military advantage. Nuclear technology and medical science have produced dangerous and unforeseen outcomes; why should agricultural science be any different? Experienced journalists will know that bland reassurance from acknowledged experts and confident ministers is likely to be suspect; the blander and more confident, the more uneasy a reporter should feel. There are always awkward questions to be asked, and complacencies to be challenged.

There have been a number of academic studies of the way the GM debate was 'framed' in the UK media,[2] and several commentators made the point that the issue quickly ceased to be the sole preserve of the science reporters: it rapidly became a political debate, an environmental argument, a social question. One suspects that there was a subtext to this observation: that some commentators felt that had it been left to reporters who understood and were familiar with the science, then it might not have become such an issue. I am not so sure. It may not be

[2] For instance, The 'Great GM Food Debate' Report 138, Parliamentary Office of Science and Technology, May 2000; also GM Crops: Good or Bad? by Sue Mayer and Andy Stirling, EMBO Reports, Vol. 5, No. 11, 2004.

enough simply to explain the science. Reporters certainly should ask questions about the purpose, direction, finance and business implications of GM research. In a media culture that – in general – prefers its science stories to be a procession of distractions, amusements, entertainments and intellectual breakthroughs, it is sometimes easy to forget that certain areas of science are *for* something: in the service of products and technologies that will be employed, manipulated or consumed; that will cut costs or create jobs, or perhaps put people out of work; that will save producers money, make businesses more profitable, and keep retail cash-registers ringing; advance national economies, create new streams of wealth for new enterprises, or sustain the cash flow and establish the dominance of multinational corporations. There can be nothing inherently sinister about GM technology: it is a scientific technique, a tool already widely used in medical and pharmaceutical research. Whether it is a good or bad thing depends on how well it is used, and how well it is used is not easily separable from public acceptance.

Shortly after the GM 'disaster' – a word used at the time by some scientists to describe the public response in 1989 – a different group of biologists set about trying to introduce yet another potentially disturbing scientific advance: the use of embryo stem cell therapy as a possible treatment for so-far incurable disorders. They enlisted the interest of the media and the public, and the backing of medical charities and patient groups; they provoked a protracted national debate, confronted criticism from the established churches and from people worried about the morality of employing human embryos for a new kind of research, and they helped politicians in two Houses of Parliament to see the strength of their arguments, and make appropriate legislative changes. Conspicuously, and before embarking on the research, they had told the public what they wanted to do and why they wanted to do it; explained who would benefit from the research; and outlined how they wanted to proceed. The lesson could not be clearer: scientists showed that they were prepared to trust the public's judgement, and the public responded by trusting the scientists to get on with the research. Reporters and broadcasters supported the debate because scientists were prepared to explain their ambitions in clear, vivid and patient language; but reporters were also in sympathy because they could see that stem cell research was a 'good story' – good in the sense that people with terrible neurodegenerative disorders might be treated, and people debilitated by spinal injury might gain hope; good, too, in the sense that by reporting all the arguments fairly and seriously, the media really could make a difference.

The contrast with the GM adventure could not be more obvious. It is not enough simply to have a good idea; you have to persuade enough people that the idea is valuable. Jonathan Swift in *Gulliver's Travels* quotes the King of Brobdingnag as saying that '*whoever could make two ears of corn or two blades of grass, to grow upon a spot of ground where only one grew before, would deserve better of mankind, and do more essential service to his country, than the whole race of politicians put together*'. Quite right, but in a democracy, royal endorsement is not enough: you have to get the public to see things the same way.

23

The environmental movement's earlier and current viewpoints and positions

Technologists are politicians under cover.

AFTER 15 YEARS THE CONTROVERSY OVER GM SOYA GOES ON

In the beginning of the 1990s, major environmental organisations in Europe, like Greenpeace, the European Environmental Bureau (EEB) and the World Wildlife Fund (WWF), had hardly any staff working on genetic engineering in agriculture and food production, except Friends of the Earth Europe. In 1996 this changed radically. That year a small fleet of rubber vessels from Greenpeace tried to prevent the first ships with Monsanto's genetically modified (GM) soya from the USA from docking in European harbours, and the first banners calling for a ban on GM food were unfolded in front of the head offices of major food manufacturers and supermarket chains.

Since then Monsanto's GM soya and other GM crops have been the focus of numerous campaigns by environmental and other non-governmental organisations (NGOs) from all over the world. Meanwhile, organic farmers worldwide and major European food producing and retailing organisations have joined their ranks out of concern over possible economic losses due to the presence of GM materials in their products. Major consumer organisations in Europe have however lost most of their interest in campaigning on GM food, as they were satisfied with the revision of European Union (EU) market approval and labelling procedures

Successful Agricultural Innovation in Emerging Economies: New Genetic Technologies for Global Food Production, eds. David J. Bennett and Richard C. Jennings. Published by Cambridge University Press. © Cambridge University Press 2013.

at the end of the 1990s. In their opinion, these EU regulations sufficiently ensure the safety of GM foods coming into the EU market and guarantee consumer choice through mandatory labelling of GM foods as well.

Nonetheless, after 15 years Monsanto's GM soya is still targeted by many (environmental) NGOs with the notable exception of the WWF. Their main tactic still is to draw (media) attention to downstream operators in the agri-food chain that utilise GM soya ingredients and are close to the consumer. In 2012 a petition to major European retailers from the campaign group Toxicsoy.org shows that Monsanto's GM soya is not only criticised as such but also criticised as a crucial element in a system of 'factory farming', in which retailers would drive the use of soya in the production of cheap meat products (Toxicsoy.org, 2012).

The petition of 2012 from Toxicsoy.org

In February 2012 the Dutch multinational food retailer Ahold received the signatures of 26000 people across Europe demanding an end to greenwash projects such as the Round Table for Responsible Soy (RTRS). Ahold agreed that the RTRS has a long way to go and that the current certification scheme cannot be called sustainable. For that reason the supermarkets have chosen not to communicate the RTRS label on the product. It will be communicated in their corporate responsibility report, though.

The petition kicked off in six countries in 2011 and targeted supermarket chains and food companies around Europe such as Ahold, Aldi, Arla, Carrefour, Colruyt, Coop, Delhaize, Marks & Spencer and Unilever. International environmental groups including Friends of the Earth International, Action Aid, Global Forest Coalition and Food & Water Europe supported the petition. Since 'Europe imports 34 million tons of GM soya every year, mainly to feed factory farmed animals', Toxicsoy.org argued that 'this system can never be called responsible and does not deserve a green label'.

According to Toxicsoy.org, the RTRS is an initiative of the WWF together with the soya industry and companies with a vested interest in soya expansion, such as the agribusiness and oil giants Monsanto, Syngenta, Cargill, BP and Shell. The Dutch food and animal feed industry is actively supporting the RTRS, and the Dutch government is providing financial support to the scheme in particular via the Sustainable Trade Initiative. The RTRS has faced strong opposition from civil society for years. Hundreds of organisations from Europe and South America have signed declarations against the RTRS. In April 2011 the German platform of environmental organisations Deutscher Naturschutzring (DNR) sent a letter to WWF and asked them to withdraw from the RTRS, stating: *'DNR cannot accept that WWF protects a failed system of agriculture and secures the profits of companies like Monsanto and BP.'*

The soya that is being certified by the RTRS is mostly Monsanto's Roundup Ready GM soya, made resistant to Monsanto's own herbicide Roundup based on glyphosate, which, according to Toxicsoy.org, has been increasingly linked to serious health impacts on humans and wildlife. Mixtures of pesticides are sprayed over large surfaces by aeroplane or large machines, causing severe health problems for the local population, pollution of water and damage to crops.

Ahold wrote in response to the petition that there are at present too few alternatives to soya imports. However, retailers like Ahold drive the use of soya by promoting cheap meat products. Instead, soya animal feed should be replaced by locally grown animal feed, and factory farming should be banned. In addition, Ahold wrote that it *'does not intend to communicate the use of certified soya to consumers via packaging'*, which, according to Toxicsoy.org, shows once more that certified 'responsible' soya has already failed as a brand.

A few months earlier, in October 2011, a global coalition of six NGOs from North and South America, eight NGOs from Europe, three NGOs from Africa and four NGOs from Asia Pacific issued a Global Citizens Report on the state of GMOs under the title *The Emperor Has No Clothes* (Navdanya International (India) and Centre for Food Safety (USA), 2011). According to the report's foreword,

> *Hans Christian Andersen's fable is an apt parody for what is happening today with GMOs in food and agriculture. The GMO Emperor Monsanto has no clothes; its promises to increase crop yields and feed the hungry have proven to be false; its genetic engineering to control weeds and pests has created super weeds and super pests. Yet the Emperor struts around hoping the illusion will last and the courtiers, not wanting to be seen as stupid, will keep applauding and pretending they see the magnificent robes of the GMO emperor. Citizens around the world can see*

the false promises and failures of GMOs, and, like the child who speaks up,
are proclaiming 'What the Emperor is telling us is not true. It is an illusion.
The GMO Emperor has no clothes.'

The 251-page report itself extensively examines a series of 'false promises'. It starts by noting that only two GM traits, herbicide tolerance and insect resistance, have so far been exploited in crops such as soya, maize and cotton. It then suggests that despite claims that GM crops would lower the level of chemical insecticides and herbicides, this has not been the case. The report cites several case studies from the USA, Argentina, Brazil and India that found increases of the use of chemicals and the emergence of herbicide-resistant weeds that need more toxic chemicals to control.

The report further points at claims from Monsanto that it could breed, through genetic engineering, crops for drought tolerance and other climate-resilient traits. But this is also considered a false promise, with reference to the US Department of Agriculture's (USDA) draft environmental risk assessment of Monsanto's drought-tolerant GM maize that suggested that *'equally comparable varieties produced through conventional breeding techniques are readily available in irrigated corn production reviews'*. Concern is further expressed about the bio-piracy from nature and farmers through 1600 patents on climate-resilient crops, while none of the traits such as drought tolerance or nitrogen fixation have been 'invented' by genetic engineering.

Another false claim from Monsanto and the biotechnology industry, according to this report, would be that GM foods are safe. Indeed, according to the report, there are enough independent studies to show that GM food can cause health damage. It notes that the biotechnology industry has attacked every scientist who has done independent research on GMOs, and provides evidence for the repeated suppression of the environmental risk assessment of GM crops and the safety assessment of GM foods by 'bad science'.

Another myth promoted by the biotechnology industry, the report claims, is the peaceful coexistence of GM, conventional and organic crops, because GM contamination via cross-pollination would be unavoidable. As illustration the report cites several cases of GM contamination of fields or shipments from several continents.

Finally, the report suggests that GMOs are intimately linked to seed patents. *'In fact, patenting of seeds is the real reason why industry is promoting GMOs'*, it says. The companies have succeeded in marketing their crops to more than 15 million farmers, largely by heavy lobbying of governments, buying up local seed companies and withdrawing conventional seeds from the market. Monopolies on seed are being established through patents, mergers and acquisitions, and cross-licensing, resulting in the control of nearly 50% of global seed sales by Monsanto, Dupont and Syngenta, the world's three largest GM companies. The combination of patents, GM contamination and the spread of monocultures means that society is rapidly losing its seed freedom and food freedom, the report argues. The refusal of the US government to label GM foods is therefore seen as one dimension of the totalitarian structures associated with the introduction of genetic engineering in agriculture and food production. '*Our biodiversity and our seed freedom are in peril. Our food freedom, food democracy and food sovereignty are at stake*', the report says, and instead of genetic engineering it advocates an 'agro-ecological' approach to agriculture and food production.

Both the Toxicsoy.org petition and the Global Citizens Report are recent examples of the current position of many (environmental) NGOs on GM crops. The NGOs not only assess GMOs from an environmental and health perspective, but also in conjunction with social perspectives, and it is clear that they have not changed their viewpoints during the last 15 years, as they see no reasons or evidence to do so.

The same goes for what in Dutch innovation politics is called the 'Golden Triangle' of government, science and industry in many developed and emerging economies which has continued to push the use of genetic engineering in agriculture and food production since its inception. The Toxicsoy.org petition, for example, brings to light the support provided to the Round Table on Responsible Soy by the Dutch government and the Dutch food and animal feed industry. Not surprisingly, the 'Golden Triangle' in many European and other countries throughout the world expressed its frustration about the (environmental) NGO campaigns against GMOs, accusing them of irresponsible behaviour in the light of global challenges such as hunger and climate change, while at the same time being concerned about the anxiety created in consumer markets. Therefore, in addition to the usual large public and private expenditures for further development of GM and other genomics-based plant breeding tools, the 'Golden Triangle' felt the need to spend large sums of money on social science research, consumer surveys and public communication. They saw this as necessary to embed genetic engineering technology in society, i.e. to increase the public acceptance of GM food. Apparently this was in vain, as after 15 years of battles in the media and political arenas many working within the 'Golden Triangle' sighed that the public debate on GMOs had failed.

A CLASH OF TWO PARADIGMS

The example of Toxicsoy.org illustrates that the opposition of NGOs against GM crops is deeply rooted in a more fundamental critique of the agro-industrial system. Since the 1970s agro-industrial systems have been criticised by many NGOs for causing environmental harm such as soil degradation, vulnerability to pests, greater dependence on agrichemicals, run-off pollution, genetic erosion and crop uniformity. Moreover, it is now also (scientifically) recognised that agro-industrial systems contribute significantly to greenhouse gas emissions, while climate change potentially destabilises agricultural systems, whose greater vulnerability may therefore need to be addressed through adaption and resilience.

These problems are diagnosed in various ways which inform agricultural research and innovation agendas (CREPE, 2011). The dominant paradigm since the 1990s has been the Life Sciences paradigm which promises to make agriculture more sustainable through greater efficiency, e.g. genetically precise changes which could protect crops from external threats and increase their productivity, while its R&D agenda favours new knowledge that can be privatised. By way of contrast, another paradigm is the Agro-Ecology paradigm, in which eco-efficiency is promised through keeping cycles as short and closed as possible, as a means to use biodiverse resources more efficiently, as for instance in organic agriculture. According to the agro-ecological diagnoses of sustainability problems, farmers' knowledge of natural resources has been displaced by laboratory knowledge and distant commodity chains, state research agendas have generally locked out agro-ecology, and incremental farmer-led improvements are not officially valued as innovation in contrast to laboratory-based biotechnology research. As a result, the state research agendas have been shifted by the 'Golden Triangle' towards specialist knowledge for agro-inputs and processing methods, while the institutional basis for disinterested science, public good training and agricultural extension services has largely been dismantled and replaced by public–private partnerships that are based on the belief that scientific research can best lead to successful innovation if it is targeted at market opportunities.

Given the dominance of the Life Sciences paradigm in setting agricultural research agendas, many environmental and other NGOs from all over the world cheered the World Agriculture Report of 2008. This report, formally the International Assessment of Agricultural Science and Technology for Development (IAASTD), puts out a call for governments and international agencies to redirect and increase their funding towards agro-ecological approaches. According to Greenpeace, the IAASTD's core message is '*the urgent need to move away from destructive and chemical-dependent industrial agriculture and to adopt environmental modern farming methods that champion biodiversity and benefit local communities*' (Greenpeace, 2008). Greenpeace acknowledged that it only presents options for actions for governments and funders to bring the much-needed paradigm shift in agricultural about, but none of these options are legally binding. Although the power of the report is considered a balanced, scientific and sobering view of the facts, Greenpeace expected that '*it will require substantial work in the coming years to alert the relevant decision makers about the report and its key findings*'.

The IAASTD report resulted from a unique collaboration initiated by the World Bank in multi-stakeholder partnerships with several UN organisations and representatives of governments, civil society, private sectors and scientific institutions. In April 2008, nearly 60 governments signed the IAASTD's final report. Notably, the USA, Canada and Australia, major grain producers and heavy users of genetic engineering, did not approve the IAASTD's report. A few months earlier, the biotechnology industry had already left the process because it felt the assessment of modern biotechnology and GM crops was unbalanced. By contrast, many NGOs applauded the report's conclusion on GM crops that '*such techniques as genetic engineering are no solution for soaring food prices, hunger and poverty*'.

Yet, it remains to be seen what impact the IAASTD report will have in practice on public and private agendas for agricultural research and innovation. For example, according to an experienced NGO campaigner for sustainable agriculture (Haerlin, 2011), its recommendations do not appear to have been taken into account in the revision of the EU's Common Agricultural Policy (CAP); and he suggests that only a citizens' revolt would help. Given the interests of the 'Golden Triangles' in developed and emerging economies, it is likely that they will continue seeking to shape agricultural research and innovation in accordance with the Life Sciences paradigm, through which agriculture becomes a biomass factory in a (knowledge-based) bio-economy for the eco-efficient production and conversion of biomass into various food, feed, fibre and fuel products.

Moreover, while several African countries have been moving towards GM crops, often as a result of funding from Monsanto, Syngenta and the US Agency for International Development (USAID), many NGOs are also deeply suspicious of philanthropic initiatives such as the

Alliance for a New Green Revolution in Africa (AGRA). This programme which operates across Africa, has received many millions of dollars in funding from billionaires, such as the Bill and Melinda Gates Foundation and Warren Buffet. While AGRA claims that it wishes to establish food security, in practice, several NGOs from Africa suggest that '*it is opening up huge new markets for the agribusiness industry by persuading millions of African farmers to become dependent on their seeds and chemicals*'. This will eventually lay the groundwork for GMOs to enter Africa on a large scale, because the Bill and Melinda Gates Foundation, AGRA's main funder, has clearly stated its belief that GMOs are part of the solution to hunger in Africa. The NGOs thereby noted that '*several former Monsanto staff work for the Gates Foundation, whose portfolio has invested in more than 23 million dollars of Monsanto stock*'.

Some quotations on GM crops from the IAASTD report listed in the Greenpeace Briefing that may explain why the biotechnology industry left the process and major grain-producers and genetic engineering users as the USA, Canada and Australia did not sign it

'*A problem-oriented approach to biotechnology R&D would focus investment on local priorities identified through participatory and transparent processes, and favor multifunctional solutions to local problems.*'

'*The impacts of transgenic plants, animals and microorganisms are currently less understood. This situation calls for broad stakeholder participation in decision making as well as more public domain research on potential risks.*'

'*Assessment of modern biotechnology is lagging behind development; information can be anecdotal and contradictory, and uncertainty on benefits and harms is unavoidable. There is a wide range of perspectives on the environmental, human health and economic risks and benefits of modern biotechnology, many of which are as yet not understood.*'

'*Regimes of intellectual property rights that protect farmers and expand participatory plant breeding and local control over genetic resources and their related traditional knowledge can increase equity.*'

'*The use of patents for transgenes introduces additional issues. In developing countries especially, instruments such as patents may drive up costs, restrict experimentation by the individual farmers or public researchers, while also potentially undermining local practices that enhance*

> food security and economic sustainability. In this regards, there is particular concern about present IPR instruments eventually inhibiting seed-saving, exchange, sale and access to proprietary materials necessary for the independent research community to conducts analyses and long-term experimentation on impacts. Farmers face new liabilities: GM farmers may become liable for adventitious presence if it causes loss of market certification and income to neighboring organic farmers, and conventional farmers may become liable to GM seed producers if transgenes are detected in their crops.'

THE PUBLIC DEBATE ON AGRICULTURAL RESEARCH AND INNOVATION HAS NEVER STARTED

Over the last 15 years most (environmental) NGOs have operated from the Agro-Ecology paradigm in their campaigns for sustainable agriculture and food production, whereas the 'Golden Triangles' of government, science and industry in developed and emerging economies have operated from the Life Sciences paradigm, thereby dominating the shaping of agricultural research and innovation. Whether the public debate on GMOs after 15 years has failed or succeeded is not the point here, nor is the question 'Who is right and who is wrong?' The issue at stake here is the 'social constitution' of a new technology.

Obvious or not, the political and economic interests of those who own and control the new technology largely determine how a new technology is used. Yet, how a new technology is socially constituted could be key to its public acceptance. For Greenpeace (2008), 'any technology placed in the hands of those who care little about possible environmental, health, or social impacts is potentially disastrous', and 'global technologies can, particularly in the long term, be of greater significance than prime ministers and presidents'. So, 'when entire national economies are adapted to take advantage of the economic opportunities offered by new technologies, the potential environmental and social consequences are of huge public importance, and clearly of importance to Greenpeace'. Thorough public scrutiny before financial or political commitments to new technologies become irreversible could be hugely beneficial, and, according to Greenpeace, 'surely a matter of democratic rights' (Arnall, 2003).

In the wake of the controversy over GMOs, public debates and stakeholder consultations about new technologies have become fashionable, but have not often been perceived as meaningful by parties that have been engaged in such exercises, thereby promoting some cynicism. If public debate or stakeholder consultation on new technologies is to mean anything, Greenpeace recommends addressing at least the following questions concerning the social constitution of a new technology:

- Who is in control?
- Where can I get information that I trust?
- On what terms is the technology being introduced?
- What risks apply, with what certainty, and to whom?
- Where do the benefits fall?
- Do the risks and benefits fall to the same people?
- Who takes the responsibility for resulting problems?

Thus, if such questions about agricultural research and innovation are not addressed in open, transparent and democratic decision-making processes, it is likely that the public controversy over genetic engineering in agriculture and food production between globalised networks of 'Golden Triangles' and NGOs will continue in both developed and emerging economies. And as long as agricultural research and innovation are shaped by the 'Golden Triangles' only, the NGOs will continue, using all their media skills, to mobilise consumers, supermarket chains and food manufacturers through criticising the official risk assessments of genetic engineering technologies and pointing at agro-ecological alternatives.

REFERENCES

Arnall, A. H. (2003). *Future Technologies, Today's Choices: Nanotechnology, Artificial Intelligence and Robotics – A Technical, Political and Institutional Map of Emerging Technologies,* a report for the Greenpeace Environmental Trust. Greenpeace, London. www.greenpeace.org.uk/MultimediaFiles/Live/FullReport/5886.pdf

CREPE (2011). *What Bio-Economy for Europe? Co-operative Research on Environmental Problems in Europe, with Support from the European Commission FP7 Science in Society Programme.* http://crepeweb.net/wp-content/uploads/2011/02/crepe_final_report.pdf

Greenpeace (2008). *The World Agriculture Report of 2008: Results and Recommendations, International Assessment of Agricultural Science and Technology for Development (IAASTD).* Greenpeace, London. www.greenpeace.org.uk

Haerlin, B. (2011). Der Paradigmenwechsel *Gen-ethischer Informationsdienst,* December, 209, 15–17.

Navdanya International (India) and Centre for Food Safety (USA) (Coordinators) (2011). *The GMO Emperor Has No Clothes: A Global Citizens Reports on the State of GMOs – False Promises, Failed Technologies.* www.navdanya.org/attachments/Latest_Publications1.pdf

Toxicsoy.org (2012). *Over 25 000 people tell Ahold: stop misleading consumers, genetically modified toxic soy is not responsible.* Toxicsoy.org, February. www.toxicsoy.org/toxicsoy/news/Artikelen/2012/2/9_Over_25.000_people_tell_Ahold__stop_misleading_consumers%2C_Genetically_Modified_toxic_soy_is_not_responsible!.html

24

Social and ethical issues raised by NGOs and how they can be understood

INTRODUCTION

The flagship of modern biotechnology is the genetically modified organism – the GM seed. GM technology has become a focus of widespread concern among non-governmental organisations (NGOs) and a considerable amount of debate has been generated as a result of this concern. Indeed, the debate has become so polarised that advocates of either extreme in the debate will, with equal conviction and authority, make entirely contradictory claims about the harm or benefit that (may) result from this technology. To some the benefits of GM agriculture are obvious, to others the dangers are equally obvious. The extreme differences in these points of view make it difficult to see a middle way through the debate – common ground is hard to find. The discussion between the two groups has become so polarised that an observer may conclude they live in different worlds – it has become virtually impossible to take an impartial point of view.

Of course genetic technology goes well beyond genetic modification. Various other kinds of genetic technology can be used in breeding and production, such as genomics, marker-assisted screening, phenotype analysis and computer modelling. Most of the authors of this book see GM agriculture as an obvious benefit to world food production. But GM agriculture is only a part of a broader technology that can contribute to world food production, and, for those to whom the dangers of GM agriculture are obvious, it is a technology that should be done without. The fact is that, of all the new and developing techniques in biotechnology, GM has become the focus of debate.

Successful Agricultural Innovation in Emerging Economies: New Genetic Technologies for Global Food Production, eds. David J. Bennett and Richard C. Jennings. Published by Cambridge University Press. © Cambridge University Press 2013.

This chapter will look at the sources of this debate and consider two of the principal social and ethical issues that are contested. I will refer to those who favour the use of GM technology as GM advocates, and to those who are against GM technology as AGM advocates. The two principal issues that I will discuss are the debate over the social economic implications of GM technology, and the debate over whether GM foods affect human health.

With regard to the social question of wealth creation and distribution, GM advocates argue that the use of GM can create crops that will contribute to the productivity and livelihood of farmers across the world, especially poor farmers in emerging economies. In reply, AGM advocates argue that GM will not benefit those who are most in need, but will benefit those who are least in need – the rich transnational corporations (TNCs). They argue that, instead of a technology that concentrates knowledge and wealth in the hands of a few, biotechnology should be used to benefit the poor and should be freely shared with them.

With regard to GM food, both GM advocates and AGM advocates wish to improve global health and welfare, but they are entirely opposed on the question of whether GM can contribute to these goals. GM advocates argue that there is no evidence of harm to human health resulting from GM foods, AGM advocates argue that there is. GM advocates say many people have been eating GM foods for years and have shown no adverse reactions, AGM advocates point to cases such as tobacco smoking or working with asbestos where similar claims were made and turned out to be wrong.

The issues that I address – economic and health impacts – are far from the only points of difference between these two extreme views. There are a variety of other contested issues between GM and AGM advocates. One of the main areas concerns the environment. This includes issues such as the development of 'super-weeds', the question of whether use of pesticides increases or decreases with GM crops, and whether GM agriculture leads to a reduction of carbon emissions. Other issues include the yield of GM crops, and how far research has advanced in producing the promised advantages of GM crops. But the economic and health impacts are important issues that matter to a world that is facing global food shortages and in which the economically deprived face starvation.

The welfare of the poor in the world is a common concern of both GM and AGM advocates. But they disagree over the implications of this GM technology for the poor in the world. GM advocates see themselves as providing a technology that can overcome the difficulties of food production in emerging economies, while the opponents see GM as a tool for exploitation in the hands of TNCs.

INTERESTED PARTIES

There are in fact *three* main parties involved in the controversy over GM. There are the TNCs with their commercial interests, there are the NGOs with their interest in global equity, and, somewhat eclipsed by the tension between these two, there is the (scientific) academic community with an interest in using GM technology to avert the growing problem of global hunger. By and large the academic community believes they have a technology that will benefit humanity, and tend to believe that those opposed to GM are resisting a technology that can alleviate global hunger and contribute to global equity. But they also believe the results of their research should be available to the public, to be used without constraint as a public good. The academic community shares with the TNCs an interest in GM technology and with the NGOs an interest in global equity.

THE RISE OF COMMERCIAL GM TECHNOLOGY

As is the case with much modern technology, GM technology was initially a product of academic research that was later spun out into the commercial world. The first commercial application of GM was in the modification of a bacterium to produce human insulin. In 1978 the first biotechnology company, Genentech (founded in California in 1977) announced that they had succeeded in modifying bacteria to produce human insulin using recombinant DNA techniques (Kevles, 1998, p. 66). At this point in time however there was no protection for the intellectual property (IP) embodied in this bacterium. In the USA there was, at the time, a long-standing policy of not allowing the patenting of products of nature, which included forms of life. This meant that the bacteria that Genentech had developed for the production of insulin could not be patented and anyone else who could engineer a bacterium to do this would be able to make use of the same process. The protection of IP characteristic of mechanical technology was not transferable to the technology of living organisms. But things were about to change.

As early as 1972 a biochemist at General Electric, Ananda Chakrabarty, had bioengineered a bacterium that would digest the oil of oil slicks. Chakrabarty had not used GM techniques, he had used traditional breeding techniques, but he had developed a living organism that was a valuable technology and he wanted to patent it – to protect it as his own intellectual property. The Patent Office initially refused to patent it on the grounds that living things were the product of nature and that products of nature could not be patented. Chakrabarty appealed against this ruling and carried his appeal to the US Supreme Court where it was heard as the case of *Diamond* v. *Chakrabarty* – Diamond being the current patent commissioner. On 16 June 1980 the Supreme Court voted by the narrow margin of five to four in favour of Chakrabarty. They ruled that Chakrabarty's organism was not a product of nature – it was a man-made living organism, and thus Chakrabarty had a right, within existing statutes, to patent his bacterium (Kevles, 1998, p. 70). This judgment meant that Genentech could now patent its GM bacterium which produced human insulin. This was widely seen as a social good – it was a source of clean, cheap insulin.

The possibility of claiming IP rights to GM organisms enabled industry to put considerable investment into research on new organisms knowing that the knowledge and technology that they generated could be protected from exploitation by their competitors. In the following two decades there was an explosive growth in the size of companies in the biotechnology industry, largely motivated by the need to amalgamate the various bits of IP – the techniques and materials generated in the research process that were needed to complete the genetic modification of organisms that were to be placed in the market.

At the same time as the biotechnology industry was consolidating its control over the IP rights, academic scientists across the world saw the possibilities of creating GM organisms to do good, including the possibility of genetically modifying crops to improve the health and nutrition of the poor in the world. Among the most celebrated, and sometimes vilified, of these scientists was Ingo Potrykus, who, along with Peter Beyer, created the rice variety known as Golden Rice, engineered to produce a precursor to vitamin A. Their motivation was, as far as I can tell, entirely altruistic. They saw Golden Rice as a solution to one of the major nutritional deficiencies in the world, vitamin A deficiency. (For an in-depth discussion of this case, see Adrian Dubock's Chapter 12 in this book.)

ECONOMIC IMPLICATIONS

AGM advocates are concerned that, in using new developments in biotechnology to improve the productivity of agriculture in the developing world, it may not be poor farmers and poor people who benefit, but the multinational biotechnology corporations, the TNCs, who benefit. The technology of GM is largely owned by the corporations and protected by IP rights which give to the producers of GM seed the right of ownership of the seed through its entire lifetime and that of its descendants. Currently, GM seeds are not sold by the corporations which hold the patents, they are exclusively licensed, and the licensee is contractually obliged to the seed company in various ways, not least to respect the seed company's ownership of succeeding generations of the seed by paying a licence fee if seed is saved and resown.

Impact of intellectual property rights

One of the central issues that concerns AGM advocates is the IP rights that it creates. This is particularly true when the seeds are not sold to the farmer and are not owned by the farmer, and this is normally the case for GM seeds. If national food security is to be attained, it will require that the seed be owned by the farmer, or at least by the nation. National food security depends on national ownership of the means of food production. This is a problem that does not affect organic agriculture, nor even conventional agriculture – it is unique to GM agriculture. The problem might be overcome by creating public licences as found in the free software movement, but so much of the technology is already owned by corporations within the biotechnology industry that it becomes difficult to see how this can be achieved.

Since 2006 Friends of the Earth (FOE) has published annual reports under the heading *Who Benefits from GM crops?* (FOE, 2011). These reports provide an analysis of developments in GM biotechnology and explain who benefits from these developments. FOE claims that they expose *'the reasons why GM crops cannot, and are unlikely ever, to contribute to poverty reduction, global food security or sustainable farming'* (FOE, 2009, p. 3). What is important to note is that the critique is aimed at GM, not at biotechnology as a whole.

An example: Golden Rice

To illustrate the role of IP ownership of GM technology I will consider the development of Golden Rice. By the time Potrykus and Beyer developed their technique for producing Golden Rice, the biotechnology industry had become a powerful force in seed production and held control of much of the technology needed to produce Golden Rice. They found that they needed a partner commercially experienced in international agricultural product development and so they turned to the Swiss agrichemical company Zeneca (now Syngenta) for help. The agri-chemical company did a deal with Potrykus and Beyer whereby the rice would be freely licensed to farmers whose annual turnover was less than $10000, but would be commercially licensed to large-scale industrial farmers. This meant that poor farmers in developing nations would bene-fit, but that the company would profit from licensing the seed to commer-cial producers. In a word, the humanitarian efforts were realised, but only through assigning control over the commercial use of Golden Rice to the company. Potrykus is reported to have been furious that the agrichemical giant was able to claim a share of his intellectual property for their own commercial purposes (Enserink, 2008). Moreover, NGO resistance to GM crops was growing rapidly at the time, and unfortunately Potrykus, des-pite his good intentions, was caught in the crossfire of the conflict that had developed between the biotechnology industry and its NGO critics.

EFFECTS ON HUMAN HEALTH

An AGM position

A recent publication summarises the current position of the AGM move-ment on the health risks of GM food. The GMO Emperor Has No Clothes is a 246-page book published in 2011 by Vandana Shiva, Debbie Barker, and Caroline Lockhart. The book (Shiva et al., 2011) is a coordinated effort on the part of an international consortium of 20 NGOs concerned with the ethical and social issues raised by GM technology. Subtitled A Global Citizens Report on the State of GMOs – False Promises, Failed Technologies, the book presents the most recent thinking of the AGM movement.

Several classic cases are cited in which evidence of health risks arise from GM food. Perhaps the most notorious case is that of Arpad Pusztai whose work on GM potatoes brought to public attention the possible impact on health of eating GM foods. He found that the rats to which he was feeding GM potatoes containing the GNA lectin gene from the snowdrop plant suffered damage to their internal organs, including their stomachs, intestines and brains, and their immune systems were compromised. On 13 January 1998 an interview was broadcast in which he discussed his findings. The consequences were far-reaching – the potential health impact of GM food was brought to the attention of the country and possibly of the whole world. The scientific establishment reacted swiftly and viciously against Pusztai, destroying his career. One explanation that has been suggested for this is that the US biotechnology

industry felt these revelations were damaging to their commercial interests and called upon the US government to put a stop to such talk (Smith,
2005). After some controversy, Pusztai's research was published in *The
Lancet*. According to Shiva *et al.*, '[Pusztai's] 1999 GM study results, co-
authored with Dr. Stanley Ewen and controversially published in the
respected British medical journal, *The Lancet*, remains the most sensitive
and rigorous GM feeding trial ever conducted' (Shiva *et al.*, 2011, p. 240).
A long-term study raising similar controversy has appeared while this
book was in production – Gilles-Eric Séralini's study of a Roundup
herbicide and a Roundup-tolerant GM maize (Séralini *et al.*, 2012).

Another example of a health issue that worries the AGM movement is the creation of food crops that produce the *Cry* (or *Bt*) toxin
produced by the bacterium *Bacillus thuringiensis* (*Bt*). The toxin acts as
an insecticide and thus renders the crop insect-resistant. In this way
Monsanto has produced an insect-resistant maize, *Bt* maize. The obvious
worry is that in eating such crops (e.g. *Bt* maize) we are also eating the
toxin, an insecticide, and, on the face of it, this appears to pose a danger
to our health. Those who defend GM argue that the toxic *Bt* protein
breaks down during digestion, but Shiva *et al.* cite a Canadian study
which showed that traces of the *Bt* toxin from Monsanto *Bt* maize
were found in the blood of 93% of women and 80% of their umbilical
cord and foetal blood in an agricultural township of Quebec, Canada
(Shiva *et al.*, 2011, p. 18). The concern with *Bt* found in the umbilical
cord and foetal blood is that the developing embryo is particularly
sensitive to disruption caused by toxins of any sort.

The argument

Though GM technology may have potential humanitarian value, there is
still a heated debate over the effects GM food may have on human and
animal health. Many GM advocates categorically assert that there is no
evidence that GM food is harmful to health, while AGM advocates offer
extensive evidence that it is harmful. In reply GM advocates claim that
this evidence is unscientific and inadequate, and they put considerable
effort into deconstructing the AGM claims, and they offer their own
evidence in support of the safety of GM food. Then in turn, AGM advocates deconstruct these claims, arguing that no real trials of GM food have
been undertaken, and that research which shows evidence of harm to
human health has been ignored or suppressed. And so the debate goes on.

At first sight it would seem that this is an empirical question and
it should be a simple matter of applying scientific methods to decide the

matter. In an ideal scientific world, the experiments used to establish each claim would be replicated to determine whether the same results obtained. But we do not live in an ideal scientific world – various non-scientific interests are at stake and these interests lead to different preferences about which claims are accepted. Indeed, in the debate between the GM and AGM advocates, each side is keen to point out the non-scientific interests that motivate the other side and to explain how these distort their understanding of the truth.

For example, GM advocates accuse their AGM opponents (the NGOs) of creating a crisis in order to raise money, or of acting out of a misplaced idealism about the nature of food and farming, or perhaps a vaguely defined 'public interest'; while AGM advocates accuse their opponents of backing a technology that contributes to the wealth of TNCs at the expense of the world's poor. TNCs are typically located in the richer developed nations of the world, so the money raised by those TNCs will mainly find its way back to these rich nations, further increasing their wealth. The consequence of this is that the governments of rich nations will support their local TNCs and will support the technology contributing to their wealth. Once we begin to focus on the different non-scientific interests involved in this dispute we can begin to understand why a simple empirical resolution of the dispute is not forthcoming.

Clearly the impact of GM food on human health is an important factor in its use, and to that end those who wish to develop GM crops are subject to varying regimes of health and safety regulations. Indeed, the need to establish health and safety regulations is a theme that recurs in many of the discussions in this book. But the existence of such regulations does not end the controversy. Rather the controversy moves on to ask whether the regulations are too stringent (GM advocates) or are insufficient (AGM advocates).

The logic of the debate

There is a certain asymmetry in the logic of the debate over health and safety issues between GM and AGM advocates. Insofar as AGM advocates wish only to establish that there are health risks from GM foods, they only need to show real cases of health damage. They do not need to show that *all* GM foods cause damage to health, only that *some* do. However on the GM side the story is different – the GM advocates wish to claim that there are no health risks, or at least there is no evidence of health risk. But such universal claims can be falsified merely by providing one instance of health damage resulting from GM food. Now, it may be that some GM foods are safe to eat and some are not. But without some adequate method for testing the safety of GM foods this will never be known. And this is a fact that particularly bothers AGM advocates – the fact that GM advocates work very hard to deconstruct their evidence of health damage and to undermine adequate testing methods. Here the operative word is 'adequate'. The GM industry will not want to wait for the kind of testing that takes place in the testing of new medicines – trials first with animals and then with humans. And given that some of the evidence for harm from GM foods appears as birth defects, there is serious concern amongst the AGM advocates that such testing be done over at least one generation of animals. The controversy is not whether to test GM foods or not, but how far they need to be tested, i.e. what methods of testing need to be employed.

For example, the AGM group GM-Free Cymru (GM-Free Wales) says the Danish government is to be applauded for paying attention to the experiences of the farmer Ib Borup Pedersen and his undertaking to study the effects of GM soya on pigs. Pedersen found that by taking GM soya out of the diet of his pigs their health increased markedly and the number of birth defects fell dramatically. This and other factors led the Danish government's Pig Research Centre to initiate a study on the use of GM soya for feeding pigs. But GM-Free Cymru raises several questions of methodology. One in particular concerns the feeding trials themselves. GM-Free Cymru argues that since the study did not start at 28 days when the piglets are approximately 7 kg in bodyweight, but are only started when they reach 30 kg, it is *quite possible – and indeed likely – that during the weight gain period between 7 kg and bodyweight, toxic effects will be triggered off by the use of GMO feed in both* [the test and the control] *groups of animals'* (GM-Free Cymru, 2012).

The issue at stake here is one that recurs in the GM/AGM debate – it is the issue of what counts as an adequate test for the safety of GM foods. In a polarised debate such as this the battle rages over a changing terrain – once it is accepted that regulation needs to be in place to determine the safety of GM food, the debate turns to the nature and extent of the regulation. The possibility of health damage has clearly been established, and one of the central concerns of organisations promoting the uptake of biotechnology in developing nations is that health and safety regulations governing this technology must be in place. In answer to the question *'Are GM foods safe?'*, the World Health Organization (WHO) does not categorically answer *'Yes'*, but rather answers that '*individual GM foods and their safety should be assessed on a case-by-case basis and that it is not possible to make general statements on the safety of all GM foods*' (WHO, 2012, question 8).

ALTERNATIVES TO POLARISATION: GM IN PERSPECTIVE

If the GM/AGM debate is ever to reach any resolution or real conclusion, there needs to be a defusing of tensions or a relocation of the discussion. Relocation of the discussion seems to be the most viable solution, and various groups and publications attempt to do this. I will look at two examples – a 2011 report published by the UK Government Office for Science, and a report published in 2008 by the International Assessment of Agricultural Knowledge, Science and Technology for Development (IAASTD).

The middle road

In January 2011, the UK Government Office for Science published a report on the *Future of Food and Farming* (GO-Science, 2011). This report argues that '*New technologies (such as the genetic modification of living organisms and the use of cloned livestock and nanotechnology) should not be excluded a priori on ethical or moral grounds*' but it puts little emphasis on these new technologies. Instead, it argues that '*much can be done today with existing knowledge…any claims that a single or particular new technology [e.g. GM] is a panacea are foolish*' (GO-Science, 2011, p. 11). The report notes that GM food crops have so far been largely a commercial product and, though their use is increasing, they still make up only a small part of worldwide food production. The report rates biotechnology highly as a source of increasing food production, but GM crops are not of central interest – rather they are included with cloned livestock and nanotechnology as *potential* sources of improvement. In this report of over 200 pages 'genetic modification' is only mentioned eight times. Much is made of what modern genetics offers in the way of developing new varieties and breeds of crops, as well as livestock and aquatic organisms, but GM is only mentioned as a last measure: '*Modern genetics offers new ways to select for desirable traits (for example, marker-assisted selection) that are far more efficient than traditional breeding. They make use of information about an organism's genome but are not restricted to species whose complete genome is known, offering the prospect of the improvement of relatively neglected species…. Yet some advances would also require, or could be done faster or more efficiently, using genetic modification or techniques such as animal cloning*' (GO-Science, 2011, pp. 90–91).

The broad view

One of the most well-developed and well-balanced analyses of the state of worldwide agriculture is provided by the International Assessment of Agricultural Knowledge, Science and Technology for Development (IAASTD) which was initiated in 2002 by the World Bank and the Food and Agriculture Organization of the United Nations. The report consists of a Global Report, a Synthesis Report, an Executive Summary (ES) of the Synthesis Report and various summaries and sub-reports (IAASTD, 2008). The Assessment aims to provide an analysis of the state of agriculture across all regions of the world and what is needed for the development and sustainability needed to feed a growing world population. It addresses the worldwide complexity and diversity of agricultural knowledge, science and technology (AKST) and asks: 'how can AKST can be used to reduce hunger and poverty, improve rural livelihoods, and facilitate equitable environmentally, socially, and economically sustainable development?' (IAASTD, 2008, Executive Summary, p. 3).

The second part of the report addresses eight different themes, from bioenergy and biotechnology to women in agriculture. Biotechnology is placed in a broad context of world agriculture – it is a technology that can contribute to agricultural science and knowledge, but it is only a part of the solution to the problem of global hunger. The report states that our use of scientific and technological discoveries has distracted us from some of the social and environmental consequences of these uses (IAASTD, 2008, Executive Summary, p. 3), and argues that the management of global agriculture must take into account local indigenous knowledge (pp. 10–11). One of the most important recognitions of the report is that modern biotechnology is only one of many contributions to sustainable development of world agriculture. A leading concern in the report is the number and needs of small-scale farmers in the world, the fact that the poor in the world are mainly rural poor and that the most effective way to alleviate the poverty and hunger of rural populations is to improve the productivity of small-scale farmers. The policies and institutional changes that the report envisages 'Should be directed primarily at…resource-poor farmers, women and ethnic minorities' (IAASTD, 2008, Executive Summary, pp. 4–5). The report argues that biotechnology may well have a place in responding to this concern, but should only be implemented once the health, environmental and social impact has been assessed. The ideal is to combine indigenous knowledge and local best practice with any acceptable biotechnology – to develop partnership between farmers and scientists. Such a partnership will both encourage and enable farmers to *'manage soils, water, biological resources, pests, disease vectors, genetic diversity, and conserve natural resources in a culturally appropriate manner'* (IAASTD, 2008, Executive Summary, p. 5).

Biotechnology in the IAASTD report is not confined to GM: *'[Biotechnology] is a broad term embracing the manipulation of living organisms and spans the large range of activities from conventional techniques for fermentation and plant and animal breeding to recent innovations in tissue culture, irradiation, genomics and marker-assisted breeding (MAB) or marker-assisted selection (MAS) to augment natural breeding'* (IAASTD, 2008, Executive Summary, p. 8).

The report recognises the current controversy over *'the use of recombinant DNA techniques to produce transgenes that are inserted into genomes'* (p. 8), that is, the controversy over GM crops. But in spite of the controversy over GM, there still remains a wide range of useful modern technologies for crop development. The IAASTD report is certainly not opposed to the use of GM as a technique – recognising as it does that these are early days and many of the points of contention are as yet undecided. Indeed, the claims and counter-claims made on behalf of this technology can both be true as long as they are not taken to be universally true. For example, the report points to the mixed results on yields of GM crops: *'data based on some years and some GM crops indicate highly variable 10–33% yield gains in some places and yield declines in others'* (IAASTD, 2008, Executive Summary, p. 8 – my emphasis).

CONCLUSIONS

GM advocates maintain that there is no evidence of GM foods causing harm to the health of people or other animals. AGM advocates maintain that there is considerable evidence of such harm. GM advocates say masses of GM food has been eaten by people and by animals with no evidence of harm. AGM advocates claim that there is a significant correlation between the introduction of GM foods and increased incidence of allergy and food sensitivity (FOE, 2006). GM advocates point to scientific work which finds no link between GM and health damage, AGM advocates point to scientific work which does find links between GM and health damage. GM advocates claim that AGM advocates are motivated by socio-political factors – opposition to TNCs and a 'green' ideology. AGM advocates claim that GM advocates are motivated by the economic interests of TNCs and by the money that is available for funding GM research. The difficulty with this situation is that dialogue and mutual respect have all but disappeared – the controversy has reached a point of mutual antagonism and incomprehension.

In my view this situation must be resolved. For a start, the techniques of GM are only a few of the many useful techniques available as a result of recent advances in biotechnology. Even if the risks to health posed by GM foods are established, there are other techniques that can be fruitfully employed in improving the yield of food plants. Given current forecasts of world population growth, improving the yield of agriculture is an imperative. Traditional breeding continues apace and the use of other techniques in the arsenal of biotechnology can improve the speed and efficiency of these traditional techniques, while genetic modification is still only a small source of such improvements.

One of the major concerns of the AGM advocates is the fact that GM gives the agrochemical industry property rights over seeds that traditional agricultural methods did not. The farmer who owns the seeds has the advantage of only having to purchase them once and then the seeds can be saved from each year's harvest to plant for the next year's harvest. Moreover the farmer can select the seeds to be kept, and maintain or even improve the preferred variety. This is a significant advantage for poor farmers who cannot afford to buy seeds each year.

In this chapter I have discussed a polarisation in the debate between GM advocates and AGM advocates. I have suggested that they are effectively living in different worlds – looking at different evidence and listening to different voices. Perhaps through a sympathetic discussion of all points of view we can achieve a closer engagement between these two worlds of discourse – perhaps we can realise the best of both worlds. It is imperative that we find a way to listen to each other and resolve these differences because, after all, we do live in the same world, a world that faces significant problems.

REFERENCES

Enserink, M. (2008). *Tough Lessons from Golden Rice.* http://fbae.org/2009/FBAE/website/news_tough-lessons-from-golden-rice.html
FOE (2006). *Could GM Foods Cause Allergies?* www.foe.co.uk/resource/briefings/gm_alergies.pdf
FOE (2009). *Who Benefits from GM Crops? – Feeding the Biotech Giants, not the World's Poor*, Executive Summary: www.foei.org/en/resources/publications/pdfs/2009/gmcrops2009exec.pdf/at_download/file
FOE (2011). *Who Benefits from GM Crops? 2006–2011.* www.foei.org/en/resources/publications/pdfs/2011/who-benefits-from-gm-crops

GM-Free Cymru (2012). *GM Soy Linked to Health Damage in Pigs: A Danish Dossier.* www.gmwatch.org/latest-listing/1-news-items/13882

GO-Science (Government Office for Science) (2011). *Foresight: The Future of Food and Farming – Challenges and Choices for Global Sustainability.* www.bis.gov. uk/assets/bispartners/foresight/docs/food-and-farming/11–546-future-of-food-and-farming-report.pdf.

IAASTD (2008). *Agriculture at a Crossroads.* The Global Report (606 pp.), the Synthesis Report (106 pp.) and various summaries and sub-reports are available in pdf format at www.agassessment.org/

Kevles, D. J. (1998). *Diamond* v. *Chakrabarty* and beyond. In A. Thackray (ed.) *Private Science: Biotechnology and the Rise of Molecular Sciences*, pp. 65–79. University of Pennsylvania Press, Philadelphia, PA.

Séralini, G.-E., *et al.* (2012). Long-term toxicity of a Roundup herbicide and a Roundup-tolerant genetically modified maize. *Food. Chem. Toxicol.* 50 Issue 11 (November 2012), http://www.sciencedirect.com/science/article/pii/ S0278691512005637.

Shiva, V., Barker, D., and Lockhart, C. (eds.) (2011). *The GMO Emperor Has No Clothes: A Global Citizens Report on the State of GMOs – False Promises, Failed Technologies.* www.navdanya.org/publications

Smith, J. M. (2005). *Seeds of Deception.* Chelsea Green Publishing, Co., White River Junction, VT. (Originally published 2003 by Yes Books.)

WHO (World Health Organization) (2012). *20 Questions on Genetically Modified Foods.* www.who.int/foodsafety/publications/biotech/20questions/en/

25

Advancing the cause in emerging economies

INTRODUCTION AND OVERVIEW: THE DIALOGUE ON THE USE OF GM CROPS IN EMERGING ECONOMIES

The deficit model in teaching is out, science in developing countries shows growing valuable assets, and only dialogue and active listening will unearth mutual understanding between local science and local agriculture. Consequently, there is no room for corporate or environmental imperialism. Genetically modified (GM) crops have their real chance in developing countries in combination with the genomes of landraces and local modern cultivars – provided those rules of dialogue are respected and as a result collaborative breeding programmes are implemented that meet the local needs of the farmers. Such programmes give African agriculture the opportunity to make use of topnotch science as done by researchers of the University of Wageningen; one example is the fruitful collaboration between Germany and Namibia. We need to search for a *combination* of the various agricultural approaches *adapted to regional and local needs;* there is no use simply campaigning, teaching and preaching about one's own convictions.

THE 'GENOMIC MISCONCEPTION': A MAJOR REASON FOR THE SLOWDOWN OF REGULATION AND COMMERCIALISATION OF GM CROPS

The contrast between natural selection and transgenesis has clearly been overestimated. There is no difference between natural hybridisation and genetic engineering on the level of the molecular

Successful Agricultural Innovation in Emerging Economies: New Genetic Technologies for Global Food Production, eds. David J. Bennett and Richard C. Jennings. Published by Cambridge University Press. © Cambridge University Press 2013.

processes; this has been emphasised and underpinned with robust science years ago.

Arber (2002) notes:

> Interestingly, naturally occurring molecular evolution, i.e. the spontaneous generation of genetic variants, has been seen to follow exactly the same three strategies as those used in genetic engineering. These three strategies are:
>
> (a) small local changes in the nucleotide sequences,
> (b) internal reshuffling of genomic DNA segments, and
> (c) acquisition of usually rather small segments of DNA from another type of organism by horizontal gene transfer.

However, at the breeding level differences are important and have been helpful for the success of the new technologies. Arber continues:

> However, there is a principal difference between the procedures of genetic engineering and those serving in nature for biological evolution. While the genetic engineer pre-reflects his alteration and verifies its results, nature places its genetic variations more randomly and largely independent of an identified goal. Under natural conditions, it is the pressure of natural selection which eventually determines, together with the available diversity of genetic variants, the direction taken by evolution. It is interesting to note that natural selection also plays its decisive role in genetic engineering, since indeed not all pre-reflected sequence alterations withstand the power of natural selection. Many investigators have experienced the effect of this natural force which does not allow functional disharmony in a mutated organism.

Unfortunately, international biosafety protocols such as the Cartagena Protocol and also the European biosafety legislation have not followed product-oriented regulation as suggested by the majority of scientific authorities; rather they have followed the regulatory path focusing on the process of transgenesis with all its negative aspects. Canada and other countries with great success have chosen the product-oriented approach.

HIGH REGULATORY COSTS AS AN INDIRECT CONSEQUENCE OF THE 'GENOMIC MISCONCEPTION'

There are numerous signals that biotechnology as a whole is a victim of growing anti-science campaigns. Many recent publications make a serious plea to lower the regulatory hurdles which are the main cause for the exorbitantly high development costs of commodity crops.

The Public Regulation and Research Initiative (PRRI), a worldwide effort of public-sector scientists involved in research and development of biotechnology for the public good, have sent an open letter to the members of the European Commission to aid them in their orientation discussion on biotechnology. PRRI (2008) expressed deep concern about the effect of the political situation in Europe affecting GM foods and crops.

The delays of approvals for GM crops run into years for certain traits, which is basically an intolerable situation. It is true that the main reason for the grotesque delays can be seen in the obscure and too-complex decision-making structures of the European regulatory system, as clearly diagnosed by independent experts for European Commission Directorate-General for Health and Consumers (Sanco).

As a result of a mistaken focus on the process of transgenesis, opponents of GM crops and food mount a detailed critique, including constructions of negative 'facts' which do not stand up to critical review and which are also contradicted by the evident success of GM crop planting worldwide. And worse: critical questions are often launched by non-specialists with no deep understanding of the science involved. Vested interests of important parties are heavily influencing the negotiations about changes in international regulatory legislation, since many have a clear interest to keep the pot on the boil. See the overall analysis of the GM debate by Arntzen *et al.* (2003):

> In the corpulent cafe societies of Europe, with their glut of good food, GM stands no chance of being accepted until there is an obvious benefit for consumers, but it is a crime for indifference and hostility to block the development of GM by and for the world's poorest.

Influential and well-funded activist groups like the Norwegian GENOK organise numerous biosafety classes in developing countries, with detrimental effects for the perception of modern biotechnology in agriculture worldwide.

A comprehensive review of the regulatory system of GM crops of the United States has been published by McHughen and Smyth (2011), a critical one on Europe by Morris and Spillane (2010). Investment in research and development is discouraged by this state of affairs, and the situation is clearly asymmetric between rich and poor countries. More recent statistics from Africa show a good correlation between R&D investment and productivity due to technical progress, rather than due to the still-lagging efficiency.

As a result (besides the often lacking research infrastructure), agricultural production in developed countries shows dramatic differences compared to the emerging economies of Africa, as illustrated by Figure 25.1 (Royal Society, 2009). A lot of work remains to be done if we want realistically to ameliorate the situation. Innovative concepts need to

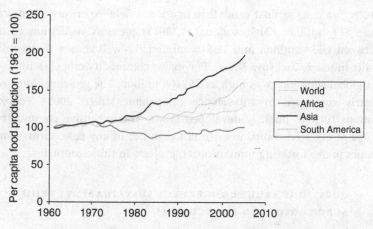

Figure 25.1 Changes in per capita agricultural production, 1961–2005. (Royal Society, 2009)

be introduced, always with the focus on local conditions and human development, including both technological and socio-economic innovation.

The trade policy of Europe is still going in the wrong direction of protectionism, which causes a lot of difficulties in developing countries as Graff and Zilberman (2004) explain:

> European policies blocking genetically engineered crops are conventionally attributed to the concerns of European consumers, but they can be attributed to the self-interests of European industry and farmers as well. Biotech policies maintained in the name of consumer interests are helping European chemical firms to slow their losses in the global crop protection market and are helping European farmers differentiate their conventional crops on environmental and safety grounds, maintain their agricultural subsidies and win new non-tariff trade protections.

In another paper Graff et al. (2009) are even more explicit:

> The analysis suggests that in Europe and in some developing countries a 'strange bedfellows' constellation of concentrated economic interests (including incumbent agrochemical manufacturers, certain farm groups, and environmental protest activists) act in rational self-interest to negatively characterize GM technology in the public arena and to seek regulations that block or slow its introduction.

As early as 1997 Guasch described precisely the dilemma between high regulatory costs and the urgent need to enhance agricultural production in the developing world but it did not help – on the contrary, the situation got worse (Guasch and Spiller, 1997).

More recent papers have documented the growing regulatory costs. Bernauer and colleagues (2011) have documented the high protection costs against vandalism in a Swiss field experiment of more than SF1 million. Kalaitzandonakes (2011) presents regulation costs between US$4 million and US$15 million for well-known transgenic traits in maize and soya beans. For major commodity crops, estimates for global adoption go as high as US$100 million. The growing costs are clearly correlated to anti-science campaigns (Miller, 2009). Poorer nations turn to publicly developed crops. The expensive commodity crops of big seed companies are not popular; in any case, most companies prefer fostering humanitarian projects in those countries.

QUESTIONS ABOUT CONCEPTS OF SUSTAINABILITY FROM AGRO-ECOLOGY TO BIO-ECONOMY

Concepts of sustainable agriculture

Concepts of sustainability are numerous, and often they are abused as defensive weapons to serve ideology. It is useful to study the original Brundtland Report, since it offers a remarkably broad-minded view. Far from being exclusively defensive and retrospective, it explicitly adds the important elements of progress and search for innovation. Sustainable development has been defined in many ways, but the most frequently quoted definition is from *Our Common Future*, the Brundtland Report (UN, 1987):

> Sustainable development is development that meets the needs of the present without compromising the ability of future generations to meet their own needs. It contains within it two key concepts: the concept of needs, in particular the essential needs of the world's poor, to which overriding priority should be given; and the idea of limitations imposed by the state of technology and social organization on the environment's ability to meet present and future needs.

Opponents of GM crops usually refer to the Brundtland Report in their attempt to preserve traditional agro-ecology, but they forget that the report also envisions a way forward which asks not only for conservation but also for the *development* and *management* of sustainable patterns of *production* and *consumption*. One should be aware of an extensive theoretical discussion on a 'principle-based approach for the evaluation of sustainability' as elaborated by the cited report – drawing an intricate philosophical map of intertwined factors of sustainability, where the main elements are justice and resilience.

A move forward to a more pragmatic, more concrete concept of sustainability is offered by the Organisation for Economic Co-operation and Development (OECD) with a focus on agriculture. The declaration of the OECD catalogues a range of concrete measures and rules in order to achieve a more sustainable agriculture. It is remarkable that the proposed indicators do not distinguish between farming with or without transgenic crops. The report contains a comprehensive list of agricultural sustainability factors, many of them implemented in modern agriculture or ready to be implemented soon.

Agriculture is at the core of renewable natural resources, including energy. Its worldwide recycling potential remains largely underexploited. Industrial agriculture is often still stuck in the petrochemical age, and organic agriculture panders too much to urban nostalgia and thus wastes its potential to contribute to the solution of the real problems on this planet in a more efficient way. The main goals of sustainable agriculture are indeed to foster renewable resources and a knowledge-based agriculture.

The controversy about agro-ecology

The eco-imperialist attitude towards farmers in the developing world is seen critically by this author. Many authors cast doubts on the frequent claims that agro-ecology-based non-GMO production strategies would be better for smallholder farmers than solutions including modern breeding, a claim which is not supported by data. The fact is that some 80% of farmers from the developing world who have adopted GM crops are smallholder farmers making considerable economic profits with the technology.

Numerous papers by Miguel Altieri offer tempting concepts of agro-ecology with some good elements and ideas, but they are not based on hard production data. Except for one publication (Altieri, 2000) with a focus on production but lacking sufficient data to allow verification, his concepts are more wishful thinking than agricultural reality. Other notorious and often cited examples of seemingly positive yield results by applying agro-ecological methods (even a doubling of yield is claimed) come from Jules Pretty and colleagues (2006). They are efficiently debunked by Phalan et al. (2007):

(a) *There is a strong selection bias towards successful projects.*

(b) *Methods used to measure changes in yields, water and pesticide use, and carbon sequestration are poorly explained, and therefore, hard to reproduce.*

(c) *Crucially, the study lacks adequate controls, thereby failing to show that it is the introduction of resource-conserving practices which is responsible for reported increases in yield and sustainability.*

(d) *The extent to which these practices provide greater net benefits to farmers than conventional techniques is unclear.*

In the answers to the critique of Phalan, Pretty *et al.* (2007) basically admit the weakness of their study, but offer the excuse of unreasonably high costs to overcome the flaws in field data gathering. Nevertheless, Altieri seems to be 100% convinced that his way is the right one, otherwise it would be hard to understand why he helps fundamentalists to occupy research areas near Berkeley, hindering biotech research with the false accusation that the projects in question are supported by corporate money.

Another sometimes cited paper from Africa, describing comparative field research with maize cultivation, shows the seemingly positive effects of the push–pull technology, basically a fascinating idea to attract and trap pest insects with the weed *Desmodium*, but a careful study of the paper shows bias (Hassanali *et al.*, 2008). It compares a traditional inefficient maize with its own push–pull technology, obtaining a favourable result. If the team had worked with a modern *Bt* maize, the result would most probably have been reversed.

Nevertheless, the hard reality today is that we urgently need to produce more – on the basis of enhancing yield dramatically, the gaps are clear, and astonishingly enough those gaps are higher in Eastern and Central Europe than in Africa (Hengsdijk and Langeveld, 2009).

Organic agriculture versus biotech-based agriculture

Organic farming generally gets a largely unjustified bonus especially in Europe, where the market for organic produce is booming, in spite of higher prices and despite of the fact that local organic farming does not cover market demands. This is why imports, even from overseas, are routine today and transport costs hardly influence retail prices. Organic is a buzzword, hardly contested in a fashionable world of wealthy Western consumers.

A striking example of a nearly unbeatable positive image of organic food is the dangerous *Escherichia coli* outbreak in northern Germany in 2011, which had only a low impact in the European press despite the fact that the event caused more than 50 deaths and that hundreds of patients suffered severe and permanent kidney damage. The outbreak was clearly related to organic farming. Nevertheless, it did not harm the organic boom at all. Highly virulent enterohaemorrhagic *E. coli* (EHEC) strains originating from human faeces introduced into organic cultures by liquid manure were the reason for dozens of uncontrollable cases, pointing to serious hygiene problems in organic cultivation. Dozens of papers clearly relate the presence of virulent EHEC to organic farming in former outbreaks. In soil, the persistence of EHEC for many months, if not years, has been monitored with hard field data. There are also many publications on recent field research: they all demonstrate the connection between the application of liquid manure and (mostly) organic farming. This tragic recent case, exacerbated by newly acquired multiple antibiotic resistance genes in the new strains, only got a minor echo in the press and the public in Europe. This does not mean that organic farming is dangerous in principle: for instance in Switzerland the hygienic rules are so strict that such deadly infection cases can be virtually excluded.

On the positive side of organic farming there is some pioneering work in developing recycling loops and on indirect effects resulting in better landscape management. There is no reason from a scientific point of view why organic methods of production should not go well together with some of the genetically improved plant varieties.

The global success of biotech-based farming has been well documented over many years by the International Service for the Acquisition of Agri-biotech Applications (ISAAA) (www.isaaa.org), with its centres in many countries and directed by Clive James. Downloading the latest reports provides a convincing picture that the success is continuing, with the exception of Europe.

Comparison of yield in organic farming and biotech-based agriculture

Organic agriculture in particular may be part of the solution but it is currently also part of the problem. The most widespread and notoriously negative side in organic farming is the low yield, documented in many long-term monitoring experiments. The paper by Badgley et al. (2007), often cited by proponents of organic farming, claims high yields, but a closer look at the data reveals cherry-picking of yield data and major statistical flaws: instead of averaging the yield data over some years, they simply added (cumulated) the results in some cases – see the comments of Avery (2007).

A blow to the slogan of 'organic farming feeding the world' comes from a meta-study published in *Nature* by Seufert et al. (2012). The main graph is convincing and does not need much comment (Figure 25.2).

The conclusions from Seufert et al. (2012) are as follows:

The results of our meta-analysis differ dramatically from previous results (Badgley et al., 2007). Although our organic performance estimate is lower than previously

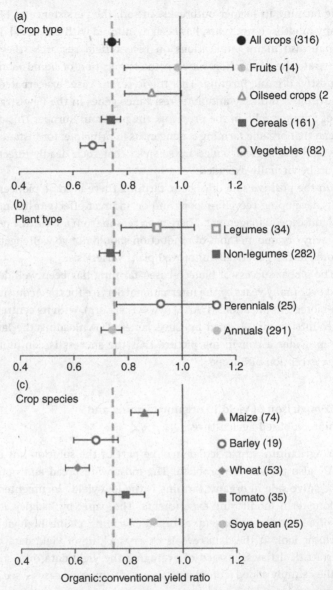

Figure 25.2 Influence of different crop types, plant types and species on organic-to-conventional yield ratios. Influence of crop type (a), plant type (b) and crop species (c) on organic-to-conventional yield ratios. Only those crop types and crop species that were represented by at least ten observations and two studies are shown. Values are mean effect sizes with 95% confidence intervals. The number of observations in each class is shown in parentheses. The dotted line indicates the cumulative effect size across all classes. (Seufert et al., 2012)

reported in developed countries (22% compared to 28%), our results are markedly different in developing countries (-43% compared to +80%). This is because the previous analysis mainly included yield comparisons from conventional low-input subsistence systems, whereas our data set mainly includes data from high input systems for developing countries.

Quite a different picture is given in a meta-study about yield from GM crops. The example of insect-resistant GM corn demonstrates a positive result globally (see Figure 4.2 in Chapter 4).

A caveat about field trials and monitoring needs to be mentioned here. When you want to find out about the reality on yield and ecology by comparing agricultural strategies, you need to establish good contacts with the farmers and their work by doing intensive fieldwork. Inevitably, there is one important factor that should also be taken into account: there is a possible learning effect and automatic amelioration of field practice when monitoring work by specialists starts. This effect could easily alter results in one or the other direction, independent of the agricultural strategy studied.

Focus on bio-economics on the search for the way forward

Balancing local food production against global agricultural trade will be a challenge, since there will be increasing divergence between food demand and supply (which is stagnating due to insufficient investment in agricultural productivity). As a consequence, there will be pressure not just to enhance local food production but also to increase the share of food that is regionally and globally traded. After all, the food-importing countries will be the ones that are most vulnerable to price shocks – and those price dynamics can be correlated with political riots.

The economic basis should be important, but local social networking and lifestyle need to be taken into account as well and protected from hidden protectionism under the false premise of import bans for GM crops.

It is shocking to discover that, according to reports, we spend US$1 billion a day on agricultural subsidies, in a very asymmetric way which results in a nearly perfect agricultural protectionism for the developed world. Considering the complexity of the global challenges in sustainable food production, we should not rely on ideology. Rather, we need a pragmatic understanding of sustainable agriculture. Sustainable agriculture must be based on efficient resource management that makes effective use of the new opportunities of the global knowledge economy and combines the best of system-oriented organic agriculture with the new tools in precision agriculture and biotechnology. From a scientific point of view there is no reason why organic methods of production should not go well together with genetically improved plant varieties.

Even though the Green Revolution was a great success, there were also detrimental effects such as the upsurge of new insect pests, growing insect resistance against widely used pesticides, negative effects on soil fertility, and a rising number of herbicide-resistant weeds. Swaminathan (2006) was one of the fathers of the Green Revolution who recognised its shortcomings:

> The initiation of exploitive agriculture without a proper understanding of the various consequences of every one of the changes introduced into traditional agriculture, and without first building up a proper scientific and training base to sustain it, may only lead us, in the long run, into an era of agricultural disaster rather than one of agricultural prosperity.

In his call for an Evergreen Revolution in 2006 (Swaminathan, 2006; Kesavan and Swaminathan, 2008) he argues that ensuring continuous productivity increases requires a rethinking of sustainable agriculture: a new emphasis on better infrastructure, crop rotation, sustainable management of natural resources as well as progressive enhancement of soil fertility and overall biodiversity. These goals can only be achieved by combining traditional *and* high technology knowledge. Detrimental effects like increasing weed and pest resistance (new resistant species moving into a huge ecological niche) are also likely to become serious problems for large-scale farmers that adopt new high-tech (GM) crops. But these are well-known problems from experience with conventional and traditional agriculture as well (Weed Science, 2012); the only difference is that these resistance problems can be addressed more quickly and more effectively with new technological means available (modern breeding measures, conservation tillage, crop rotation, mixed cropping, etc.). It is obvious that many opponents of GM crops want to take advantage of the negative news related to soya bean farming in Brazil and Argentina, but those scare stories are based on flawed science and target genetic engineering as a breeding method in an unjustified way. All those negative arguments and the rebuttals, related to soya bean farming, are summarised in Ammann (2012).

A comprehensive overview of how sustainability could be organised is offered by Reed *et al.* (2006). The good thing about this scheme is

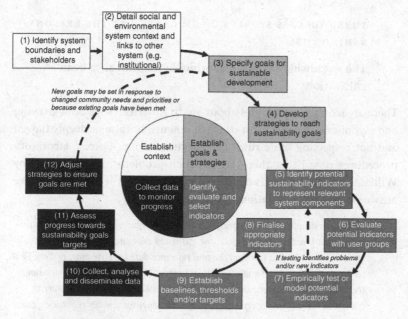

Figure 25.3 Adaptive learning process for sustainability indicator development and application. (Reed *et al.*, 2006)

that it is open-ended and conceived as a learning process, thus having the near-automatic capability of adaptation to local needs (Figure 25.3).

On a more theoretical level, but in a comparable process spirit Phillis *et al.* (2010) have chosen an approach using fuzzy logic, followed by a recent publication within the same framework giving a truly holistic picture including corporate structures:

> *Many people believe that our society is at the crossroads today because of societal and environmental problems of scales ranging from the local to the global. Such problems as global warming, species extinction, overpopulation, poverty, drought, to name but a few, raise questions about the degree of sustainability of our society. To answer sustainability questions, one has to know the meaning of the concept and possess mechanisms to measure it. In this paper, we examine a number of approaches in the literature that do just that. Our focus is on analytical quantitative approaches. Since no universally accepted definition and measuring techniques exist, different approaches lead to different assessments. Despite such shortcomings, rough ideas and estimates about the sustainability of countries or regions can be obtained. One common characteristic of the models herein is their hierarchical nature that provides sustainability assessments for countries in a holistic way. Such models fall in the category of system of systems. Some of these models can be used to assess corporate sustainability.*

THREE SUCCESS STORIES OF GM CROPS AND THE REASONS BEHIND THEM

The worldwide success of herbicide-tolerant soya bean cultivation

There are many negative tales about South American soya bean growing and glyphosate toxicity, most of them downright false or deeply flawed and not respecting basic rules in experimentation. Correct laboratory procedures reveal that there are no such problems. A new report by Williams *et al.* (2012) needs no further comments – it is a comprehensive answer to all those allegations. From the summary:

> *These data demonstrated extremely low human exposures as a result of normal application practices. Furthermore, the estimated exposure concentrations in humans are >500-fold less than the oral reference dose for glyphosate of 2 mg/kg/d set by the U.S. Environmental Protection Agency (U.S. EPA 1993). In conclusion, the available literature shows no solid evidence linking glyphosate exposure to adverse developmental or reproductive effects at environmentally realistic exposure concentrations.*

Green (2012) gives an overview on the success story of the herbicide-tolerant soya bean; the graphs need no further comment. Compared to insect resistance and combined resistance the success of GM soya beans has been overwhelming (James, 2011) (Figure 25.4). The environmental impact is overall clearly reduced due to conservation tilling and the low toxicity of glyphosate (Figure 25.5).

The introduction of a virus-resistant GM bean in Brazil

Another recent source on the Brazilian viral-resistant bean has been published on the website of Biofortified (www.biofortified.org) by De Souza (2011). It documents well an impressive success of Brazilian researchers with their own research and development and approval of a highly useful new virus-resistant transgenic bean for domestic use.

> *Why are virus-resistant beans so important:*
>
> *Beans are highly nutritious and one of the most important legumes consumed by over 500 million people in Latin America and Africa. In Brazil it is regularly an indispensable item of the everyday diet, often combined with rice and eaten by all social classes in all parts of the nation. They are found in a great variety of types with different sizes, colors and tastes consumed throughout the country.*
> *Perhaps, the most typical Brazilian dish is the 'feijoada', a black beans stew. The local consumption is around 16 kg per person every year. Given its high*

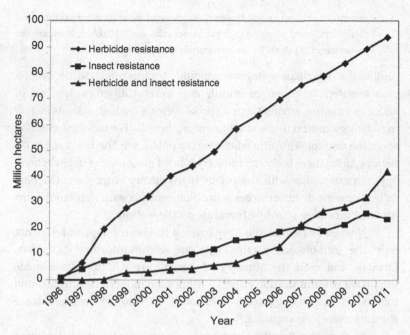

Figure 25.4 Adoption of herbicide-resistant and insect-resistant crops globally. (James, 2011)

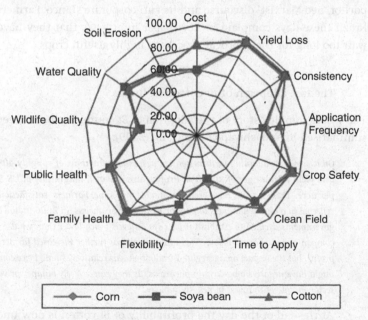

Figure 25.5 Percentage of 1176 corn, soya bean and cotton growers reporting various glyphosate-resistant crop characteristics that are very important. (Hurley *et al.*, 2009)

protein (15 to 33%) content besides B vitamins and minerals as iron, calcium and phosphorus, beans provide a high nutritional value meal. Moreover, beans are the major source of protein for the economically disadvantaged.

AnBio, the Brazilian non-governmental organisation (NGO) dealing with biosafety including genetically engineered (GE) crops, has lots of activities running, among them a special website for high schools including a biology contest (www.anbiojovem.org.br/obb). For more information about the multiple efforts in educating the public, see Traynor *et al.* (2007). Besides AnBio there is also an industry-funded group active in biotechnology communication with the public (www.cib.org.br/sec_executiva.php) called Conselho de Informações sobre Biotecnologia with numerous activities and providing scientific literature on the website.

Numerous local media have covered the bean's approval. It is not only the anti-biotech groups who are communicating. Scientists, farmers, and even the Ministry of Agriculture (CTNBio) have made important communication efforts – using such modern communication tools such as twitter (http://twitter.com/#!/CTNBio) to counterbalance the anti-science propaganda.

In fact, the Brazilian transgenic crops developed by Embrapa broke the jaw of those who always accuse genetic engineering of serving the multinational large producers. They lost the easy ideological support of 'neo-Marxist' discourse and its anti-corporate stance. Farmers in Brazil these days complain in newspaper interviews that they have to wait too long for the approval process of highly useful crops.

The fast adoption of *Bt* cotton in India

The whole complex story of the adoption of *Bt* cotton has been recently summarised by Sadashivappa and Qaim (2009):

On average, Bt-adopting farmers realize pesticide reductions of roughly 40%, and yield advantages of 30–40%. Profit gains are at a magnitude of US $60 per acre. These benefits have been sustainable over time. Farmers' satisfaction is reflected in a high willingness to pay for Bt seeds. Nonetheless, in 2006 Indian state governments decided to establish price caps at levels much lower than what companies had charged before. This intervention has further increased farmers' profits, but the impact on aggregate Bt adoption was relatively small. Price controls might have negative long-term implications, as they can severely hamper private sector incentives to invest in new technology.

At the end of the day the profitability of *Bt* cotton is now uncontested; see early comments of Müller-Jung (2007).

The connection between suicides of Indian farmers and the introduction of GE cotton in India has been thoroughly refuted by Gruere and Sengupta (2011). This does not hinder activists like Vandana Shiva from continuing with the same old and cheap propaganda linking GE crops with the sad tradition of farmer suicides in India, which started decades before the introduction of GE crops and the beginning of the activities of multinational seed companies.

A new perspective has opened since 2006 for the production of cotton seed oil for human consumption and seed meal for feed, made possible thanks to the detoxification (gossypol) successfully done by modern breeding including genetic engineering.

From ISAAA come the following highlights (James, 2011):

> India celebrated the 10th anniversary of Bt cotton, with plantings exceeding 10 million hectares for the first time, reaching 10.6 million hectares, and occupying 88% of the record 12.1 million hectare cotton crop. The principal beneficiaries were 7 million small farmers growing, on average, 1.5 hectares of cotton. India enhanced farm income from Bt cotton by US$9.4 billion in the period 2002 to 2010 and US$2.5 billion in 2010 alone. Thus, Bt cotton has transformed cotton production in India by increasing yield substantially, decreasing insecticide applications by ~50%, and through welfare benefits, contributed to the alleviation of poverty of 7 million small resource-poor farmers and their families in 2011 alone.

REFERENCES

Altieri, M. A. (2000). Enhancing the productivity and multifunctionality of traditional farming in Latin America. *International Journal of Sustainable Development and World Ecology* **7**, 50–61.

Ammann, K. (2012). *Glyphosate Information*, revised version. www.ask-force.org/web

Arber, W. (2002). Roots, strategies and prospects of functional genomics. *Current Science* **83**, 826–828.

Arntzen, C. J., Coghlan, A., Johnson, B., Peacock, J., and Rodemeyer, M. (2003). GM crops: science, politics and communication. *Nature Reviews Genetics* **4**, 839–843.

Avery, A. (2007) 'Organic Abundance' Report: Fatally Flawed. Hudson Institute, Center for Global Food Issues, Churchville, VA.

Badgley, C., Moghtadera, J., Quineroa, E., *et al.* (2007). Organic agriculture and the global food supply (including rebuttals from Kenneth Cassman and Jim Hendrix). *Renewable Agriculture and Food Systems* **22**, 86–108.

Bernauer, T., Tribaldos, T., Luginbühl, C., and Winzeler, M. (2011). Government regulation and public opposition create high additional costs for field trials with GM crops in Switzerland. *Transgenic Research* doi 10.1007/s11248-011-9486-x.

De Souza, L. (2011). Brazilian virus-resistant beans: a homemade, high potential benefit-driven development from the public sector. www.biofortified.org/2011/10/brazilian-virus-resistant-beans/#more-7511

Graff, G., and Zilberman, D. (2004). Explaining Europe's resistance to agricultural biotechnology. *UPDATE, Agricultural and Resource Economics* **7**, 4.

Graff, G., Hochman, G., and Zilberman, D. (2009). The political economy of agricultural biotechnology policies. *AgBioForum* **12**, 34–46.

Green, J. M. (2012). The benefits of herbicide-resistant crops. *Pest Management Science* **68**, 1323–1331.

Gruere, G., and Sengupta, D. (2011). Bt cotton and farmer suicides in India: an evidence-based assessment. *Journal of Development Studies* **47**, 316–337.

Guasch, J.L., and Spiller, P. (1997). *Managing the Regulatory Process: Design, Concepts, Issues*. International Fund for Agricultural Development, Rome.

Hassanali, A., Herren, H., Khan, Z. R., Pickett, J. A., and Woodcock, C. M. (2008). Integrated pest management: the push–pull approach for controlling insect pests and weeds of cereals, and its potential for other agricultural systems including animal husbandry. *Philosophical Transactions of the Royal Society B* **363**, 611–621.

Hengsdijk, H., and Langeveld, J. W. A. (2009). *Yield Trends and Yield Gap Analysis of Major Crops in the World*. Wettelijke Onderzoekstaken Natuur & Milieu, Wageningen, Netherlands.

Hurley, T. M., Mitchell, P. D., and Frisvold, G. B. (2009). Characteristics of herbicides and weed-management programs most important to corn, cotton, and soyabean growers. *AgBioForum* **12**, 269–280.

James, C. (2011). *Global Status of Commercialized Biotech/GM Crops: 2011*, ISAAA Brief No. 43. International Service for the Acquisition of Agri-biotech Applications, Ithaca, NY.

Kalaitzandonakes, N. (2011). *The Economic Impacts of Asynchronous Authorizations and Low Level Presence: An Overview*. International Food & Agricultural Trade Policy Council, Washington, DC.

Kesavan, P. C., and Swaminathan, M. S. (2006). From green revolution to ever-green revolution: pathways and terminologies. *Current Science* **91**, 145–146.

Kesavan, P. C., and Swaminathan, M. S. (2008). Strategies and models for agricultural sustainability in developing Asian countries. *Philosophical Transactions of the Royal Society B* **363**, 877–891.

McHughen, A., and Smyth, S. (2011). Regulation of genetically modified crops in North America: American overview. In C. Wozniak and A. McHughen (eds.) *Regulation of Agricultural Biotechnology: The United States and Canada*, pp. 35–56. Springer, New York.

Miller, H. I. (2009). The human cost of anti-science activism. *Policy Review* **154**, 65–78.

Morris, S., and Spillane, C. (2010). EU GM crop regulation: a road to resolution or a regulatory roundabout? *European Journal of Risk Regulation* **4**, 359–369.

Mueller-Jung, J. (2007). Wie verpackt man eine Kulturrevolution in Watte? (How to wrap up a cultural revolution in cotton wool?) *Frankfurter Allgemeine Zeitung*, 13 November.

Phalan, B., Rodrigues, A. S. L., Balmford, A., Green, R. E., and Ewers, R. M. (2007). Comment on 'Resource-conserving agriculture increases yields in developing countries'. *Environmental Science and Technology* **41**, 1054–1055.

Phillis, Y. A., Kouikoglou, V. S., and Manousiouthakis, V. (2010). A review of sustainability assessment models as system of systems. *IEEE Systems Journal* **4**, 15–25.

Pretty, J. N., Noble, A.D., Bossio, D., *et al.* (2006). Resource-conserving agriculture increases yields in developing countries. *Environmental Science and Technology* **40**, 1114–1119.

Pretty, J. N., Noble, A.D., Bossio, D., *et al.* (2007). Reply to Phalan *et al. Environmental Science and Technology* **41**, 1056–1057.

PRRI (2008). *PRRI Open Letter to EC Members.* www.isaaa.org/kc/cropbiotechupdate/article

Reed, M. S., Fraser, E. D. G., and Dougill, A. J. (2006). An adaptive learning process for developing and applying sustainability indicators with local communities. *Ecological Economics* **59**, 406–418.

Royal Society (2009). *Reaping the Benefits: Science and the Sustainable Intensification of Global Agriculture,* Policy Document No.11/09. Royal Society, London.

Sadashivappa, P., and Qaim, M. (2009). *Bt* cotton in India: development of benefits and the role of government seed price interventions. *AgBioForum* **12**, 172–183.

Seufert, V., Ramankutty, N., and Foley, J. A. (2012). Comparing the yields of organic and conventional agriculture. *Nature* **485**, 229–232.

Swaminathan, M. S. (2006). An evergreen revolution. *Crop Science* **46**, 2293–2303.

Traynor, P. L., Adonis, M., and Gil, L. (2007). Strategic approaches to informing the public about biotechnology in Latin America. *Electronic Journal of Biotechnology* **10**, 1–9.

UN (1987). Our Common Future, Chapter 2: Towards Sustainable Development. From A/42/427. *Our Common Future: Report of the World Commission on Environment and Development.* United Nations, Geneva.

Weed Science (2012). *International Survey of Herbicide Resistant Weeds.* Funded and Supported by the Herbicide Resistance Action Committee (HRAC), the North American Herbicide Resistance Action Committee (NAHRAC), and the Weed Science Society of America (WSSA).

Williams, A. L., Watson, R. E., and Desesso, J. M. (2012). Developmental and reproductive outcomes in humans and animals after glyphosate exposure: a critical analysis. *Journal of Toxicology and Environmental Health B* **15**, 39–96.

Index

418

Printed in the United States
By Bookmasters